PROTON-TRANSFER REACTIONS

R. P. Bell, F.R.S.

PROTON-TRANSFER REACTIONS

Edited by

EDWARD CALDIN

University Chemical Laboratory
University of Kent at Canterbury

and

VICTOR GOLD

F.R.S.

Department of Chemistry
King's College, London

SPRINGER-SCIENCE+BUSINESS MEDIA, B.V.

© Springer Science+Business Media Dordrecht 1975
Originally published by Chapman and Hall in 1975
Softcover reprint of the hardcover 1st edition 1975

ISBN 978-0-412-12700-7   ISBN 978-1-4899-3013-2 (eBook)
DOI 10.1007/978-1-4899-3013-2

Library of Congress Catalog Card Number 75-5805

CONTENTS

EDITORS' ACKNOWLEDGEMENTS

The editors gladly acknowledge the help they have received from all the authors, and from the publishers, Chapman and Hall.

E.F.C.
V.G.

PREFACE

The study of proton-transfer reactions is a continual source of inspiration for enquiry into reaction mechanisms and in physical chemistry generally. These processes are chemically simple and amenable both to accurate measurements and to quantitative theoretical analysis. Much of the credit for the widespread interest in them must go to Professor R. P. Bell, and to his outstanding theoretical and experimental contributions over more than four decades. His books, especially *Acid-Base Catalysis* (1941) and *The Proton in Chemistry* (1959, 1973), have had a world-wide effect in educating more than one generation of chemists, and he has won the affection and respect of numerous colleagues and collaborators.

The present volume, by a group of Professor Bell's fellow-workers and friends, is intended as a companion to *The Proton in Chemistry* and aims to build on the foundations laid in that book. It discusses some key issues in this central area of chemistry in a series of cognate studies.

In gratefully acknowledging our debt to Ronnie Bell, we dedicate this book to him on the occasion of his retirement from his Chair at the University of Stirling.

May 1975

E. F. Caldin	E. Grunwald
V. Gold	F. J. Kézdy
W. J. Albery	A. J. Kresge
E. M. Arnett	E. S. Lewis
M. L. Bender	C. J. Liddiard
B. Capon	R. A. More O'Ferrall
J. E. Crooks	G. D. Morgan
D. Eustace	B. H. Robinson
R. J. Gillespie	R. W. Taft

CONTRIBUTORS

John Albery — Physical Chemistry Laboratory, South Parks Road, Oxford, England.

Edward M. Arnett — Department of Chemistry, University of Pittsburgh, Pittsburgh, Pennsylvania 15213, U.S.A.

Myron L. Bender — Department of Chemistry, Northwestern University, Evanston, Illinois 60201, U.S.A.

Brian Capon — Chemistry Department, The University, Glasgow G12 8QQ, Scotland.

John E. Crooks — Department of Chemistry, King's College, Strand, London WC2R 2LS, England.

Daniel Eustace — Department of Chemistry, Brandeis University, Waltham, Massachusetts 02154, U.S.A.

Ronald J. Gillespie — Department of Chemistry, McMaster University, Hamilton, Ontario, L8S 4MI, Canada.

Victor Gold — Department of Chemistry, King's College, Strand, London WC2R 2LS, England.

Ernest Grunwald — Department of Chemistry, Brandeis University, Waltham, Massachusetts 02154, U.S.A.

Ferenc J. Kézdy — Department of Biochemistry, University of Chicago, Chicago, Illinois 60637, U.S.A.

A. Jerry Kresge — Scarborough College, University of Toronto, West Hill, Ontario, Canada.

Edward S. Lewis — Department of Chemistry, Rice University, Houston, Texas 77001, U.S.A.

Colin J. Liddiard — Department of Chemistry, King's College, Strand, London WC2R 2LS, England.

Gareth D. Morgan — Department of Chemistry, King's College, Strand, London WC2R 2LS, England.

Rory A. More O'Ferrall — Department of Chemistry, University College, Belfield, Stillorgan Road, Dublin 4, Eire.

Brian H. Robinson — University Chemical Laboratory, Canterbury, Kent CT2 7NH, England.

Robert W. Taft — Department of Chemistry, University of California, Irvine, California 92664, U.S.A.

1

Ronald J. Gillespie

PROTON ACIDS, LEWIS ACIDS, HARD ACIDS, SOFT ACIDS AND SUPERACIDS

1.1 Introduction

Although the terms acid and base have been in use since the early days of chemistry, the meaning of these terms has undergone considerable modification through the centuries and continues even today with the recent introduction of new terms such as hard acid, soft acid and superacid. It is the purpose of this chapter to consider some of the classic definitions of the terms acid and base and also some of the more recent developments of our ideas concerning acids and bases [1].

The terms acid and base originated as descriptions of two different types of substance with which were associated characteristic chemical and physical properties. The term acid was applied to substances which have a sour taste, dissolve many other substances, effervesce with chalk and change the colour of certain plant dyes, while the term base (or alkali) was used for substances which have the property of forming salts with acids and in general of neutralizing or reversing the effects of acids. Naturally attempts were soon made to formulate more precise definitions based on various theoretical interpretations of acid-base behaviour. Consequently there have been a number of controversies between the proponents of various definitions of the terms acid and base, from the time of Lavoisier's oxygen theory and the rival hydrogen theory due to Davy up to more recent times when the relative merits of the Brønsted and the Lewis theories have been hotly disputed. Much of this discussion has, however, not been too profitable as no one definition of an acid or a base can be said to be more correct than another although, certainly, some are more useful than others.

1.2 The Arrhenius definition

The first really clear definition of acids and bases resulted from the work of Ostwald and Arrhenius on electrolytic dissociation in water. In 1887 Arrhenius defined acids as substances that give rise to the hydrogen ion in water and bases as substances that give rise to the hydroxide ion in water. In the following years a very large amount of important quantitative work on the dissociation of acids and other related ionic equilibria in water was carried out. However the restriction to water as a solvent was clearly a severe limitation and it was not clear, for example, whether anhydrous hydrogen chloride or a solution of hydrogen chloride in benzene should be regarded as acids or if only the aqueous solution is an acid. It was also not clear if the ammonia molecule NH_3 should be regarded as a base or if only the corresponding hydroxide NH_4OH has basic properties. The inconvenience of restricting definitions of acids and bases to the aqueous solvent system only, was clearly realized, for example, by Franklin in his early work on liquid ammonia as a solvent [2]. He pointed out that, although acids in this solvent might be defined as substances that give rise to the proton, it was necessary to define bases as substances that give the amide ion NH_2^- rather than the hydroxide ion OH^-, neutralization in liquid ammonia being represented by the equation:

$$H^+ + NH_2^- \longrightarrow NH_3 \qquad (1)$$

It has subsequently become clear that the free proton is never present in significant concentrations in any solvent and that it is always solvated by at least one solvent molecule. Assuming that one solvent molecule is bound more strongly than the others, the proton in water is written as H_3O^+, in liquid ammonia as NH_4^+ and in ethanol as $C_2H_5OH_2^+$, etc.

1.3 The Brønsted definition

The difficulties associated with the Arrhenius definition were largely overcome by the definitions proposed by Brønsted [3], according to which an acid is a species having a tendency to lose a proton and a

2

base is a species having a tendency to add a proton. These definitions can be represented by the equation

$$A \rightleftharpoons B + H^+ \qquad (2)$$

where A and B are called a conjugate acid-base pair. The symbol H^+ represents a free proton, and, since this species cannot exist in anything more than an infinitesimal concentration in any solution, the ionization of an acid can occur to an appreciable extent only in the presence of a suitable base that is capable of accepting a proton. All actual acid-base reactions are of the general form

$$A_1 + B_2 \rightleftharpoons A_2 + B_1 \qquad (3)$$

where $A_1 - B_1$ and $A_2 - B_2$ are two conjugate acid-base pairs. One of these acid-base pairs is often derived from the solvent. For example water, which has both acid and base properties, gives rise to the two conjugate acid-base pairs $H_3O^+ - H_2O$ and $H_2O - OH^-$. Thus the ionization of an acid HA in water is represented by the equation

$$HA + H_2O = H_3O^+ + A^- \qquad (4)$$

and the ionization of a base by the equation

$$B + H_2O = BH^+ + OH^- \qquad (5)$$

In sufficiently dilute solutions in which the concentration of water is effectively constant and the activities of the ionic species may be replaced by their concentrations, the dissociation constant of an acid HA is given by

$$K_c = [H_3O^+][A^-]/[HA]$$

It is not necessary to give a separate definition of base strength since this is conveniently given for any base by the dissociation constant of its conjugate acid.

Acids defined in this manner may be more specifically called proton acids, proton donors or Brønsted acids and the corresponding bases may be called proton acceptors or Brønsted bases.

1.3.1 The effect of the solvent on acid-base behaviour

The solvent has a profound effect on acid-base behaviour. Thus, for example, acetic acid behaves as a weak acid in solution in water:

$$CH_3CO_2H + H_2O \rightleftharpoons H_3O^+ + CH_3CO_2^- \qquad (6)$$

whereas in solution in sulphuric acid it behaves as a strong base:

$$CH_3CO_2H + H_2SO_4 \rightarrow CH_3CO_2H_2^+ + HSO_4^- \qquad (7)$$

Sulphuric acid is a strongly acidic solvent in which a very large number of substances exhibit basic behaviour, *i.e.* are protonated by the solvent, but very few exhibit acidic behaviour, *i.e.* are capable of protonating the solvent. An example of an acid of the sulphuric acid solvent system is disulphuric acid $H_2S_2O_7$, which is partially ionized in solution in sulphuric acid according to the equation:

$$H_2S_2O_7 + H_2SO_4 \rightleftharpoons H_3SO_4^+ + HS_2O_7^- \qquad (8)$$

Liquid ammonia provides an example of a solvent that is more basic than water. Very many substances exhibit acidic behaviour in liquid ammonia, including, for example, triphenylmethane:

$$(C_6H_5)_3CH + NH_3 \rightarrow NH_4^+ + (C_6H_5)_3C^- \qquad (9)$$

while rather few exhibit basic properties; hydride ion H^- is one of the few strong bases in the liquid ammonia solvent system.

It is clear that the strength of an acid can be measured only relative to some suitable base which is generally chosen to be the solvent. The most commonly quoted acid strengths (*i.e.* dissociation constants) are, of course, those that refer to water as a solvent.

It is also clear that the range of acidity or basicity that can be studied in a given solvent is limited by the acid-base properties of the solvent. In general for any solvent HS the strongest acid that can exist in the solvent is the conjugate acid of the solvent, H_2S^+, any other intrinsically stronger acid HA being completely converted to the solvent cation H_2S^+:

$$HA + HS \rightarrow H_2S^+ + A^- \qquad (10)$$

Similarly any base that is appreciably stronger than the conjugate base of the solvent S^- is essentially quantitatively converted to the solvent anion S^-:

$$B + HS \rightarrow BH^+ + B^- \qquad (11)$$

The solvent is said to exert a levelling effect on the strengths of acids and bases. Thus, for example, many of the common mineral acids are fully ionized in water; they all behave as strong acids and their strengths cannot be distinguished.

1.3.2 *Autoprotolysis equilibria in amphiprotic solvents*

Many of the best and most widely used solvents for acids and bases, such as water, have both acidic and basic properties. They are described as amphoteric or amphiprotic solvents, and they undergo an autoprotolysis or self-ionization reaction in which one molecule acts as a base and another as an acid, for instance:

$$H_2O + H_2O \rightleftharpoons H_3O^+ + OH^- \tag{12}$$

$$H_2SO_4 + H_2SO_4 \rightleftharpoons H_3SO_4 + HSO_4^- \tag{13}$$

The autoprotolysis reaction

$$HS \rightleftharpoons H_2S^+ + S^- \tag{14}$$

is characterized by the autoprotolysis constant K_{ap}. When, as is usually the case, the solvent is only very slightly dissociated the activity of the solvent is practically the same as it would have been in the hypothetical undissociated solvent and thus may be given the value of unity, and the activities of the ionic species may be replaced by their concentrations, so that

$$K_{ap} = [H_2S^+][S^-] \tag{15}$$

where H_2S^+ and S^- are the conjugate acid and the conjugate base of the solvent respectively, *i.e.* they are the characteristic cation and anion of the solvent. Some values of $pK_{ap} = -\log K_{ap}$ are given in Table 1. For the pure solvent the concentrations of the characteristic

Table 1

Values of the autoprotolysis constant (ionic product) for various amphiprotic solvents $K_{ap}/mol^2\,l^{-2}$ at 25°C (except as otherwise indicated)

Solvent HS	Cation H_2S^+	Anion HS^-	$pK_{ap} = -\log K_{ap}$
Sulphuric acid	$H_3SO_4^+$	HSO_4^-	2.9
Formic acid	$HCO_2H_2^+$	HCO_2^-	6.2
Hydrogen fluoride	H_2F^+	F^-	10.7 (0°C)
Water	H_3O^+	OH^-	14.0
Acetamide	$CH_3CONH_3^+$	CH_3CONH^-	14.6
Acetic acid	$CH_3CO_2H_2^+$	$CH_3CO_2^-$	15.2
Methanol	$CH_3OH_2^+$	CH_3O^-	16.7
Formamide	$HCONH_3^+$	$HCONH^-$	16.8
Acetonitrile	CH_3CNH^+	CH_2CN^-	19.5
Ammonia	NH_4^+	NH_2^-	27.7
Hydrogen sulphide	H_3S^+	HS^-	32.6

solvent ions are given by:

$$[H_2S^+] = [S^-] = K_{ap}^{\frac{1}{2}} \tag{16}$$

or

$$-\log [H_2S^+] = -\log [S^-] = \tfrac{1}{2}pK_{ap} \tag{17}$$

1.3.3. The nature of the hydrogen ion

Although it was a natural consequence of the Brønsted theory that the hydrogen ion, in aqueous solution, was written as the hydronium ion H_3O^+, it was not until much later that definite evidence was obtained for this ion and for its structure. The first conclusive evidence was provided by Richards and Smith [4] who measured the proton magnetic resonance spectrum of the solid hydrates $HNO_3 \cdot H_2O$, $HClO_4 \cdot H_2O$ and $H_2SO_4 \cdot H_2O$, and showed that the structures were ionic and that the H_3O^+ ion has a rather flat pyramidal shape. It is to be expected, however, that this ion might be further hydrated in aqueous solution, and a variety of evidence strongly suggested that the most important hydrated species is $H_3O^+(H_2O)_3$ which has the structural formula 1. Recently X–ray crystallographic [5] and vibrational spectroscopic evidence [6] has been obtained not only for $H_9O_4^+$ but also for $H_7O_3^+$ and $H_5O_2^+$ which have the analogous hydrogen bonded structures 2 and 3. In fact several different forms of these cations are known due to differences in the hydrogen bonds which are in some cases symmetrical and in other cases unsymmetrical, depending on the particular anion with which they are associated.

Information on the nature of the conjugate acids of other solvents is much more sparse. Obviously there is no doubt about the ammonium ion in liquid ammonia, although information on its further solvation is lacking. Recent work has provided spectroscopic evidence for the angular H_2F^+ ion 4 in solid HF–SbF$_5$ mixtures at low temperatures [7]. The structures of the solvated proton in sulphuric and fluorosulphuric acids are presumed to be 5 and 6 but there is no direct evidence for this. The conjugate acid of acetic acid has been shown by nmr spectroscopy [8] to have the structure 7 in solution in fluorosulphuric acid, but this does not, of course, necessarily prove that the structure is the same in solution in acetic acid.

1.3.4 Measurements of acidity; pH

Addition of an acid to a solvent HS causes the concentration of the characteristic solvent cation, i.e. the solvated proton H_2S^+, to increase to a value greater than its initial concentration $K_{ap}^{\frac{1}{2}}$, while the addition of a base to such a solvent reduces the H_2S^+ concentration to a value lower than $K_{ap}^{\frac{1}{2}}$. Thus it is convenient to measure the acidity of a solution in a given solvent by the concentration of the solvated proton, i.e. the solvent conjugate acid, or by some property related to the concentration of the solvated proton.

In practice the acidity is measured by the pH which in dilute aqueous solution may be written in terms of the concentration of the hydrated proton as $pH = -\log[H^+(\text{hydrated})]$. The solvated proton is usually written as H_3O^+ but is certainly further hydrated and, as discussed above, it seems that the most likely formula is $H_9O_4^+$. However lack of knowledge of the exact nature of the hydrated proton is of no importance for solutions that are sufficiently dilute that the proton is solvated to the maximum extent possible and does not change to any significant extent with increasing concentration. In more concentrated solution the degree of hydration decreases and $H_9O_4^+$ is replaced by less hydrated species such as $H_7O_3^+$ and $H_5O_2^+$ and for this reason, and also because of the increasing importance of interionic effects, the pH is no longer given by $-\log[H^+(\text{hydrated})]$.

In another solvent the activity of the proton will, in general, be greatly different from that in water. Thus the solvated proton in liquid ammonia is NH_4^+, while the solvated proton in sulphuric acid is the $H_3SO_4^+$ cation, and solutions of NH_4^+ in liquid ammonia or of $H_3SO_4^+$ in sulphuric acid have vastly different acidities from solutions of H_3O^+ in water. Comparison of the acidities of such widely different systems is a difficult problem that is discussed later. We simply point out here that an approximate pH scale may be defined for each solvent HS as follows

$$pH(HS) = -\log[H_2S]^+$$

In the pure solvent $pH = \frac{1}{2}pK_{ap}$, and this is the pH value in a neutral solution in this solvent. Acidic media in this solvent are those for which $pH < \frac{1}{2}pK_{ap}$ and basic solutions are those for which $pH > \frac{1}{2}pK_{ap}$. If we limit the discussion to dilute solutions, and if we make the arbitrary assumption that the limit of dilute solutions is 1M, then the minimum value of pH in an acid solution is zero and the maximum value of pH for a basic solution is pK_{ap}. Thus the range of pH, *i.e.*, of acidity that is available in a given solvent, is $\Delta pH = pK_{ap}$; in other words, the greater the extent of dissociation of the solvent, the smaller is the range of acidity available in the solvent. This is understandable when it is remembered that the greater the concentration of the characteristic solvent ions in the pure solvent the less can their concentrations be varied by addition of an acid or a base. The value of the autoprotolysis constant reflects the properties of the solvent as both an acid and a base; thus although sulphuric acid is a very strong acid, it is also relatively a good base, in contrast to hydrogen fluoride which although also a good acid is a relatively poor base, and thus the extent of autoprotolysis of hydrogen fluoride is considerably less than that of sulphuric acid.

During the course of an acid-base titration the pH varies rapidly in the vicinity of the end-point, and for a strong-acid strong-base titration the pH changes over essentially the whole range accessible in the solvent. Thus titrations can be carried out much more easily in a solvent with a low autoprotolysis constant, such as water, than in a solvent with a high autoprotolysis constant, such as sulphuric acid.

The range of acidity $\Delta pH(HS) = pK_{ap}(HS)$ available in a number of solvents is shown in Figure 1. The various solvents are also compared on an absolute basis. Each solvent is characterized by two pK values, pK' and pK'', such that if an acid-base system has $pK(H_2O) < pK'$ then on dissolving in this solvent it is essentially completely converted to base + solvent cation, whereas if $pK(H_2O) > pK''$ the acid-base system is essentially completely converted to acid + solvent anion. Evidently $pK' - pK'' = \Delta pH(HS) = pK_{ap}(HS)$. The horizontal scale also represents the Hammett acidity function H_0 (see Section 8). In addition to a number of amphiprotic solvents this Figure also illustrates the acidity range available in a solvent such as diethyl ether which has no acid properties, and a solvent such as hexane which has neither acid nor base properties.

8

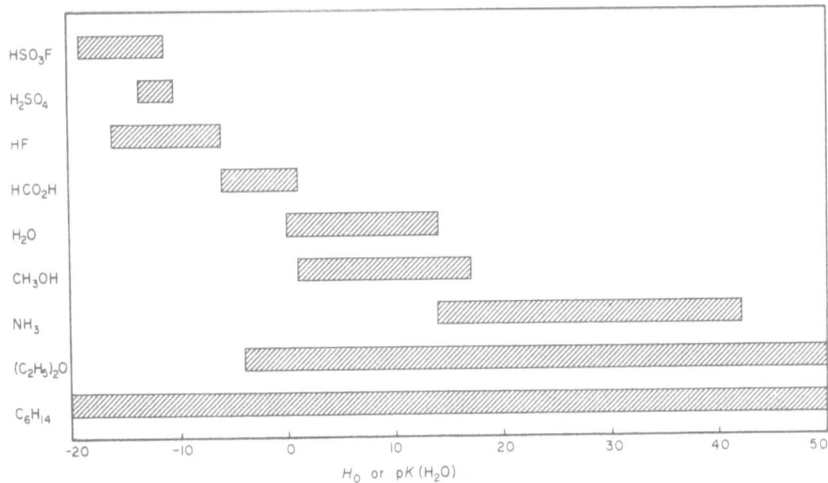

Figure 1. Comparison of acidity ranges (pH(HS)) available in different solvents at 25°C. (Adapted from R. P. Bell, *The proton in Chemistry*, Cornell University Press (1959), and B. Tremillon, *La Chimie en Solvants Non-aqueux*, Presses Universitaires de France (1971).)

As diethyl ether has no acidic properties, it has no levelling effects on the strengths of bases. As hexane has neither acidic nor basic properties, there are no limitations on the range of acid base strengths which can be studied in this solvent. However, such a solvent is far from being an ideal solvent in which to study acid-base properties, since it is non-polar and has a low dielectric constant, which makes it a poor solvent for electrolytes which are in general very incompletely ionized in it. Thus, although a complete range of acid-base strengths is in principle available in a solvent such as hexane, there is the considerable disadvantage that no single scale of acid strengths can be set up, because acids are not converted to a common solvated proton and thus the extent of ionization of an acid depends on the base that is present. There are further complications associated with the very incomplete ionization of the conjugate acid-base pair that is formed by the proton transfer, and by the formation of associated (polymeric) hydrogen-bonded complexes.

1.4 The Lewis definition

Lewis pointed out that some of the characteristic properties of protonic acids and bases, such as their ability to change the colours of suitable indicators, their ability to displace a weaker acid or base from its salts, and their ability to act as catalysts, are often possessed by non-protonic substances. For example, pyridine can be titrated with boron trichloride in an inert solvent such as chlorobenzene with crystal violet as an indicator:

$$C_5H_5N + BCl_3 \rightarrow C_5H_5N \colon BCl_3 \tag{18}$$

Lewis suggested that in such a reaction pyridine is behaving as a base and that boron trichloride may be regarded as an acid. He proposed that in general an acid may be defined as an electron-pair acceptor and a base as an electron-pair donor. Neutralization is then the formation of a covalent coordinate bond between the acid and the base, as for instance in the reactions:

$$R_3N + BCl_3 \rightarrow R_3N \colon BCl_3 \tag{19}$$

$$2H_3N + Ag^+ \rightarrow Ag(NH_3)_2{}^+ \tag{20}$$

and the product of the reaction is termed an acid-base complex.

At first sight this definition does not bear much resemblance to the Brønsted definition. According to Lewis an acid-base reaction or a neutralization is the formation of a donor-acceptor bond, whereas in proton acidity there is a transfer of a proton from the acid to the base. In a typical proton-transfer reaction there is a competition for the proton by two bases, for instance A^- and B in the reaction:

$$HA + B \rightleftharpoons A^- + HB^+ \tag{21}$$

However, there is, in fact, a strictly analogous competition between two bases (donor molecules) X and Y for a Lewis acid, e.g.,

$$AlCl_3 \colon X + Y \rightleftharpoons AlCl_3 \colon Y + X \tag{22}$$

Thus it may be seen that the proton is simply one special example of a Lewis acid or acceptor molecule.

Much less quantitative work has been done on Lewis acid-base reactions than on proton transfer reactions, and there are good reasons for this. In a given solvent all proton acids give rise to the same species, the solvated proton; a relative scale of strengths of acid may be drawn up in terms of the concentration of the solvated proton formed by the ionization of a series of acids, and this scale is independent of the nature of the bases with which the acids might react in this solvent. A similar scale of base strengths can be established. This is, however, not the case for Lewis acids and bases, as the strengths of a series of Lewis acids may depend on the particular base chosen as a basis for comparison. There have, nevertheless, been some attempts to establish scales of acidity and basicity for Lewis acids and bases. For example, Gutman has defined a *donor number* for a large number of solvents as a measure of basicity towards the Lewis acid $SbCl_5$ [9]. This donor number is obtained from the enthalpy change for the reaction of a basic solvent with $SbCl_5$. Some examples are given in Table 2. However, there is no reason to suppose that this is a unique order and it must, therefore, be of limited usefulness.

Table 2

Donor number DN_{SbCl_5} of certain solvents
Data from Reference 8

Solvent	DN_{SbCl_5}
Sulphuryl chloride	0.1
Thionyl chloride	0.4
Acetyl chloride	0.7
Nitromethane	2.7
Nitrobenzene	4.4
Acetic anhydride	10.5
Selenium oxychloride	12.2
Acetonitrile	14.1
Sulpholane	14.8
Ethylene carbonate	16.4
Acetone	17.0
Ethyl acetate	17.1
Water	18.0
Diethyl ether	19.2
Tetrahydrofuran	20.0
Dimethylformamide	26.6
Pyridine	33.1

1.5 Comparison of the Lewis and Brønsted definitions

It should be emphasized at the outset that it is not intended to demonstrate that one definition is better or more correct than the other. Nevertheless, it is the author's opinion that there would be advantages to restricting the use of the terms acid and base to the Brønsted definition, and to using the terms acceptor and donor or electron-pair acceptor and electron-pair donor for Lewis acids and bases. However, it seems that 'Lewis acid' is now deeply entrenched in chemistry and is used by most chemists.

The reasons for suggesting that it would be preferable to restrict the terms acid and base to the Brønsted definitions are that proton transfer reactions are, to a very large extent, rapid and reversible, and quantitative scales of acid and base strengths can be set up for any given solvent. Moreover, many of the common inorganic solvents such as water, ammonia, alcohols and sulphuric acid exhibit Brønsted acid-base behaviour. On the other hand Lewis acid-base reactions are often not rapid or reversible; no general acidity or basicity scales can be set up for Lewis acids and bases; and the majority of Lewis acids react with protonic solvents such as water and NH_3 to increase the concentration of solvated protons, i.e., they are effectively converted to Brønsted acids, and thus their properties as Lewis acids cannot be studied in these important solvents.

Acceptor molecules or Lewis acids are, nevertheless, important in a discussion of Brønsted acids. Indeed, in some cases the addition of a Lewis acid to a solvent is the only known way in which the concentration of the solvated proton can be increased, i.e., in which an acid solution in the Brønsted sense can be obtained. The Lewis acid disturbs the autoprotolysis equilibrium of the protonic solvent by combining with the characteristic anion, thus diminishing its concentration and accordingly raising the concentration of the characteristic cation; for instance:

$$2HS + SbF_5 \rightleftharpoons SbF_5S^- + H_2S^+ \tag{23}$$

A common example of this is the increased acidity produced by the addition of a number of metal ions to water, e.g., Al^{3+} and Zn^{2+}, because of the formation of species such as $ZnOH(H_2O)_3{}^+$, which diminishes the OH^- concentration and thus increases the H_3O^+ concentration. An interesting example is provided by the solvent HSO_3F which is so weakly basic that there are no known simple protonic acids HA that are capable of increasing the concentration of $H_2SO_3F^+$. However, the Lewis acid SbF_5 can do this by combining with the solvent anion SO_3F^-:

$$SbF_5 + 2HSO_3F \rightleftharpoons H_2SO_3F^+ + SbF_5SO_3F^-$$

Various oxides also behave as Lewis acids in that they exhibit a considerable affinity for water or the OH^- ion. In this sense they behave as Lewis acids and they accordingly enhance the concentration of H_3O^+ in water, e.g.:

$$SO_3 + OH^- \rightarrow HSO_4{}^- \tag{24}$$

This is equivalent to

$$SO_3 + 2H_2O \rightarrow H_3O^+ + HSO_4{}^- \tag{25}$$

The free acid H_2SO_4 is not formed in the presence of excess water as it is a strong acid. It is interesting to note, however, that even when the ionization is incomplete, i.e. when the oxide behaves as a weak acid, there is in some cases little or no evidence for the formation of the free un-ionized acid, for instance in the reactions:

$$CO_2 + H_2O \rightleftharpoons H_3O^+ + HCO_3^- \qquad (26)$$

$$SO_2 + H_2O \rightleftharpoons H_3O^+ + HSO_3^- \qquad (27)$$

Lewis acids behave similarly in many other solvents. Thus SO_3 gives H_2SO_4 in water, $H_2S_2O_7$ in H_2SO_4 and HSO_3F in HF. In HF various fluorides such as AsF_5 and SbF_5 give acid solutions, for example:

$$SbF_5 + 2HF \rightarrow SbF_6^- + H_2F^+ \qquad (28)$$

Boric acid ionizes in water in a similar fashion:

$$B(OH)_3 + OH^- \rightarrow B(OH)_4^- \qquad (29)$$

the overall reaction being:

$$B(OH)_3 + 2H_2O \rightarrow B(OH)_4^- + H_3O^+ \qquad (30)$$

Although in this case, and in the analogous reaction of BF_3 and SbF_5 with HF, the formation of an intermediate Brønsted acid is in principle possible, there is no evidence for the existence of such species.

$$B(OH)_3 + 2H_2O \rightleftharpoons [B(OH)_3OH_2 + H_2O] \rightleftharpoons B(OH)_4^- + H_3O^+$$
$$(31)$$

$$BF_3 + 2HF \rightleftharpoons [HBF_4 + HF] \rightleftharpoons BF_4^- + H_2F^+ \qquad (32)$$

The existence of acids such as $HB(OH)_4$, HBF_4 and $HSbF_6$ in the undissociated form must remain an open question for the present. It may be noted, however, that the frequently-cited objection that these acids are unlikely to exist because they contain 2-coordinated fluorine is not valid, because 2-coordinated fluorine is known in H_2F^+ and in many fluorine-bridged species such as $Sb_2F_{11}^-$.

1.6 The solvent-system definition

Since acceptor molecules such as $B(OH)_3$ and SbF_5 can enhance the acidity of water and HF respectively, although they are not simple Brønsted acids, there is considerable merit in defining an acid as any substance which increases the concentration of the characteristic cation of a solvent, and a base as any species which increases the concentration of the characteristic solvent anion. These are the solvent-system definitions of acid and base, and are very useful for amphiprotic and related solvents.

An analogous situation is encountered with bases such as triphenyl carbinol and nitric acid, which ionize in sulphuric acid in the following manner:

$$(C_6H_5)_3OH + H_2SO_4 \rightarrow (C_6H_5)_3C^+ + H_3O^+ + 2HSO_4^- \quad (33)$$

$$HNO_3 + 2H_2SO_4 \rightarrow NO_2^+ + H_3O^+ + 2HSO_4^- \quad (34)$$

These substances cannot be regarded as Brønsted bases in the normal sense of simple proton acceptors, since there is no evidence for the existence, under the experimental conditions in which the above ionizations are observed, of the conjugate acids $(C_6H_5)_3COH_2^+$ and $H_2NO_3^+$. It might seem desirable to have a special name for such bases, but no reasonable name has been suggested. If, however, one accepts the solvent-system definition of a base as a species that increases the concentration of the characteristic anion of a solvent, no new name is needed.

The solvent-system definition is also particularly useful for extending the concepts of acid and base to non-protonic non-aqueous systems. Many such systems are considered to undergo a self-ionization reaction analogous to autoprotolysis but involving the transfer of an anion such as F^-, Cl^-, O^{2-}, etc., rather than the transfer of a proton. In a few cases conductivity and other measurements provide substantial support for the existence of the postulated equilibrium, while in other cases there is no evidence at all for such a self-dissociation equilibrium. Nevertheless, the assumption that the equilibrium exists in a given solvent has proved useful for systematizing the chemistry in that solvent.

In the case of BrF_3 there is indeed good experimental evidence for the self-ionization [10]:

$$2BrF_3 \rightleftharpoons BrF_2^+ + BrF_4^- \quad (35)$$

This is a fluoride-ion transfer reaction and the ions BrF_2^+ and BrF_4^- are the characteristic solvent cation and anion respectively. Thus SbF_5 is regarded as an acid of this system, since it ionizes in the following manner:

$$SbF_5 + BrF_3 \rightarrow BrF_2^+ + SbF_6^- \quad (36)$$

while F^- is a base, since it produces the BrF_4^- anion:

$$BrF_3 + F^- \rightarrow BrF_4^- \quad (37)$$

There are a number of solvents in which chloride-ion transfer is believed to be important, e.g. $POCl_3$, $SeOCl_2$, $SOCl_2$, $NOCl$ and $COCl_2$. A self-ionization involving such a chloride ion transfer appears to be reasonably well established in the case of $SeOCl_2$:

$$2SeOCl_2 \rightleftharpoons SeOCl^+ + SeOCl_3^- \tag{38}$$

with pK (self-dissociation) = 9.7 [11]. Acids of this system are substances such as $SbCl_5$ which give rise to the $SeOCl^+$ cation:

$$SeOCl_2 + SbCl_5 \rightarrow SeOCl^+ + SbCl_6^- \tag{39}$$

and bases are substances such as pyridine which give rise to the $SeOCl_3^-$ ion:

$$C_5H_5N + 2SeOCl_2 = C_5H_5N:SeOCl^+ + SeOCl_3^- \tag{40}$$

Acidity in these systems may be measured in terms of the concentration of the solvated chloride ion $SeOCl_3^-$. One may define a pCl^- scale by analogy with the pH scale, so that:

$$pCl^- \simeq -\log [SeOCl_3^-]$$

Considering a range of concentrations of self-dissociation ions between $[SeOCl_3^-] = 1$ M and $[SeOCl^+] = 1$ M, we see that the acidity scale extends over approximately 10 pCl^- units, *i.e.* $\Delta pCl^- = pK_{sd} = 10$. Figure 2 depicts the range of acidity and basicity available in the solvent $SeOCl_2$ and lists a number of representative acid-base pairs.

By a further extension of the solvent-system concept, the definitions of acid and base may be extended to molten salts. Thus the tetra-

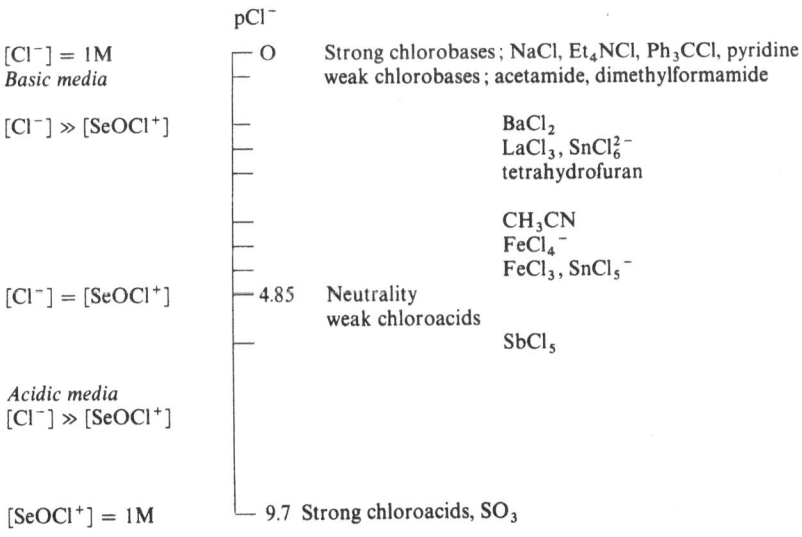

Figure 2. pCl^- scale for the solvent $SeOCl_2$. (Adapted from B. Tremillon, *La Chimie en Solvants Non-aqueux*, Presses Universitaires de France (1971).)

chloraluminate ion in molten $NaAlCl_4$ undergoes the following self-dissociation:

$$2AlCl_4^- = Al_2Cl_7^- + Cl^- \qquad (41)$$

Here $Al_2Cl_7^-$ is the characteristic acidic ion and Cl^- the characteristic basic ion. This is again the chloride-ion transfer reaction. Addition of $AlCl_3$ increases the acidity of the solution while the addition of Cl^- increases the basicity. In molten sulphates an analogous equilibrium is set up:

$$2SO_4^{2-} = S_2O_7^{2-} + O^{2-} \qquad (42)$$

Here the characteristic acid ion is $S_2O_7^{2-}$ and the basic ion is O^{2-}

1.7 Hard and soft acids and bases

One of the important differences between Brønsted acids and Lewis acids that has been referred to above is the apparent impossibility of giving any unique order for the strengths of Lewis acids, as may be done for a series of proton acids in a suitable solvent such as water. For example, it is well known that the proton is a much stronger acid towards the hydroxide ion than is the silver ion, but towards ammonia the silver ion is a much stronger acid than the proton. It would seem that any attempt to classify Lewis acids and bases according to a single order of relative strength is doomed to failure. Nevertheless a certain amount of progress has been made during the last ten years by the development of the idea of hard and soft acids and bases [12]. These terms describe two classes of acids and bases which exhibit contrasting behaviour and within one class there is perhaps a more consistent pattern of acid-base behaviour. Many chemists appear to have found the hard and soft acid-base concept a useful one, although it has engendered considerable discussion and controversy. The names *hard* and *soft* were coined by Pearson [13] following a suggestion by Busch, as descriptions for two classes of acids which had been first recognized in the special case of metal ions by Ahrland, Chatt and Davies [14]. These authors had pointed out that many metals form much more stable complexes with a ligand containing a nitrogen atom than with the corresponding ligand in which nitrogen is replaced by phosphorus or one of the other elements of Group V. These metals were called class (a) metals. On the other hand many metals form much more stable complexes with ligands containing P, As, etc. rather than with the corresponding ligand with nitrogen; these were called class (b) metals. Similar consistent behaviour was found for ligands containing the elements of groups VI and VII, *i.e.* a class (a) metal is consistently found to form more stable complexes with ligands containing N, O or F as the donor atom than with the corresponding ligands containing the heavier elements of these groups, whereas for class (b) metals the converse holds. Pearson [13] extended this idea into a general classification of acids and bases as class (a) or *hard* and class (b) or *soft*. The distinguishing properties of a hard acid are that it has a small size, a high positive charge, and no valence electrons that are easily removed or distorted, *i.e.* it has a low polarizability. A soft acid is one which has an acceptor atom of large size, small or zero positive charge, and several valence electrons that are easily distorted or removed, *i.e.* it has a high polarizability. A soft base is similarly defined as one in which the valence electrons are easily distorted or removed, *i.e.* are polarizable. A hard base has the opposite properties, *i.e.* it is one that holds on to its valence electrons rather strongly. Examples are given in Table 3.

The usefulness of these definitions is that there is a general principle governing the interactions of hard and soft acids and bases, namely that hard acids prefer to coordinate with hard bases and soft acids prefer to coordinate with soft bases. Clearly these are qualitative definitions and a qualitative principle, and there are many intermediate cases and exceptions. Nevertheless the ideas are useful in that they bring a certain limited order into the otherwise varied and rather incomprehensible behaviour of Lewis acids and bases.

Many examples of the usefulness of the concept could be quoted. It is interesting to note, for example, that many years ago Berzelius pointed out that some metals occur naturally as oxide and carbonate ores while others occur as sulphides. We see now that the metal ions that are hard acids such as magnesium, aluminium and calcium occur in combination with the hard bases carbonate and oxide, while the soft metal ions such as copper, mercury and lead occur in combination with the soft base sulphide. The hard acid BF_3 forms a stable complex with the hard base F^- but not with the soft base CO, whereas the soft acid BH_3 forms a strong complex with CO. It may also be noted that the hard acid *par excellence* is, of course, the proton H^+ and there is no evidence, even in the strongest superacid medium known [15], that the proton forms a stable complex with the soft base CO, although the latter forms many stable complexes with soft metal ions.

There has been much discussion of the theory underlying the hard and soft acid-base concept. It is clear that a number of factors contribute to hardness and softness, and their relative importance is

Table 3

Classification of Lewis acids and bases according to the hard and soft acid-base concept

---

Acids

Hard	Soft
H^+, Li^+, Na^+, K^+	Cu^+, Ag^+, Au^+, Tl^+, Hg^+
Be^{2+}, Mg^{2+}, Ca^{2+}, Sr^{2+}, Mn^{2+}	Pd^{2+}, Cd^{2+}, Pt^{2+}, Hg^{2+}, CH_3Hg^+,
	$Co(CN)_5^{2-}$, Pt^{4+}, Te^{4+}
Al^{3+}, Sc^{3+}, Ga^{3+}, $In3^+$, La^{3+}	Tl^{3+}, $Tl(CH_3)_3$, BH_3, $Ga(CH_3)_3$
N^{3+}, Cl^{3+}, Gd^{3+}, Lu^{3+}	$GaCl_3$, GaI_3, $InCl_3$
Cr^{3+}, Co^{3+}, Fe^{3+}, As^{3+}, CH_3Sn^{3+}	RS^+, RSe^+, RTe^+
$Si4^+$, Ti^{4+}, Zr^{4+}, Th^{4+}, U^{4+}	I^+, Br^+, HO^+, RO^+
Pu^{4+}, Ce^{3+}, Hf^{4+}, WO^{4+}, Sn^{4+}	
UO_2^{2+} $(CH_3)_2Sn^{2+}$, VO^{2+}, MoO^{3+}	I_2, Br_2, ICN, etc.
$BeMe_2$, BF_3, $B(OR)_3$	trinitrobenzene, etc.
$Al(CH_3)_3$, $AlCl_3$, AlH_3	chloranil, quinones, etc.
RPO_2^+, $ROPO_2^+$	tetracyanothylene, etc.
RSO_2^+, $ROSO_3^+$, SO_3	O, Cl, Br, I, N, Ro^-, Ro_2^-
I^{7+}, I^{5+}, Cl^{7+}, Cr^{6+}	$M°$ (metal atoms)
RCO^+, CO_2, NC^+	bulk metals
HX (hydrogen bonding molecules)	CH_2, carbenes

Borderline

Fe^{2+}, Co^{2+}, Ni^{2+}, Cu^{2+}, Zn^{2+}, Pb^{2+}, Sn^{2+}, Sb^{3+}, Bi^{3+}, Rh^{3+}, Ir^{3+}, $B(CH_3)_3$, SO_2, NO^+, Ru^{2+}, Os^{2+}, R_3C^+, $C_6H_5^+$, GaH_3

Bases

Hard	Soft*
H_2O, OH^-, F^-	R_2S, RSH, RS^-
$CH_3CO_2^-$, PO_4^{3-}, SO_4^{2-}	I^-, SCN^-, $S_2O_3^{2-}$
Cl^-, CO_3^{2-}, ClO_4^-, NO_3^-	R_3P, R_3As, $(RO)_3P$
ROH, RO^-, R_2O	CN^-, RNC, CO
NH_3, RNH_2, N_2H_4	C_2H_4, C_6H_6
	H^-, R^-

Borderline

$C_6H_5NH_2$, C_5H_5N, N_3^-, Br^-, NO_2^-, SO_3^{2-}

---

* The symbol R stands for an alkyl or aryl group.

not entirely clear and certainly varies from one case to another. Among factors that have been considered are the following. (1) The ionic or covalent nature of the acid-base interactions; hard-acid hard-base interactions may be thought of as being primarily ionic, while soft-acid soft-base interactions are thought of as being primarily covalent. Such an idea was put forward a long time ago to account for the difference in properties of NaCl and AgCl, for example. (2) Soft-acid soft-base interaction is often enhanced by π-bond formation or back-donation of loosely-held electrons of the acid to empty d orbitals or π^*-molecular orbitals on the base. (3) Soft-acid soft-base interactions may be stabilized relative to hard-acid hard-base interactions by the van der Waals interactions that may be significant between the large atoms of typical soft acids and bases. We are clearly far from a good understanding of the concept of hardness and softness, although this does not detract from the usefulness of the concept as a method of classification. One should, however, be alert to the danger that naming a phenomenon inhibits further creative thought about it; a name has a tendency to become accepted as an explanation. It is quite possible that, as we do begin to understand the phenomena that are at present described as hardness and softness, the need for these special terms will disappear.

It seems fitting to conclude this discussion with a slight paraphrase of the concluding paragraphs of R. P. Bell's classic book *Acids and Bases* [1]. It seems preferable to confine the term *acid* to proton acids and to denote Lewis acids as *electron-pair acceptors*, which explicitly describes their functions. This usage does not automatically imply the qualitative resemblances stressed by G. N. Lewis, but it is quite natural that proton-donors and electron-acceptors should often produce similar effects. The nomenclature of bases offers less difficulty, since a proton-acceptor can always donate an electron-pair to Lewis acids: however, it might be advisable to use the term *base* only in contexts involving the transfer of a proton, since it is only here that the correspondence of acid-base pairs and the quantitative aspects of acid-base strength are applicable. In other contexts the term *electron-pair donor* is more appropriate.

In conclusion it should be stressed that these questions of nomenclature are concerned only with convenience and consistency, and not with any fundamental differences in interpretation. It is, therefore, misleading to attach much scientific importance to controversies about acid-base definitions. The chief importance of new definitions lies in their stimulating effect on experimental work. Just as the Brønsted–Lowry definition initiated many investigations of acid-base equilibria and kinetics in different solvents, so the Lewis definition has led to much valuable work on the reactions of acceptor molecules, which will retain its importance even if it is not found convenient to describe them as acids.

1.8 Superacids

In recent years there has been a very considerable increase in our knowledge of very highly acidic protonic systems. Concentrated aqueous solutions of acids such as sulphuric acid and hydrochloric acid have been known as highly acidic media since the early days of chemistry, but today we have available a number of systems with acidities some 10^6 to 10^{10} times as great as that of concentrated (98 per cent) H_2SO_4 or even 100 per cent H_2SO_4. It has proved convenient to call these systems, which have acidities greater than that of 100 per cent H_2SO_4, *superacids*. These superacid systems are in general non-aqueous, since the acidity of any aqueous system is limited by the fact that the strongest possible acid is the hydronium ion H_3O^+. The interest in such superacid systems arises not only from the important but nevertheless rather academic problem of trying to prepare the most highly proton-donating medium possible, but also from the new areas of chemistry which have been opened up as a consequence of the use of these media for carrying out reactions.

Before describing some of these systems, it will be convenient to discuss how their acidity may be measured on an absolute scale – a problem that has not been entirely satisfactorily solved. We might conveniently describe the variation of the acidity of solutions in H_2SO_4 in terms of the concentration of the characteristic cation or solvated proton $H_3SO_4^+$, *i.e.* in terms of $pH(H_2SO_4)$; this, however, does not give the acidity of 100 per cent H_2SO_4 or of the solutions in H_2SO_4 on any absolute scale. Ideally one would like to be able to continue the scale of $pH(H_2O) = -\log[H_3O^+]$ into the region of 100 per cent H_2SO_4, but clearly this is impossible. This simple relationship implies, among other things, that the intrinsic acidity of the hydrated proton is constant, and this implies that its extent of hydration remains constant. However, this is certainly not the case, since the extent of hydration decreases from $H_3O(H_2O)_3^+$ to H_3O^+ as the acid concentration is increased and finally at low water concentrations the proton is present in the form of undissociated H_2SO_4 and in the vicinity of 100 per cent H_2SO_4 as $H_3SO_4^+$. Even ignoring the changing extent of hydration of H_3O^+, it is clear that the concentration of H_3O^+ must go through a maximum in the region of the molar composition $H_2O:H_2SO_4 = 1:1$, and it diminishes to a very small value in the vicinity of 100 per cent H_2SO_4. Since it is well known that the acidity of sulphuric acid (as measured by its effect on indicators, *i.e.* its ability to protonate weak bases, or its catalytic activity, for example) does not go through a similar minimum but increases steadily from dilute aqueous solution up to 100 per cent H_2SO_4, it is clear that the acidity is not simply measured by the concentration of H_3O^+.

Hammett first proposed that the acidity of such concentrated aqueous acid systems, and indeed in principle any acid system, could be measured by an acidity function defined in the following manner:

$$H_0 = pK_{BH^+} - \log [BH^+]/[B] \qquad (43)$$

where $K_{BH}{}^+$ is the ionization constant, referred to water as the solvent, of the conjugate acid BH^+ of a suitable basic indicator B, and $[BH^+]/[B]$ is the ionization ratio, which can usually be conveniently measured spectroscopically. Values of H_0 were first obtained by Hammett in the 1930's for the $H_2SO_4-H_2O$ system, using a series of primary anilines as indicators. By starting measurements in a sufficiently dilute aqueous solution, the pK of the strongest base of the series was determined by conventional methods. The pK's of the series of successively weaker primary aniline bases were then determined by measuring the ionization ratios of two indicators at the same solvent composition and making use of the relation:

$$pK_{B_1H^+} - pK_{B_2H^+} = \log [B_1H^+]/[B_1] - \log [B_2H^+]/[B_2] \quad (44)$$

Hammett's original work has subsequently been corrected and amplified by several groups of workers using the primary anilines as bases.

The basic assumption of the Hammett method is that the activity coefficient ratio f_{BH^+}/f_B is independent of the nature of B at any given solvent composition. This seems to be a reasonably good assumption for a closely related series of bases, such as the primary anilines, but may not be true for bases of different structural types. This has led to the proposal of many different types of acidity function, and to much controversy as to the meaning and usefulness of the Hammett and related acidity functions. This subject has been extensively reviewed elsewhere [16]. Although it now appears that the concept of the acidity function is perhaps not as generally useful as was originally believed, at the present time we do not have any more reliable method of assessing the acidities of superacid media.

The H_0 values for several H_2O-HX systems are shown in Figure 3 [17]. It may be seen that 100 per cent H_2SO_4, with $H_0 = 11.9$, is approximately 10^{12} times as acidic as a 1 molar solution of sulphuric acid in water. It would appear the $HClO_4$ should give the most highly acidic $HX-H_2O$ system, but nobody has yet had the courage to complete these measurements up to 100 per cent $HClO_4$!

With the possible exception of the $HClO_4-H_2O$ system, higher acidities than that of 100 per cent H_2SO_4 can be obtained only in non-aqueous systems, and these may be somewhat arbitrarily

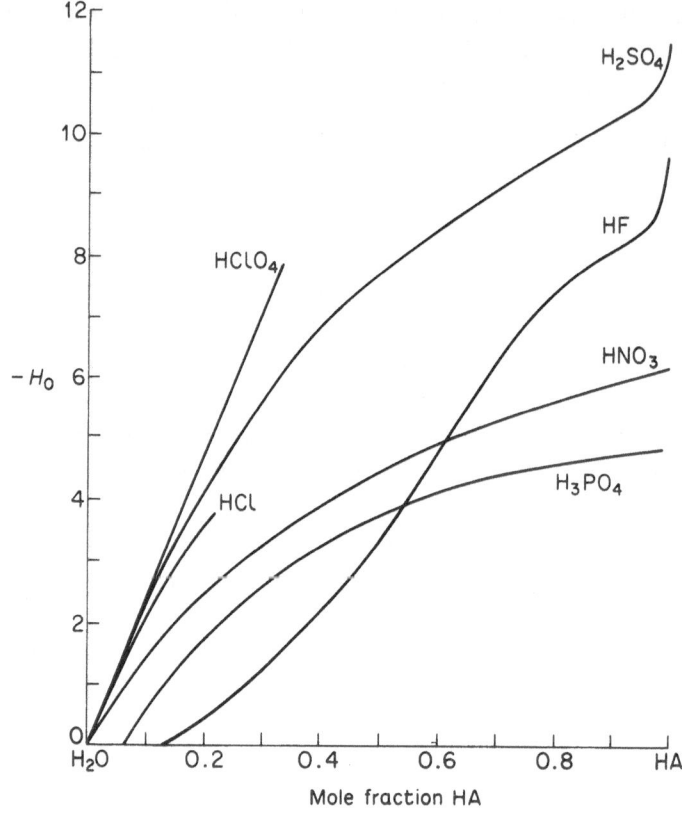

Figure 3. Hammett acidity function (H_0) values for some acid–water systems.

described as superacid systems. In extending the H_0 measurements to superacid media, it was necessary to use another set of bases, namely aromatic nitro-compounds, since the primary anilines used by Hammett and other workers for the aqueous acid systems are all too basic. A satisfactory overlap of these nitro-indicators with the weakest of the Hammett indicators, namely 2,4,6-trinitroaniline, was demonstrated [17], [18]. Nevertheless it might be objected that the values obtained are not true H_0 values. Although this may in fact be true, no other measure of the acidities of these systems is available at the present time, and the results do appear to give a consistent picture of the relative acidities of a number of superacid systems.

1.8.1 *Sulphuric acid*

The acidity of 100 per cent H_2SO_4 can, in principle, be increased by the addition of any solute which increases the $H_3SO_4^+$ concentration.

There are only a very few simple acids HA which are capable of ionizing in sulphuric acid according to the equation:

$$HA + H_2SO_4 \rightleftharpoons H_3SO_4^+ + A^- \tag{45}$$

These include $H_2S_2O_7$ and the higher polysulphuric acids $H_2S_nO_{3n+1}$, HSO_3F and HSO_3Cl.

Sulphuric acid oleums (solutions of SO_3 in H_2SO_4) were probably the first superacid systems to be studied. They have been shown to contain disulphuric acid $H_2S_2O_7$ and higher polysulphuric acids, such as $H_2S_3O_{10}$ and $H_2S_4O_{13}$, which behave as acids of the sulphuric acid system, ionizing according to equations such as:

$$H_2S_2O_7 + H_2SO_4 \rightleftharpoons H_3SO_4^+ + HS_2O_7^- \tag{46}$$

The variation of the acidity function in the H_2SO_4–SO_3 system is shown in Figure 4 [18].

Fluorosulphuric acid and chlorosulphuric acid are weaker than $H_2S_2O_7$ and consequently cause a slower increase in the acidity than is caused by $H_2S_2O_7$ (Figure 4). However, as the composition HSO_3F is approached the value of H_0 rises rapidly, and the acidity of 100 per cent HSO_3F exceeds that of $H_2S_2O_7$. This is to be attributed to the fact that the autoprotolysis constant of HSO_3F is considerably smaller

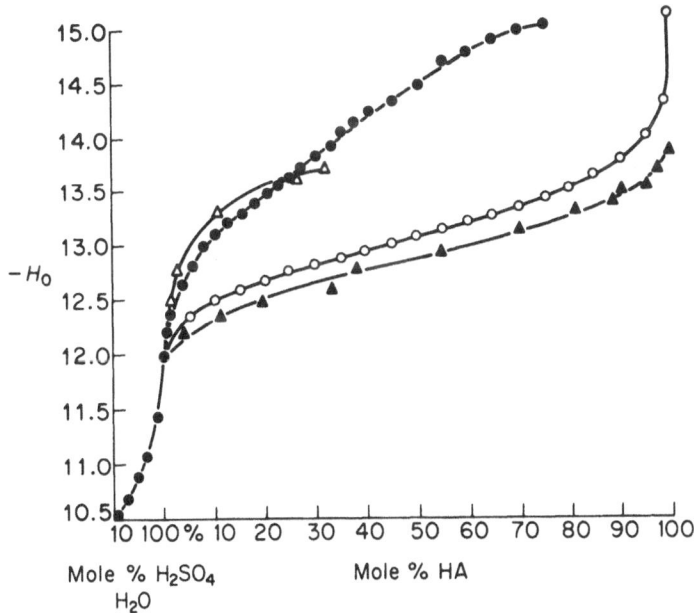

Figure 4. Hammett acidity function (H_0) values for the systems: $\triangle$ H_2SO_4 – $HB(HSO_4)_4$; $\bullet$ H_2SO_4 – $H_2S_2O_7$; $\bigcirc$ H_2SO_4 – HSO_3F; and $\blacktriangle$ H_2SO_4 – HSO_3Cl.

23

than that of $H_2S_2O_7$, and thus the concentration of $H_2SO_3F^+$ rises very rapidly in the vicinity of the composition 100 per cent HSO_3F. It is interesting that boric acid, a weak acid of the aqueous system, gives rise to a strong acid of the sulphuric acid solvent system. When boric acid is dissolved in oleum it is converted into the trihydrogen sulphate of boron:

$$B(OH)_3 + 3SO_3 \rightarrow B(SO_4H)_3 \qquad (47)$$

This is a strong acceptor and it accepts a hydrogen sulphate ion from the solvent:

$$B(HSO_4)_3 + HSO_4^- \rightarrow B(HSO_4)_4^- \qquad (48)$$

Thus the overall reaction of boric acid with oleum may be written:

$$B(OH)_3 + 3SO_3 + 2H_2SO_4 \rightarrow H_3SO_4^+ + B(HSO_4)_4^- \quad (49)$$

The ionization of $B(HSO_4)_3$ in sulphuric acid is extensive and it behaves as an essentially strong acid. Whether or not any un-ionized $B(HSO_4)_3$ is present as the acid $HB(HSO_4)_4$ is an open question. There is similarly no evidence for the existence of the acids $HB(OH)_4$ or HBF_4 in solutions of $B(OH)_3$ in water or BF_3 in HF. As $HB(HSO_4)_4$ is a strong acid, the value of H_0 rises more rapidly than for $H_2S_2O_7$ (Figure 4). However, it is not possible to reach very high values of the acidity, because of the separation of insoluble complex polysulphato-boric acids from concentrated solutions [18].

1.8.2 Fluorosulphuric acid

Pure 100 per cent HSO_3F has a considerably higher acidity than that of 100 per cent H_2SO_4 and this may be further increased by increasing the concentration of $H_2SO_3F^+$. It is not surprising that no simple acids are known which are capable of ionizing according to the equation:

$$HA + HSO_3F \rightarrow H_2SO_3F^+ + A^- \qquad (50)$$

There are, however, a number of acceptor molecules, such as AsF_5 and SbF_5, which can increase the concentration of $H_2SO_3F^+$ by forming a complex with the fluorosulphate anion:

$$MF_5 + SO_3F^- \rightarrow MF_5SO_3F^- \qquad (51)$$

leading to the overall reaction:

$$MF_5 + 2HSO_3F \rightleftharpoons MF_5(SO_3F)^- + H_2SO_3F^+ \qquad (52)$$

Among the various pentafluorides that have been studied, SbF_5 causes the largest and most rapid increase in the value of H_0 (Figure 5), and this is consistent with the results of conductivity measurements which show that it is more extensively ionized than the other penta-fluorides [19–22]. As the concentration rises above 1 per cent SbF_5 the increase in the acidity becomes quite slow, and this is at least in part due to the formation of polymeric species such as $(SbF_5)_2SO_3F^-$:

$$SbF_5(SO_3F)^- + SbF_5 \rightarrow (SbF_5)_2SO_3F^- \qquad (53)$$

in place of the continued formation of $H_2SO_3F^+$.

Among the pentafluorides, SbF_5 is the strongest acceptor known, but still more acidic solutions have been prepared by the addition of SO_3 to solutions of SbF_5 in HSO_3F. The SO_3 reacts with SbF_5 to form fluoride fluorosulphates such as $SbF_4(SO_3F)$, $SbF_3(SO_3F)_2$ and $SbF_2(SO_3F)_3$. These are progressively stronger acids than SbF_5, and $SbF_2(SO_3F)_3$ appears to be a strong acid ionizing essentially completely according to the equation:

$$SbF_2(SO_3F)_3 + 2HSO_3F \rightarrow H_2SO_3F^+ + SbF_2(SO_3F)_4^-$$
$$\qquad (54)$$

For a 7 mole per cent solution of $SbF_2(SO_3F)_3$, which has the highest concentration studied, $-H_0$ has a value of 19.3; this is the highest value of H_0 so far recorded (Figure 5) [19].

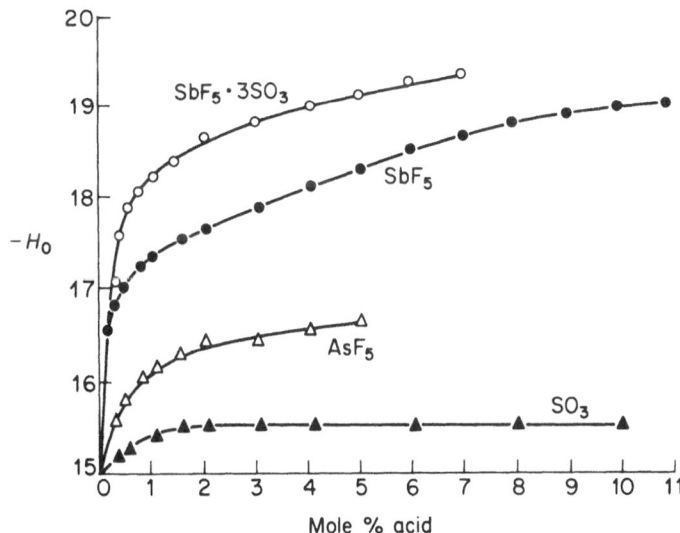

Figure 5. Hammett acidity function (H_0) values for acids of the fluorosulphuric acid solvent system.

1.8.3 *Hydrogen fluoride*

Although hydrogen fluoride has the disadvantage that it cannot be handled in glass apparatus, it has considerable potential as a superacid medium. Because HF has a rather low autoprotolysis constant, the concentration of H_2F^+ is markedly affected by small traces of impurities such as water and other basic substances that give F^-; consequently the H_0 value for the 100 per cent acid has not yet been determined with certainty, but it is approximately -11. Antimony pentafluoride is again an acid of considerable strength, ionizing in solution in HF according to the equations:

$$2HF + SbF_5 \rightarrow H_2F^+ + SbF_6^- \tag{55}$$

$$2HF + 2SbF_5 \rightarrow H_2F^+ + Sb_2F_{11}^- \tag{56}$$

$$2HF + 3SbF_5 \rightarrow H_2F^+ + Sb_3F_{16}^- \text{ etc.} \tag{57}$$

Conductivity and cryoscopic measurements have shown that the extent of ionization of the pentafluorides decreases in the series $SbF_5 > AsF_5 > NbF_5 > PF_5$ [23].

Although the value of H_0 for 100 per cent HF is lower than that for HSO_3F, the lower autoprotolysis constant means that H_0 increases more rapidly on the addition of an acid such as SbF_5; this is illustrated in Figure 6. However, the limited number of H_0 measurements that have been made indicate that H_0 does not reach the values attained in the HSO_3F—SbF_5 and HSO_3F—SbF_5—SO_3 systems, which remain the most acidic known [24].

It is not the purpose of this chapter to discuss all the interesting chemistry that has been carried out in superacid systems, since this has been extensively reviewed elsewhere [21, 22, 25, 26]. It will suffice to mention that these systems have been extensively used for studying the protonation of many organic molecules, for obtaining stable solutions of carbonium ions that had hitherto only been postulated as reactive intermediates, and for preparing many new cationic inorganic species. It is noteworthy perhaps that a 1:1 mixture of SbF_5 and HSO_3F has such remarkable ability to give stable solutions of a wide variety of carbonium ions, even those as intrinsically reactive as $CH(CH_3)_2^+$, that it has become known, particularly among organic chemists, as 'magic acid' [27]. Although this name is perhaps useful in drawing attention to the remarkable and useful properties of this medium for carbonium ion chemistry it would seem unwise to take this as a precedent for naming these systems. Thus although one could somewhat facetiously suggest that the still more highly

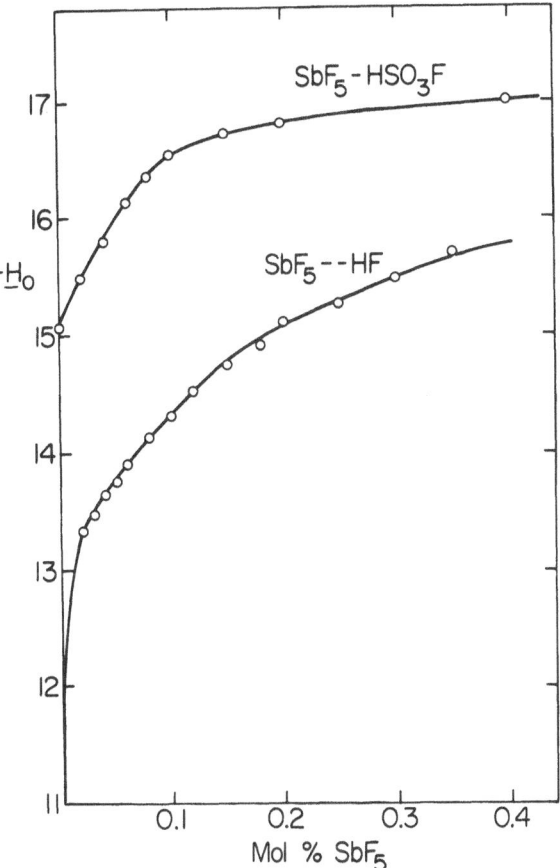

Figure 6. Comparison of H_0 values for the $HSO_3F - SbF_5$ and $HF - SbF_5$ systems.

acidic $SbF_5—SO_3—HSO_3F$ mixture might perhaps be called 'super-natural' acid, further extension of this nomenclature would seem to present difficulties. It would, moreover, be unwise to assume that this system represents the most highly acidic system that could be prepared.

It is interesting to note that even the most highly acidic medium yet prepared, namely $HSO_3F—SbF_5—SO_3$ is incapable of protonating a number of simple molecules that potentially, at least, have basic properties, such as CO, CO_2, SO_2 and xenon, and some of which, *e.g.* CO and SO_2, are known to form stable complexes with certain Lewis acids [14]. Although it might not seem reasonable at first sight to consider methane as a potential base, it is known to form CH_5^+ in the mass spectrometer, and it has been argued that a reported deuterium exchange between CH_4 and DSO_3F proceeds through the formation of CH_4D^+ Nevertheless there is no evidence of the formation of any detectable amount of CH_5^+ in solutions of CH_4 in $HSO_3F—SbF_5—SO_3$. Consequently at the present time all claims that this species is a stable intermediate in reactions in superacid media must be treated with reserve. The most that can be said is that it may be a very unstable intermediate or transition state.

REFERENCES

[1] R. P. Bell, *The Proton in Chemistry*, Chap. 2, 3, 4, Chapman and Hall, London (1973); *Acids and Bases: Their Quantitative Behaviour*, Chapman and Hall, London (1969).

[2] E. C. Franklin, *The Nitrogen Systems of Compounds*, Reinhold, New York, N.Y. (1935).

[3] J. N. Brønsted, *Rec. Trav. Chim.*, **42**, 718 (1923).

[4] R. E. Richards and J. A. S. Smith, *Trans Faraday Soc.*, **47**, 1261 (1951).

[5] Y. K. Yoon and G. B. Carpenter, *Acta. Cryst.*, **12**, 17 (1959); F. S. Lee and G. B. Carpenter, *J. Phys. Chem.*, **63**, 279 (1959); J. O. Lundgren and I. Olovsson, *Acta. Cryst.*, **B23**, 966, 971 (1967); I. Olovsson, *J. Chem. Phys.*, **49**, 1063 (1968); I. Olovsson and J. O. Lundgren, *J. Chem. Phys.*, **49**, 1068 (1968); J. O. Lundgren, *Acta. Cryst.*, **B26**, 1544 (1970); D. Mootz and J. Fayos, *Acta. Cryst.*, **B26**, 2046 (1970); J. Lundgren, *Acta. Cryst.*, **B28**, 475 (1972); J. O. Lundgren and P. Lundin, *Ibid.*, **B28**, 486 (1972); J. Almlof, *Acta. Cryst.*, **B28**, 481 (1972); J. O. Lundgren and J. M. Williams, *J. Chem. Phys.*, **58**, 788 (1973).

[6] A. S. Savoie and P. A. Giguere, *J. Chem. Phys.*, **41**, 2698 (1964); M. Fournier, G. Mascherpa, D. Rosselet, and J. Potier, *C.R. Acad. Sci.*, **269**, 279 (1969); J. Roziere and M. Fournier, *C.R. Acad. Sci.*, **270**, 729 (1970); E. Chemouni, M. Fournier, J. Roziere and J. Potier, *J. Chim. Phys.*, **67**, 517 (1970); A. S. Gilbert and N. Sheppard, *J. Chem. Soc. Chem. Commun.*, 337 (1972).

[7] M. Couzi, J. C. Cornut, and P. V. Huong, *J. Chem. Phys.*, **56**, 427 (1972).

[8] T. Birchall and R. J. Gillespie, *Can. J. Chem.*, **43**, 1045 (1965); M. Brookhart, G. C. Levy, and S. Winstein, *J. Amer. Chem. Soc.*, **89**, 1735 (1967); G. A. Olah and A. N. White, *J. Amer. Chem. Soc.*, **89**, 3591 (1967).

[9] V. Gutmann, *Coord. Chem. Reviews*, **2**, 239 (1967); *Coordination Chemistry in Non-Aqueous Solutions*, Springer-Verlag, Vienna, New York (1968).

[10] A. G. Sharpe, Chapter 7 in *Non-Aqueous Solvent Systems*, Ed. T. C. Waddington, Academic Press, London and New York (1965).

[11] J. Devynik and B. Tremillon, *J. Electroanal. Chem.*, **23**, 241 (1969); **30**, 443 (1971).

[12] *Hard and Soft Acids and Bases*, ed. R. G. Pearson, Dowden Hutchinson and Ross, Strousburg, Pennsylvania (1973).

[13] R. G. Pearson, *J. Amer. Chem. Soc.*, **85**, 3533 (1963).

[14] S. Ahrland, J. Chatt and N. R. Davies, *Quart. Rev. Chem. Soc.* (London), **12**, 265 (1958).

[15] R. J. Gillespie and G. P. Pez, *Inorg. Chem.*, **8**, 1233–1235 (1969).

[16] C. H. Rochester, *Acidity Functions*, Academic Press, London (1970).

[17] R. J. Gillespie and T. E. Peel, *Adv. Phys. Org. Chem.*, **9**, 1 (1971).

[18] R. J. Gillespie, T. E. Peel and E. A. Robinson, *J. Amer. Chem. Soc.*, **93**, 5083 (1971).

[19] R. J. Gillespie and T. E. Peel, *J. Amer. Chem. Soc.*, **95**, 5173 (1973).

[20] R. J. Gillespie, K. Ouchi and G. P. Pez, *Inorg. Chem.*, **8**, 63.

[21] R. J. Gillespie, *Accounts of Chem. Research*, **1**, 202 (1968).

[22] R. J. Gillespie, *Endeavour*, **32**, 3 (1973).

[23] P. A. W. Dean, R. J. Gillespie, R. Hulme and D. A. Humphreys, *J. Chem. Soc.*, 341 (1971).

[24] R. J. Gillespie and J. Liang, unpublished observations.

[25] R. J. Gillespie and J. Passmore, *Adv. Inorg. Chem. and Radiochem.*, **17**, 49 (1975).

[26] R. J. Gillespie and J. Passmore, *MTP International Review of Science, Inorganic Chemistry Series II*, Vol. 3, ed., V. Gutmann, Butterworths, pp. 121–136.

[27] G. Olah, *Chem. in Britain*, 261 (1972).

2

Robert W. Taft

GAS-PHASE PROTON-TRANSFER EQUILIBRIA

2.1 Introduction	2.5 Acid strengths
2.2 Regarding experimental methods	2.6 Ion-molecule attachment
2.3 Base strengths	equilibria
2.4 Hydrogen-atom transfer equilibria between cation radicals and saturated cations	2.7 Summary

2.1 Introduction

Acid-base equilibria have played a predominant role in structural theories of chemistry [1–8]. Yet, with the exception of limited studies in the gas phase with Lewis acids, such as $B(CH_3)_3$ [7, 8], the interpretation of the observations has been clouded by the uncertain role of the solvent. Recent advances in high pressure mass spectrometry [9,10] and pulsed ion cyclotron resonance spectroscopy [11–16] have made possible the study of proton-transfer equilibria in the gas phase with a precision which allows determination of even small effects of changes in molecular structure.

The process represented by equation 1 may be regarded as the simplest of all processes in which an electron-pair bond is formed with conservation of the opposed spins of the electron pair.

$$B:_{(g)} + H^+_{(g)} \rightleftharpoons B:H^+_{(g)} \tag{1}$$

The simplicity follows of course from the small nuclear size of the acid and the presumption that only the single $1s$ orbital is utilized in bonding by hydrogen. The proton-transfer equilibrium represented by equation 2 is therefore the prototype of all Lewis acid-base processes:

$$B_0H^+_{(g)} + B_{(g)} \rightleftharpoons BH^+_{(g)} + B_{0(g)}; \delta_R \Delta G^0_{(g)} = -RT \ln K_{(2)} \tag{2}$$

and gives the structural effect of the general base B relative to the standard of reference B_0. The negative of the standard enthalpy change

31

for equation 1 is called the proton affinity of B. Accordingly, the negative of the standard enthalpy change for equation 2 gives the proton affinity of B relative to B_0. The relative base strength of B compared to the reference base B_0 may be given by the standard free energy change for equation 2, or (by analogy to the convention in aqueous solutions) as units of $\log_{10} K_{(2)}$, defined by equation 3, in which square brackets denote the equilibrium concentrations for equation 2:

$$K_{(2)} = [BH^+][B_0]/[B_0H^+][B] \qquad (3)$$

The role of solvent is evaluated quantitatively by the quantity $K_{(s)}/K_{(g)}$, where $K_{(g)}$ is the equilibrium constant for equation 2 and $K_{(s)}$ is the corresponding equilibrium constant for the proton-transfer equilibrium in dilute solution written formally as:

$$B_0H^+_{(s)} + B_{(s)} \rightleftharpoons BH^+_{(s)} + B_{0(s)}$$

Equation 2 has been arbitrarily written for neutral bases. Extensive studies with both neutral and anionic bases have now been carried out. The convention of representing a neutral proton acceptor as a base B and a neutral proton donor as an acid HA is herein followed. Consequently, the acidity of HA relative to a reference acid HA_0, is given by equation 4:

$$HA_{(g)} + A^-_{0(g)} = A^-_{(g)} + HA_{0(g)}$$
$$K_{(4)} = [A^-][HA_0]/[A_0^-][HA]; \delta_R \Delta G^0 = -RT \ln K_{(4)} \qquad (4)$$

The present chapter is concerned primarily with measured molecular structural effects on reactions 2 and 4 in the gas phase. These have been obtained only very recently from direct equilibrium-constant determinations. Work in this area is still in a very active state, so that this chapter serves as a preliminary progress report. Useful comparison can now be made of structural effects on equation 2 with the following related topics: (1) proton-transfer equilibria in condensed phases; (2) other Lewis acid-base equilibria in the gas phase; (3) theoretical calculations of proton-transfer energetics; (4) hydrogen-atom transfer equilibria between cation radicals and saturated cations; (5) hydrogen-bonded complex formation in hydrocarbon solvents; and (6) gas-phase equilibria for attachment of neutral molecules to cations and anions. Each of these topics is considered at least briefly.

In the companion chapter by Professor E. M. Arnett, detailed consideration is given to the thermodynamics of proton-transfer equilibria in condensed phases, including thermodynamic quantities for the solution of cations from the gas phase. In the present chapter, the standard free energy changes for proton-transfer equilibria in the gas phase are compared with those for the corresponding reactions in water and (where literature data permit) in acetonitrile. The objective of this simple comparison is two-fold. First, the magnitude of the solvent effect problem is quantitatively defined. Second, the comparison appears to succeed in differentiating and evaluating certain of the major structural effects observed in gas-phase proton-transfer equilibria. Similarly, in the presentation of topics (3), (4), (5), and (6) above, no attempt has been made to include or discuss all available data, but rather to provide points of interest and significance which have become available and which illustrate problems needing further study and analysis. Some additional considerations and definitions for the discussion of topics (4), (5), and (6) are as follows.

The standard enthalpy change for the reverse of equation 1 is the heterolytic bond dissociation energy for BH^+ (equal by convention to the proton affinity of B). The homolytic bond dissociation energy for BH^+, *i.e.* the standard enthalpy change for reaction 5, has been obtained [17, 18] for some bases from the proton affinity and the

$$BH_{(g)}^+ \rightleftharpoons B_{(g)}^+ + \cdot H_{(g)} \qquad (5)$$

adiabatic ionization potentials of B and H. Relative values of the homolytic bond dissociation energies of BH^+ are of interest since no change in charge is involved in reaction 6. Instead the only formal change is the electron-unpairing process in this hydrogen-atom transfer equilibrium. Electronic delocalization within the formally

$$B_0 H_{(g)}^+ + \cdot B_{(g)}^+ \rightleftharpoons BH_{(g)}^+ + \cdot B_{0(g)}^+ \qquad (6)$$

unsaturated cations $\cdot B_0^+$ and $\cdot B^+$, relative to their saturated counterparts $B_0 H^+$ and BH^+, apparently provide the principal driving force of this equilibrium.

Gaseous 'solvation' of cations and anions occurs by the attachment of neutral molecules. Although energies of attachment for more than one molecule to a given ion have been measured [9], this chapter will be concerned only with preliminary studies of equilibria involving single attachments [19–22]. Such equilibria probably frequently involve acid-base reactions of the hydrogen-bonding type [22].

33

For present purposes we will distinguish two types of attachment exchange reactions for both cations and anions:

$$A_0^- \cdots HA_{1(g)} + HA_{(g)} \rightleftharpoons A^- \cdots HA_{1(g)} + HA_{0(g)} \qquad (7)$$

$$B_0H^+ \cdots B_{1(g)} + B_{(g)} \rightleftharpoons BH^+ \cdots B_{1(g)} + B_{0(g)} \qquad (8)$$

and:

$$A_1^- \cdots HA_{0(g)} + HA \rightleftharpoons A_1^- \cdots HA_{(g)} + HA_{0(g)} \qquad (9)$$

$$B_1H^+ \cdots B_{0(g)} + B_{(g)} \rightleftharpoons B_1H^+ \cdots B_{(g)} + B_{0(g)} \qquad (10)$$

Equilibria 7 and 8 correspond to the proton-transfer reactions 4 and 2, respectively, for which a single common 'solvent' molecule has been attached to the anions or cations. Equilibria 9 and 10 involve displacement of one ion-'solvating' molecule by another. Equilibria 10 are of interest, for example, to compare with corresponding hydrogen-bonded complexes formed from neutral proton donors in hydrocarbon solvent [23, 25] as in reaction 11

$$B_1H \cdots B_{0(s)} + B_{(s)} \rightleftharpoons B_1H \cdots B_{(s)} + B_{0(s)} \qquad (11)$$

Insight into the role of hydrocarbon solvents and, particularly, of the charge on the proton donor in hydrogen-bonding may be provided by these comparisons.

2.2 Regarding experimental methods

This chapter is concerned with quantitative equilibrium constants obtained for proton transfer equilibria at near room temperatures. There have been numerous reports of qualitative orders of base and acid strengths [10, 17, 26–33, 52, 53] which have been valuable as guide-lines for subsequent quantitative studies as well as for isolating a number of important structural features. The quantitative results are required, however, for analysis and separation of molecular structural effects. The first report of equilibrium constant determinations for gaseous proton-transfer reactions is due to Bowers, Aue, Webb, and McIver [14].

Quantitative studies of gas-phase base strength have been reported from four laboratories. (1) Kebarle and his associates at the University of Alberta utilizes a high pressure (~ 1 Torr) mass-spectrometric method [9, 34, 35] and obtain results at about 600 K (to avoid the higher aggregates that form at these pressures at room temperatures). (2) Bowers, Aue, and their associates at the University of California, Santa Barbara, make use of room-temperature ion cyclotron resonance

Table 1

Gas-phase base strengths by different methods. Standard free energy changes for the reactions:

$$\text{(i)} \quad MeNH_{3(g)}^+ + B_{(g)} \rightleftharpoons BH_{(g)}^+ + MeNH_{2(g)}$$

$$\text{(ii)} \quad MeOH_{2(g)}^+ + B_{(g)} \rightleftharpoons BH_{(g)}^+ + MeOH_{(g)}$$

	$-\delta_R \Delta G_{i(g)}^0/\text{kcal mol}^{-1}$		
B	(a) 600 K	(b) 300 K	(c) 300 K
(i)			
NH_3	-10.8	—	-9.1
$C_6H_5NH_2$	-1.9	—	-2.4
$MeNH_2$	(0.0)	(0.0)	(0.0)
Me_2NH	7.5	6.7	6.4
C_5H_5N	7.8	6.5	6.9
$c\text{-}C_6H_{11}NH_2$	8.7	—	7.2
Me_3N	12.5	11.3	10.9
Et_3N	—	17.8	17.6
$(n\text{-}Pr)_3N$	—	20.2	19.6
(ii)			
MeCHO	3.1	—	2.8
EtOH	4.1	4	4.4
MeCN	5.1	5	4.7
HCO_2Me	5.6	—	5.4
EtCHO	5.7	—	5.7
Me_2O	7.9	7	8.2
HCO_2Et	9.0	—	9.0
Me_2CO	12.3	12	11.7

(a) P. Kebarle *et al.* [34, 35, 93]; method (1).
(b) D. H. Aue *et al.* [16, 44]; method (2).
(c) R. W. Taft, J. L. Beauchamp *et al.* [41–43, 106]; methods (3), (4).

(ICR) pressure plots at moderate pressures [14, 16, 36] ($\sim 10^{-4}$ Torr). (3) Beauchamp and his associates at California Institute of Technology use a low-pressure ($\sim 10^{-6}$ Torr) trapped ICR drift cell method at room temperature [12]. (4) McIver, Taft, and their associates at the University of California, Irvine, utilize pulsed ICR measurements with a trapped-ion cell analyser [11, 13, 15, 18, 37–40] at room temperature and at low pressures ($\sim 10^{-6}$ Torr). Some typical results from the four laboratories for gas-phase amine base strengths relative to methylamine and for oxygen base strengths relative to methanol are summarized in Table 1. While the agreement between the values is not always within the precision measures reported by each group, there nevertheless is agreement between widely different methods.

Table 2

Acid strengths by different methods

Reaction	$-\Delta G^0$/kcal mol^{-1} at 298 K	
	(a)	(b)
Alcohols and Alkynes		
$^-$OEt + i-PrOH $\rightleftharpoons$ EtOH + i-PrO$^-$	1.9 [39, 40]	1.6 [48, 49]
$^-$OMe + HC$\equiv$CH $\rightleftharpoons$ MeOH + HC$_2^-$	5.3 [39, 40]	4.8 [48, 49]
$^-$OEt + MeOH $\rightleftharpoons$ EtOH + MeO$^-$	-2.9 [39, 40]	-3.1 [48, 49]
$^-$SH + HCN $\rightleftharpoons$ CN$^-$ + H$_2$S	1.4 [15]	1.6 [49]
MeO$^-\cdots$HOMe + EtOH $\rightleftharpoons$ MeOH + EtO$^-\cdots$HOMe	1.2 [37]	1.2 [49]

(a) McIver *et al.*; low-pressure ICR.
(b) Bohme *et al.*; flowing afterglow.

The results by the latter two methods have been found repeatedly to give standard free energy changes agreeing to ±0.1–0.2 kcal or better, and are favoured if discrepancies exist. The results of part (ii) of Table 1 provide a number of examples of excellent agreement of values at 300 and 600 K. The agreement indicates that simple gas phase proton-transfer equilibria involve little or no entropy effects (other than for rotational symmetry numbers).

Quantitative determinations of gas-phase acid strengths have been reported from three laboratories: (1) Kebarle and his associates utilizing their high-pressure mass-spectrometric method [9, 45, 47]; (2) Bohme and his associates of York University utilizing a flowing-afterglow method [10, 31, 48, 49] for room-temperature determinations of rate and equilibrium constants; and (3) McIver and his associates utilizing their low-pressure pulsed ICR method [15, 37–40]. Table 2 gives typical results on acid strengths obtained by the low-pressure pulsed ICR and the flowing-afterglow methods. Agreement is within the reported precision measures of ±0.2 kcal mol^{-1}.

There are four general concerns with all methods: (a) have the excited ions relaxed to thermal? (b) is the temperature of the experiment well defined? (c) are the ionic species of equations 2 and 4 rapidly interconverted? (d) are there sampling errors due to instrumental discrimination between ions?

In the low-pressure ICR measurements, double-resonance ejection experiments are conducted which show that the ionic species are very rapidly interconverted compared to the time scale of the equilibrium constant determinations. In the latter, ions are in fact 'trapped' in the cell for periods of time as long as are required to achieve constant abundances of the two ionic species in the proton-transfer equilibrium. In typical experiments at 10^{-6} Torr, an equilibrium concentration ratio of the ions appears to prevail within 50 ion-molecule collisions per ion, and this concentration ratio is observed up to and beyond 1000 collisions per ion, so confirming the equilibrium condition. That is, ion intensities are monitored over cell residence times of 200–1000 ms, a time scale which exceeds those of ordinary mass spectrometric operations by a factor of about 10^5. Consequently, residence times are such as to allow the initial excited ions to relax to thermal either by radiative emission or by collisions with neutral bases which are present in great excess.

The low-pressure ICR results have been successfully tested by four classical means of establishing that the equilibrium condition prevails. (1) The equilibrium constants hold (to the precision indicated) for substantial variations in the ratio of the pressures of the neutral acid or base molecules; in effect, the steady state is shown to be an equilibrium one by obtaining the same result from either side. (2) The equilibrium constant is independent of the total pressure. (3) The same value for the equilibrium constant is obtained from the ratio of the forward and reverse rate constants gotten by selective ion-ejection experiments [15]. (4) For a series of acids and bases, the standard free energy change is found to be dependent only upon the initial and final states, with no dependence on route (for instance, the value for $1 \rightarrow 3$ is the same as for $1 \rightarrow 2 \rightarrow 3$). This is illustrated for base strengths in Figure 1, which shows results for typical stair-step overlapping sequence. It should be noted that to maintain precision, intervals in the standard free energy change are always < 3.0 kcal mol^{-1}. Consequently, in none of the proton-transfer equilibria measured is there sufficient reaction exothermicity to inhibit the thermalized condition. Finally, the excellent agreement generally found between the low-pressure ICR measurements and the results by the other methods offer compelling evidence for the attainment of thermal equilibrium conditions.

All methods identify ions by mass, not by structure. Consequently it has been necessary throughout to make reasonable assumptions about the structures of the ions. The assumed structures are supported by several indirect experimental results (such as hydrogen-deuterium exchange, or structural effects such as those discussed in Section 2.3.3) and by theoretical calculations where they have been applied (*cf.* Section 2.3.6). No one line of evidence in itself is necessarily compelling, but at the present time there is no evidence to suggest general errors in the gross structural assumptions.

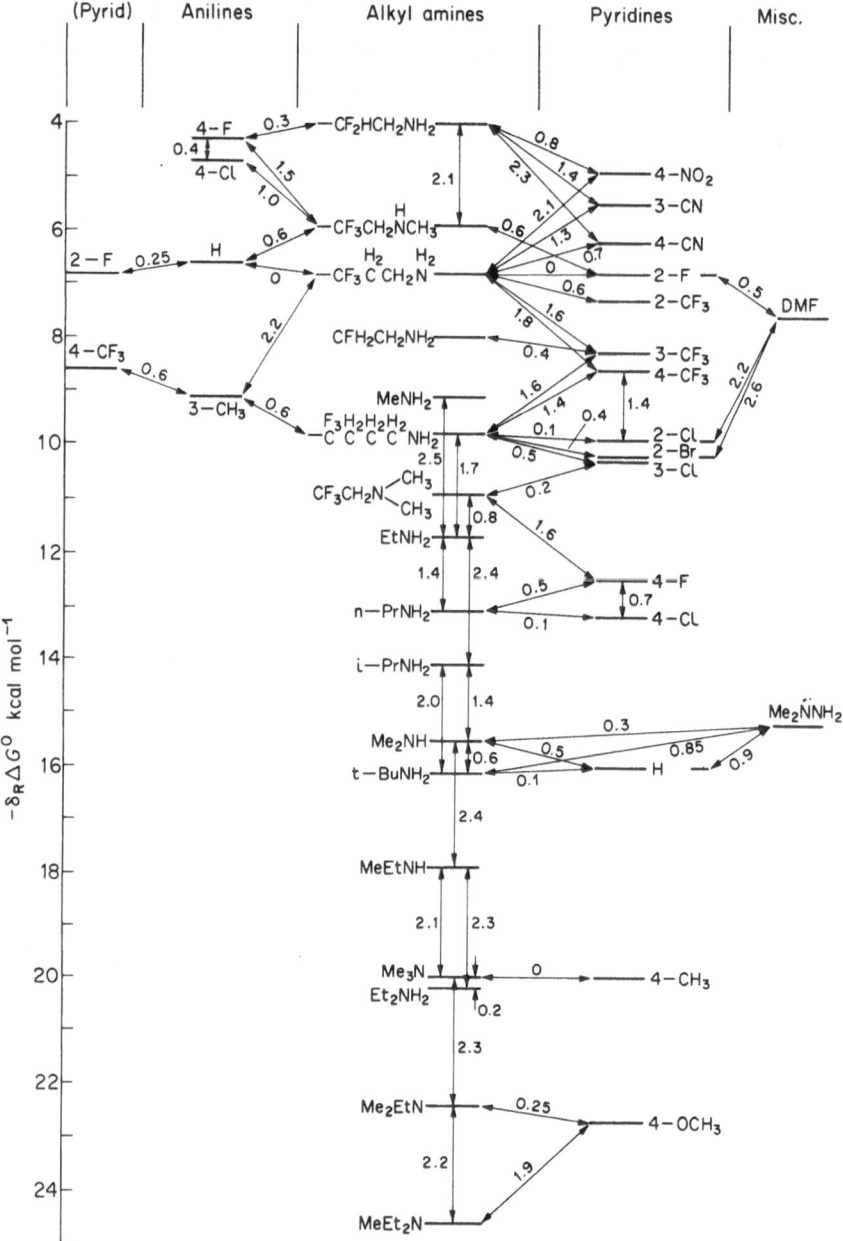

Figure 1. Gas-phase proton-transfer reactions; base strengths relative to ammonia, for typical step sequences in standard free energy changes, at 298 K.

ordinate $= -\delta_R \Delta G^0_{i(g)}$, kcal mol^{-1}

abscissa = none

2.3 Base strengths

2.3.1 Basic organic functional groups with hydrogen and hydrocarbon substituents

The simple hydrides are known [17, 29] to increase in proton affinity in the sequences: $CH_4 < NH_3 > OH_2 > HF$; $PH_3 > H_2S > HCl$; $NH_3 > PH_3 > AsH_3$; $H_2S > H_2O$. However, quantitative proton-transfer equilibrium constants connecting these hydrides have not yet been obtained. A similar situation prevails for organic functional groups, for instance $H_2CO < HCO_2Me < Me_2CO < HCO \cdot NH_2$ [32, 33]; and $MeCN < MeCH\colon NH < MeCH_2 \cdot NH_2$ [44]. Nonetheless, these sequences provide only crude bench marks predicting base strength, since even relatively simple substituents introduce such large effects that these orders are generally inapplicable. For example, although PH_3 is substantially less basic than NH_3 [17, 106], PMe_3 is more basic than NMe_3 in gaseous proton-transfer equilibria [28, 50].

Consequently, it is useful to tabulate gas-phase base strengths obtained from proton-transfer equilibria for organic functional groups with comparable hydrogen and hydrocarbon substituents. Table 3 gives a current summary, taking NH_3 as the reference base, i.e. equation 2 is utilized in the form:

$$NH_{4(g)}^{+} + B_{(g)} \rightleftharpoons NH_{3(g)} + BH_{(g)}^{+}$$

The standard free energy change for each base at 298 K is listed as $-\delta_R \Delta G_{i(g)}^0/\text{kcal mol}^{-1}$, so that base strength increases with the value of this quantity. Also given in Table 3 are corresponding values for the proton-transfer equilibria in water and in acetonitrile, i.e. $-\delta_R \Delta G_{i(H_2O)}^0$ and $-\delta_R \Delta G_{i(MeCN)}^0$ respectively. Finally, the last column in Table 3 gives values for the transfer of a p-fluorophenol molecule between hydrogen-bonded complexes in CCl_4 at 298 K, i.e. equation 8 is used in the form:

$$p\text{-FC}_6\text{H}_4\text{OH} \cdots \text{NH}_{3(s)} + B_{(s)} \rightleftharpoons p\text{-FC}_6\text{H}_4\text{OH} \cdots B_{(s)} + \text{NH}_{3(s)};$$

$$\Delta G^0 = -\delta_R \Delta G_{HB}^0$$

An estimated value of $31\ \text{M}^{-1}$ for the formation constant of the hydrogen-bonded complex of $p\text{-FC}_6\text{H}_4\text{OH}$ with NH_3 in CCl_4 at 25°C has been used to obtain the listed values from the pK_{HB} values of References [23–25].

Table 3

Base strengths of organic functional groups with hydrogen and hydrocarbon substituents

$$NH_4^+ + B \rightleftharpoons NH_3 + BH^+$$

Values of $\delta_R \Delta G^0$ in kcal mol^{-1} at 298 K

Increasing positive value indicates greater base strength

B	Gas phase		Aqueous solution		Acetonitrile solution		H-bonding in CCl$_4$	
	$-\delta_R \Delta G^0_{(ig)}$	Ref.	$-\delta_R \Delta G^0_{(H_2O)}$	Ref.	$-\delta_R \Delta G^0_{(MeCN)}$	Ref.	$-\delta_R \Delta G^0_{HB}$	Ref.
Alcohols								
MeOH	−18.2	[106]						
EtOH	−13.7	[106]	(−15.3)	[63]				
Aldehydes								
MeCHO	−15.4	[106]						
EtCHO	−12.5	[106]						
n-PrCHO	−10.7	[106]						
Azo-compound								
Me$_2$N$_2$	3.6	[31]					−0.03	[23–25]
Primary aliphatic amines								
HC≡C—CH$_2$NH$_2$	6.7	[54]	−1.50	[64]				
MeNH$_2$	9.1	[42]	1.92	[64]	2.60	[78]		
H$_2$C=CH—CH$_2$NH$_2$	11.3	[54]	0.38	[64]				
EtNH$_2$	11.8	[42]	1.96	[64]				
n-PrNH$_2$	13.0	[54]	1.81	[64]	2.64	[78]		
i-PrNH$_2$	14.1	[42]	1.94	[64]	2.39	[78]		

n-BuNH₂	13.5	[16]	1.90	[64]	2.45	[78]	0.83	[23–25]
i-BuNH₂	14.0	[44]	1.66	[64]	1.99	[78]		
s-BuNH₂	15.2	[44]	1.80	[64]				
t-BuNH₂	16.1	[42]	1.87	[64]	2.28	[78]		
c-C₆H₁₁NH₂	16.3	[41]	1.82	[64]				
t-C₅H₁₁NH₂	17.4	[44]	2.02	[64]				
Secondary aliphatic amines								
(HC≡CCH₂)₂NH	11.7	[55]	−4.30	[64]				
△NH	11.2	[44]	−1.79	[64]				
Me₂NH	15.5	[42]	2.09	[64]	3.09	[78]		
MeEtNH	17.9	[56]						
◇NH	18.0	[14]	2.79	[64]				
(H₂C=CHCH₂)₂NH	19.3	[55]	−0.01	[64]				
⬠NH	20.1	[14]	2.81	[64]	4.24	[78]		
Et₂NH	20.2	[42]	2.42	[64]	3.11	[78]		
⬡NH	21.2	[14]	2.56	[64]	3.35	[78]		
(n-Pr)₂NH	22.2	[55]	2.39	[64]				
Tertiary aliphatic amines								
(HC≡CCH₂)₃N	15.0	[54]	−8.43	[64]	1.56	[78]	−1.09	[23–25]
(CH₃)₃N	20.0	[42]	0.75	[64]				
(CD₃)₃N	20.3	[57]	1.00	[65]				
Me₂EtN	22.4	[56]	1.15	[66]				

Table 3—Continued

B	Gas phase		Aqueous solution		Acetonitrile solution		H-bonding in CCl$_4$	
	$-\delta_R \Delta G^0_{i(g)}$	Ref.	$-\delta_R \Delta G^0_{i(H_2O)}$	Ref.	$-\delta_R \Delta G^0_{i(MeCN)}$	Ref.	$-\delta_R \Delta G^0_{HB}$	Ref.
MeEt$_2$N	24.6	[56]	1.63	[66]			−0.27	[23–25]
(H$_2$C=CH—CH$_2$)$_3$N	24.7	[54]	−1.32	[64]			0.56	[23–25]
Et$_3$N	26.7	[42]	2.01	[64]	2.72	[78]	−0.07	[23–25]
(n-Pr)$_3$N	28.7	[43]	1.93	[64]				
Cyclic tertiary aliphatic amines								
N-phenylpyrrolidine	19.3†	[60]		[71]				
N-phenylpiperidine	21.8†	[60]		[71]				
diazabicyclooctane	23.5^c	[51]	0.90^c	[72]			0.95^c	[23–25]
N-Me-pyrrolidine	24.3	[44]	1.66	[64]				
benzoquinuclidine	26.0†	[60]	−1.96	[70]				
quinuclidine	27.1†	[20]	2.60	[64]			1.54	[23–25]
Anilines								
C$_6$H$_5$NH$_2$	6.70	[41]	−6.34	[67]	−8.05	[78]		
m-MeC$_6$H$_4$NH$_2$	8.90	[59]	−6.18	[67]				
p-MeC$_6$H$_4$NH$_2$	9.20	[59]	−5.67	[67]	−7.11	[78]		
C$_6$H$_5$NHMe	12.90	[34]	−5.94	[68, 69]				
C$_6$H$_5$NMe$_2$	19.50	[34]	−5.70	[68, 69]				
Diamines								
Me$_2$NNH$_2$	15.20	[61]	−2.90	[73]				
(Me$_2$NCH$_2$)$_2$	30.30	[44, 58]	−0.14	[74]				
Carboxylic acid amides								
HCO·NHMe	1.70†	[62]	(−14.8)	[63]			0.76	[23–25]
HCO·NMe$_2$	7.60†	[58, 62]	(−14.7)	[63]			1.20	[23–25]
MeCO·NMe$_2$	11.70†	[62]	(−13.1)	[63]				

Carboxylic acid esters

Compound				
HCO$_2$Me	−12.80 [106]	(−18.70) [63]		−0.56 [23–25]
MeCO$_2$Me	−5.40 [106]			
MeCO$_2$Et	−2.80† [106]			
Me$_2$O	−10.00 [106]	*Ethers*		
1,4-dioxane	−8.80 [106]	(−17.3[c]) [63]		−1.05[c] [23–25]
MeOEt	−6.70 [106]			
THF	−4.10 [106]	(−15.4) [63]		−0.33 [23–25]
Et$_2$O	−3.20 [58]	(−15.9) [63]		−0.67 [23–25]
i-PrOEt	−0.40 [58]			
(i-Pr)$_2$O	2.30 [58]			
Me$_2$CO	−6.50 [106]	*Ketones* (17.40) [63]		−0.44 [23–25]
(CH$_3$CO)$_2$CH$_2$	4.00 [59]			
MeCN	−13.50 [106]	*Nitrile* (−26.10) [63]		−0.82 [23–25]
PH$_3$	−14.00 [106]	*Phosphines* (−31.60) [75]		
MePH$_2$	0.10 [50]	−7.27 [75]		
Me$_2$PH	11.90 [50]	−0.80 [75]		
Me$_3$P	21.20 [50]		−5.62 [78]	
C$_5$H$_5$N	16.00 [41]	*Pyridines* −5.50 [76]		0.52 [23–25]
2-MeC$_5$H$_4$N	19.50 [58]	−4.41 [76]		
3-MeC$_5$H$_4$N	18.50 [58]	−4.80 [76]		0.72 [23–25]
4-MeC$_5$H$_4$N	20.00 [58]	−4.34 [76]		
2,6-di-t-BuC$_5$H$_3$N	26.40† [106]	−7.74 [110]		

Table 3—Continued

B	Gas phase		Aqueous solution		Acetonitrile solution		H-bonding in CCl_4	
	$-\delta_R \Delta G^0_{i(g)}$	Ref.	$-\delta_R \Delta G^0_{i(H_2O)}$	Ref.	$-\delta_R \Delta G^0_{i(MeCN)}$	Ref.	$-\delta_R \Delta G^0_{HB}$	Ref.
	Sulphides							
MeSH	−14.50	[106]						
Me₂S	−2.90	[106]					−1.88	[23]
Et₂S	1.90	[106]						
			Sulphoxides					
Me₂SO	7.00	[59, 62]	−15.00	[77]			1.40	[23–25]
			Olefins					
△CH=CH₂	−5.6	[108]						
C₆H₅CH=CH₂	−2.0	[107]						
C₆H₅CMe=CH₂	3.4	[107]						
(C₆H₅)₂C=CH₂	4.5	[107]	−26.8	[109]				
△CMe=CH₂	5.6	[107]						

Notes to Table 3

[a] Unless otherwise stated, the precision is ±0.2 kcal mol⁻¹ or less.

[b] Values of $\delta_R \Delta G^0_{i(H_2O)}$ given in parentheses are crude values based upon indicator or distribution methods [63].

[c] † indicates that, because of uncertainties in pressure measurement for base, the uncertainty is estimated as ±0.5 kcal mol⁻¹.

[d] c indicates that a statistical correction has been applied.

Table 4 lists some typical values of $\log(K_{(aq)}/K_{(g)})$ obtained from values of $-\delta_R \Delta G_i^0$. Solvent water is shown to give rise to medium effects on proton transfer equilibria of 1 to 25 orders of magnitude, reflecting truly major consequences [107]. On the other hand, for certain special types of proton-transfer equilibria (as discussed in the following section) quite small solvent effects are observed (<1 log unit). Figure 2 is a plot of $-\delta_R \Delta G_{i(g)}^0$ against $-\delta_R \Delta G_{i(aq)}^0$ to show that for structural effects there is *in general* little resemblance between gas-phase and aqueous proton-transfer equilibrium. A similar conclusion applies for results (where available) in the solvent acetonitrile; the values of $-\delta_R \Delta G_{i(MeCN)}^0$ resemble rather closely $-\delta_R \Delta G_{i(aq)}^0$ values

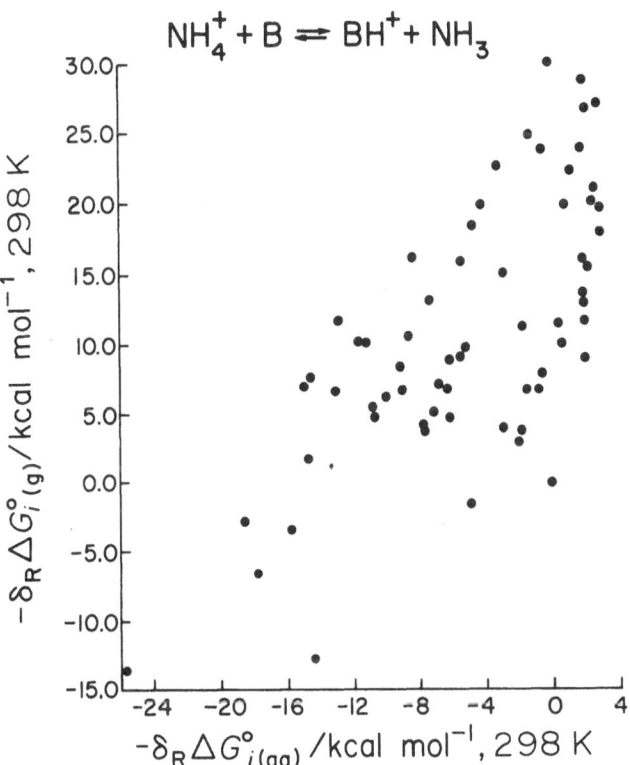

Figure 2. Comparison of gas-phase and aqueous base strengths, values of ΔG^0 in gas phase plotted against ΔG^0 in aqueous solution, for reactions; at 298 K.

$$\text{ordinate} = -\delta_R \Delta G_{i(g)}^0, \text{kcal mol}^{-1}$$

$$\text{abscissa} = -\delta_R \Delta G_{i(aq)}^0, \text{kcal mol}^{-1}$$

$$B + NH_4^+ \rightleftharpoons BH^+ + NH_3$$

45

Table 4

Some typical solvent effects of water on proton-transfer equilibria

$$NH_4^+ + B \rightleftharpoons BH^+ + NH_3$$

B	$\log K_{(g)}$	$\log K_{(aq)}$	$\log (K_{(aq)}/K_{(g)})$
EtOH	−10.00	−11.20	−1.20
MeCN	−9.90	−19.10	−9.20
Me_2CO	−4.80	−12.80	−8.00
Et_2O	−2.30	−11.70	−9.40
$CF_3CH_2NH_2$	−1.00	−3.65	−2.70
$(C_6H_5)_2C{=}CH_2$	3.30	−19.60	−22.90
$C_6H_5NH_2$	4.90	−4.64	−9.50
$(CH_3)_2SO$	5.10	−11.00	−16.10
$MeNH_2$	6.70	1.41	−5.30
$MeCONMe_2$	8.60	−9.60	−18.20
C_5H_5N	11.80	−4.03	−15.80
Me_3N	15.00	0.55	−14.40
$2,6\text{-di-t-BuC}_5H_3N$	19.40	−5.70	−25.10
$(Me_2NCH_2)_2$	22.20	−0.10	−22.30
Me_3P	15.50	−0.60	−16.10

[78]. Further, there is in general very little resemblance between structural effects on gas-phase proton-transfer equilibria and the corresponding equilibria for transfer of *p*-fluorophenol between hydrogen-bonded complexes in carbon tetrachloride solution. This latter result is shown in Figure 3, which shows a plot of values of $-\delta_R \Delta G_{i(g)}^0$ against the corresponding $\delta_R \Delta G_{HB}^0$ values from Table 3.

2.3.2 *Aliphatic Amines*

In contrast to solution results, the gas-phase base strengths show generally substantially larger effects of molecular structure (Table 3). Equally notable is the contrast in the regularity of substituent effects. The gas phase results possess a remarkable regularity. Table 5 lists values of $-\delta_R \Delta G_{i(g)}^0$ for successive substitution of H in NH_3 by propargyl, methyl, allyl, ethyl, and n-propyl substituents (the base strength increases in this order). The results are decisively non-additive, but all of these substituents exhibit effects in the ratio $1.00:1.72 \pm 0.02:2.22 \pm 0.02$. That is, the effect of the second substitution is 86 per cent of the additive value and that of the third substitution is 74 per cent of the additive value. The high precision with which the data are described in this manner is indicated in Table 5 by the agreement between observed values for R_2NH and R_3N members and the values (given in parentheses) calculated by the use of these 'saturation' factors.

46

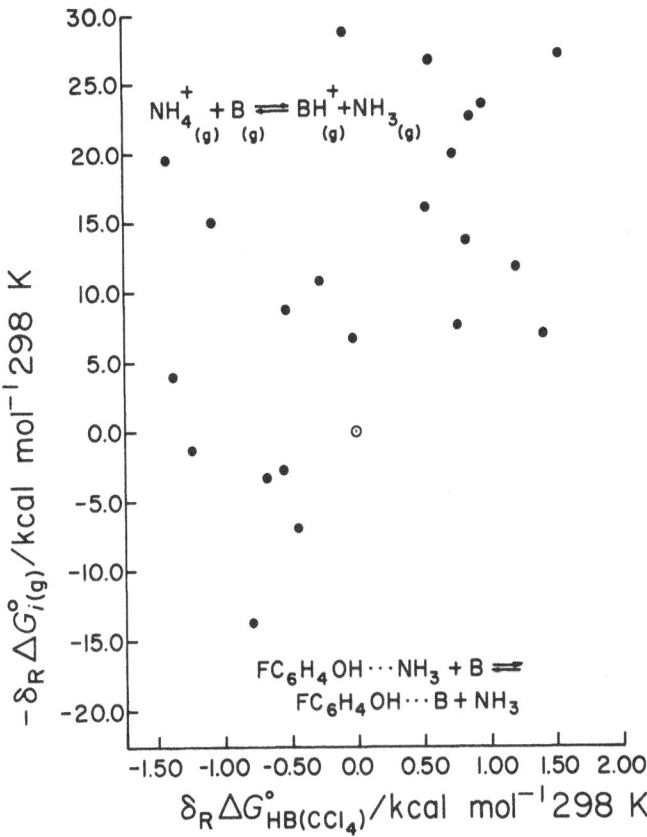

Figure 3. Comparison of gas-phase base strengths (relation to ammonia) with the hydrogen-bond formation with *p*-fluorophenol in carbon tetrachloride.

$$\text{ordinate} = -\delta_R \Delta G^0_{i(g)}, \text{kcal mol}^{-1}$$

$$\text{abscissa} = -\delta_R \Delta G^0_{HB(CCl_4)}, \text{kcal mol}^{-1}$$

Table 5

Effects of successive symmetrical substitution
$-\delta_R \Delta G^0_{i(g)}/\text{kcal mol}^{-1}$ at 298 K

R	RNH_2	R_2NH^a	R_3N^b
$HC\equiv CCH_2$	6.70	11.70 (11.50)	15.00 (14.90)
CH_3	9.10	15.50 (15.60)	20.00 (20.20)
$H_2C=CHCH_2$	11.30	19.30 (19.40)	24.70 (25.10)
C_2H_5	11.80	20.20 (20.30)	26.70 (26.20)
$n\text{-}C_3H_7$	13.00	22.20 (22.40)	28.70 (28.90)

[a] Value in parenthesis calculated as $(1.72) (\delta_R \Delta G^0_{i(g)}[RNH_2])$.
[b] Value in parenthesis calculated as $(2.22) (\delta_R \Delta G^0_{i(g)}[RNH_2])$.

Table 6

Effects of successive unsymmetrical substitution

Base	$-\delta_R \Delta G^0_{i(b)obs}$/kcal mol^{-1}	$-\delta_R \Delta G^0_{i(g)calc}$/kcal mol^{-1}
C_6H_5NHMe	12.90	13.60[a]
MeNHEt	17.90	18.00[a]
$C_6H_5NMe_2$	19.50	18.40[b]
C_6H_5N (cyclopentane ring)	19.50	19.60[b]
C_6H_5N (cyclohexane ring)	21.80	20.60[b]
Me_2NEt	22.40	22.20[b]
Et_2NMe	24.60	24.20[b]

[a] Calculated as $0.86 \, \Sigma \, R_1 + R_2$.
[b] Calculated as $0.74 \, \Sigma \, R_1 + R_2 + R_3$.

Table 6 gives values of $-\delta_R \Delta G^0_{i(g)}$ observed for a number of unsymmetrically substituted amines. These values are compared with those calculated assuming the same factors apply as for symmetrical substitution. For alkyl substituents, agreement is quite satisfactory. For N-substituted anilines, the agreement is relatively crude, as might be expected on the basis of the effects of the N-substituents on the C_6H_5-N conjugation.

For successive methyl substitution of the CH_3 hydrogens of CH_3NH_2, *i.e.* for the series $MeNH_2$, $EtNH_2$, i-$PrNH_2$ and t-$BuNH_2$, the 'saturation' factors on gas-phase base strengths are less severe. The increments in $\delta_R \Delta G^0_{i(g)}$ values from Table 1 are 2.7, 5.0, and 7.0 kcal mol^{-1}, *i.e.* a ratio of 1.00:1.85:2.60. The latter two numbers correspond to 93 per cent of the additive value for the second substitution and 86 per cent of additive for the third substitution.

The contrast between gas-phase and solution results, especially in the behaviour exhibited by saturated and unsaturated substituents, forms the basis for further analysis as presented in section 2.3.3.

2.3.3 *Hybridization-inductive, polarization and resonance effects*

Taft and coworkers [41] have formulated the rule that the conjugate acids of hydrocarbon-substituted nitrogen bases which have the same number of acidic protons (NH^+), and the same number of carbon atoms with similar frameworks, but which differ in the degree of down-chain hydrogenation, involve nearly the same base strengths towards the proton in aqueous solution as in the gas phase. More specifically, the standard free energy change for proton-transfer between a pair of nitrogen bases which meet these criteria is found to be about 20 per cent less in aqueous solution than that observed in the gas phase.

This rule has been correlated with the very similar heats of solution of the gaseous conjugate acids of the same type, as is fully discussed in the chapter by E. M. Arnett. It has been further suggested that the 20 per cent reduction factor arises primarily because of somewhat stronger hydrogen bonds to water formed by the more acidic unsaturated conjugate acid (such as $C_6H_5NH_3^+$) as well as the somewhat stronger hydrogen bonds formed by water to the more basic saturated nitrogen base (*i.e.* $C_6H_{11}NH_2$). The consequence of these hydrogen-bonding interactions may be viewed as a somewhat reduced (~ 80 per cent) apparent proton transfer in aqueous solution [19] as compared to the gas-phase reference. Presumably greater reduction factors will be found in more basic solvents.

Taft and coworkers formulated this rule on the basis of the sizeable structural effects involved in the following three proton-transfer equilibria:

$$\triangleright NH_2^+ + Me_2N = \triangleright NH + Me_2NH_2^+; \; -\Delta G_{(g)}^0 = 4.3; \; -\Delta G_{(aq)}^0 = 3.88 \qquad (12)$$

$$\langle \bigcirc \rangle NH^+ + \langle \quad N \rangle^{Me} = \langle \bigcirc \rangle N + \langle \quad N \rangle^{Me^+}_{H} \; : \; -\Delta G_{(g)}^0 = 7.9; \; -\Delta G_{(aq)}^0 = 7.16 \qquad (13)$$

$$\langle \bigcirc \rangle - NH_3^+ + \langle S \rangle - NH_2$$

$$= \langle \bigcirc \rangle - NH_2 + \langle S \rangle - NH_3^+ \; ; \; -\Delta G_{(g)}^0 = 9.6; \; -\Delta G_{(aq)}^0 = 8.16 \qquad (14)$$

Table 3 provides a number of further examples in support of the rule, and Table 7 summarizes the more important of these.

The results of the equilibria 12 and 13, as well as those of Table 7 for the tripropargyl, triallyl, and tri-n-propyl amines may all be

Table 7

Small solvent effects (Δ) of water on certain proton-transfer equilibria

1. Unsaturation effects in the aliphatic series

$$BH^+ + (n\text{-}Pr)_3N = (n\text{-}Pr)_3NH^+ + B$$

BH^+	$-\Delta G^0_{(g)}/\text{kcal mol}^{-1}$	$-\Delta G^0_{i\,(aq)}/\text{kcal mol}^{-1}$	$\Delta/\text{kcal mol}^{-1}$
$(HC{\equiv}C{-}CH_2)_3NH^+$	13.70	10.36	3.30
$(H_2C{=}CHCH_2)_3NH^+$	4.00	3.25	0.70

2. Secondary β-deuterium isotope effect

$$(CH_3)_3NH^+ + (CD_3)_3N = (CH_3)_3N + (CD_3)_3NH^+$$

$$-\Delta G^0_{(g)} = 0.312 \pm 0.030 \,\text{kcal mol}^{-1};$$

$$-\Delta G^0_{(aq)} = 0.253 \pm 0.003 \,\text{kcal mol}^{-1};$$

$$\Delta = 0.06 \,\text{kcal mol}^{-1}$$

Further examples from Table 3 which follow the rule are allylamine relative to n-propylamine and diallyamine relative to di-n-propyl amine. Although tripropargylamine relative to tri-n-propyl amine follows the rule reasonably well, dipropargylamine relative to di-n-propyl amine and propargylamine relative to n-propylamine are exceptional. The discrepancy clearly lies with the $\delta_R\Delta G^0_{i(aq)}$ values for diproparylamine and proparylamine. Although the gas phase base strengths for all three propargylamines follow the general relationship described in Section 3.2.2, the aqueous base strengths show anomalous effects of successive substitution of the propargyl substituent for hydrogen. The possibility is suggested of $-C{\equiv}C-H{\cdots}OH_2$ hydrogen-bonding solvation *of the conjugate acids* (which decreases in extent from the primary to the tertiary amine).

accounted for by hybridization theory [16, 44, 54]. That is, the nitrogen of pyridine involves proton acceptance into an orbital having greater *s* character (formally sp^2) than does the nitrogen of N-methyl pyrrolidine (more nearly sp^3). Consequently, the pyridine is less basic [6]. Additionally, the introduction of unsaturated hydrocarbon substituents having carbon orbitals of increasing *s* character reduces base strength by inductive relay. For example, base strength decreases in the sequence n-propyl > allyl > propargyl; in fact, the decrease is essentially linear with the percentage *s* character of $\beta - \gamma$ C—C bonds (25 per cent, 33 per cent and 50 per cent respectively). It appears at the present writing that these are the major structural effects involved in the cited equilibria. Stabilization by polarization [86] of saturated and unsaturated molecular cavities of the conjugate (BH^+) appears to be essentially the same for equal carbon content and similar frameworks. This conclusion is based upon the failure of solvent to differentiate in a major way between stabilities of the unsaturated and saturated conjugate acids. That is, the relatively small effect of aqueous solvent (or acetonitrile solvent – *cf.* Table 3) on the ΔG^0 values for these equilibria is taken to indicate the absence of an appreciable polarization effect, as well as the absence of any very appreciable solvent dependence upon the hybridization-inductive effects.

Subsequent discussion (Section 2.3.5) of the results for polar substituents provides an indication that a formal 'fall-off' factor of crudely $\frac{1}{2}$ per atom applies to down-chain transmission of the effects of these substituents on the gaseous proton transfer equilibria. Using this as a rough basis for analysis, the value $-9.6 \, \text{kcal mol}^{-1}$ for equilibrium 14 may be attributed to about $-4.0 \, \text{kcal mol}^{-1}$ for the hybridization-inductive effect (i.e., about half the $\Delta G^0_{(g)}$ value for equilibrium 13) and about $-5.5 \, \text{kcal mol}^{-1}$ for the extra resonance stabilization of aniline, i.e., the stabilization energy associated with the delocalization of π-electronic charge from the NH_2 group into the phenyl ring. This figure is indeed in excellent agreement with Wepster's analysis of aqueous proton-transfer equilibria [70, 81], and also accords well with theoretical calculation by the STO-3G method [83] of the energy of isomerization of a hypothetical aniline molecule with the NH_2 orthogonal to the phenyl ring to a nearly coplanar aniline molecule.

This assignment has been further confirmed by Taagepera, Mitsky, Wepster, and Taft [60], using the molecular bases designed by Wepster in his brilliant classical studies of steric inhibition of resonance (mesomerism) [70]. There is little or no delocalization of π-electronic charge from the nitrogen lone-pair into the phenyl ring of benzo-quinuclidine **1**. On the other hand, N-phenyl piperidine **2** or N-phenyl-pyrolidine **3** should involve delocalization to extents similar to that in aniline. Since the conjugate acids of **1** and **2** involve the

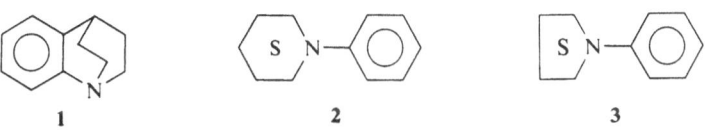

same number of carbon atoms (differing in atomic composition by only two down-chain hydrogen atoms), both should be essentially equally stabilized by polarization effects of their hydrocarbon moieties. However, some relief of ring strain is expected for protonation of **2** (six-membered rings favour an increase in coordination number from 3 to 4) [70a]. For **3**, ring strain is increased upon protonation (five-membered rings are unfavourable for CN increasing from 3 to 4) [70a]. Comparing the gas-phase base strengths of pyrrolidine and piperidine (Table 3), it may be estimated that a CH_2 group polarization effect favours **1** over **3** by $1.3\,\mathrm{kcal\,mol^{-1}}$. Thus, the gas-phase base strength of **3**, corrected for this polarization effect, is $5.4\,\mathrm{kcal\,mol^{-1}}$ less than that of **1** (Table 3). The observed gas-phase base strength of **2** is $4.2\,\mathrm{kcal\,mol^{-1}}$ less than **1**. The average of these two figures ($-4.8\,\mathrm{kcal\,mol^{-1}}$) should be largely free of I-strain effects and should be a measure of the extra resonance stabilization of the free bases **2** and **3**. The agreement of this figure with that obtained for the above analysis of (ii) and (iii) is strongly confirming. From data in 50 per cent aqueous ethanol, the average value of the standard free energy change for proton transfer from the conjugate acids of **2** and **3** to **1** is $-3.6\,\mathrm{kcal\,mol^{-1}}$ (average of -2.4 and $-4.7\,\mathrm{kcal}$, respectively). This figure indicates the approximate decrease of 25 per cent in the resonance effect in the aqueous solvent compared to the gas phase.

The polarization effect may be evaluated approximately by the empirical observation that polar solvents cause the effect nearly to disappear in $\Delta G^0_{(s)}$. Formally, bulk solvent takes over in large part from the polarizable substituent the role of stabilizing charge [27] (*cf.* also section 2.5). The predominant mechanism appears to be charge dispersal via hydrogen-bonding by BH^+ to bulk solvent. Thus, for example, the comparison of $\Delta G^0_{(g)}$ and $\Delta G^0_{(aq)}$ for equilibrium 15 places the cationic polarization stabilizing effect of a cyclohexyl (or phenyl) group as about $7\,\mathrm{kcal\,mol^{-1}}$ greater than that of a methyl group.

$$\left\langle\; S \;\right\rangle\!-NH_3{}^+ + CH_3NH_2 \rightleftharpoons CH_3NH_3{}^+ + \left\langle\; S \;\right\rangle\!-NH_2;\, \Delta G^0_{(g)} = 7.2;$$

$$\Delta G^0_{(aq)} = -0.1\,\mathrm{kcal\,mol^{-1}}$$

As an illustration of the application of these figures for the resonance (R), hybridization-induction (I), and polarization effects (P), we may consider equilibrium.

$$\left\langle\bigcirc\right\rangle\!-NH_3{}^+ + CH_3NH_2 \rightleftharpoons \left\langle\bigcirc\right\rangle\!-NH_2 + CH_3NH_3{}^+ \tag{16}$$

$$\Delta G^0_{(g)(calc)} \cong R + I - P \cong -5.5 - 4.0 + 7.0 \cong -2.5\,\mathrm{kcal\,mol^{-1}}$$
$$\Delta G^0_{soln(calc)} \cong R + I \cong -5.5 - 4.0 \cong -9.5\,\mathrm{kcal\,mol^{-1}}$$
$$\Delta G^0_{(g)(obs)} = -2.4\,\mathrm{kcal\,mol^{-1}}$$
$$\Delta G^0_{(aq)(obs)} = -8.26;\, \Delta G^0_{(MeCN)(obs)} = -10.65\,\mathrm{kcal\,mol^{-1}}$$

Finally, we may consider equilibrium 17.

$$\Delta G^0_{(aq)(calc)} \cong I \cong -4.5; \Delta G^0_{(aq)(obs)} = -4.56 \text{ kcal-mol}^{-1}$$
$$\Delta G^0_{(g)(calc)} \cong I + P' \cong -4.5 + P'; \Delta G^0_{(g)(obs)} = -0.2 \text{ kcal-mol}^{-1}.$$

Thus equating $\Delta G^0_{(g)(calc)}$ and $\Delta G^0_{(g)(obs)}$ gives 4.3 kcal mol^{-1} for the polarization effect P' of the fused benzo group.

2.3.4 Steric strains in boron trimethyl complexes

The standard enthalpy changes for gaseous proton transfer from NH_4^+ to amines (reaction 2) show no evidence of steric effects (cf. Tables 3, 5, 6). Consequently, comparison of structural effects on this reaction with the corresponding standard enthalpy change for transfer of the Lewis acid BMe_3 between ammonia and the amines in the gas phase [7, 8] reveals the strain energies of the latter complexes. Table 8 summarizes the available data, which are plotted in Figure 4. Points for all straight-chain primary amines, NH_3, and for ethyleneimine and trimethyleneimine define a shallow curve of reasonable precision. Deviations to the left of this curve in Figure 4 may be ascribed to steric strain in the BMe_3 adduct. The strain energies so obtained are listed in Table 8 and appear to be reliable to about ± 1 kcal. The strain energies for several homomorphic hydrocarbons obtained by Cox and Pilcher [84] from deviations from heats of atomization treatments are also shown in Table 8. The accord between corresponding strain energies clearly supports Brown's concept of equal strain in homomorphs [85].

Not shown in Figure 4 are data for Me_2PH and Me_3P (Table 8). Points for these phosphines also deviate from the correlation line of Figure 4. However, the deviations are probably not measures of strain energies in their BMe_3 adducts, but instead are associated with different bonding considerations for phosphorus and nitrogen (cf. the subsequent discussion in Section 4). The slope of the line defined by these two points (~ 2) is at least nearly the same as for the correlation line of Figure 4.

Table 8

Steric strain energies for trimethyl Boron adducts

All values in kcal mol^{-1}

B	$\delta_R\Delta H^0_{i(g)}$	$\delta_R\Delta H^0_{(BMe_3)(g)}$	S.E.[c]	H.S.E.[d]
$CF_3CH_2NH_2$	-1.6	0.3	—	
NH_3	$(0.0)^a$	$(0.0)^b$	—	
$HCF_2CH_2NH_2$	3.8	1.4	—	
$FCH_2CH_2NH_2$	7.8	3.0	—	
$MeNH_2$	8.9	3.8	—	
 $\triangleright NH_2$	10.8	3.8	—	
$EtNH_2$	11.6	4.2	—	
$i\text{-PrNH}_2$	13.9	3.6	2.6	3.8
Me_2NH	15.1	5.5	1.4	
C_5H_5N	15.2	3.2	3.8	2.0
$t\text{-BuNH}_2$	15.9	-0.8	8.2	8.0
 ⌐—NH	17.6	8.7	—	
2-MeC_5H_4N	18.6	~-1.8	~11	
4-MeC_5H_4N	19.2	5.6	3.8	
Me_3N	19.2	3.8	5.5	4.0
 ⬠NH	19.4	6.6	2.9	
Et_2NH	19.8	2.5	7.2	
 ⬡NH	20.7	5.9	4.3	
Et_3N	25.9	(-3.8)	~17.0	
 ⬡N	26.3	6.1	~7.0	
MePH	11.7	-2.4		
Me_3P	20.8	2.7		

[a] The value for heat of dissociation of NH_4^+ is 207.0 kcal.
[b] The value for heat of dissociation of H_3NBMe_3 is 13.8 kcal.
[c] The strain energy of BMe_3 adduct, obtained from deviations to left of correlation curve of Fig. 4.
[d] The strain energy of homomorphic hydrocarbon, *i.e.*, $\geqslant B-N\leqslant$ replaced by $\geqslant C-C\leqslant$ as obtained in reference 48.

54

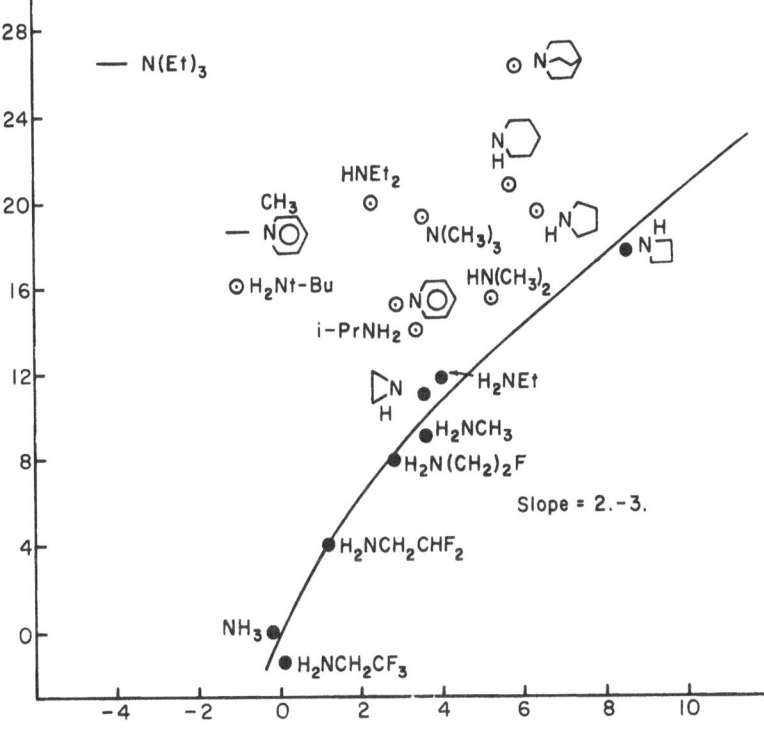

Figure 4. Comparison of ΔH^0 in the gas phase for proton-transfer with ΔH^0 for exchange of BMe$_3$ between nitrogen basic and ammonia (*cf.* Table 8).

$$\text{ordinate} = \delta_R \Delta H^0_{i(g)}, \text{ kcal mol}^{-1}$$
$$\text{abscissa} = \delta_R \Delta H^0_{(BMe_3)(g)}, \text{ kcal mol}^{-1}$$

2.3.5 *Polar substituent effects*

The effects of polar substituents (generally heteroatom) on proton transfer equilibria in the gas phase have been determined for several aliphatic series amines [86], 4-substituted pyridines [87] and a few 4-substituted anilines [59]. Data for the latter series have been included in Table 10. The results for the other series are summarized in Table 9.

Series 1 of Table 9 shows that successive substitution of β-fluorine for hydrogen substantially decreases base strength both in the gas phase and in aqueous solution. Gas-phase effects are roughly twice as large as those in aqueous solutions, but there are irregularities in the gas-phase results. The first and second substitutions of fluorine reduces base strength by 3.8 and 4.0 kcal mol^{-1}, respectively, whereas the third fluorine substitution reduces it by 5.4 kcal mol^{-1}. In aqueous solution, successive substitution of fluorine decreases base strengths in the regular pattern 1.00:1.90:2.68. The contrasting behaviour suggests that internal chelation of the type:

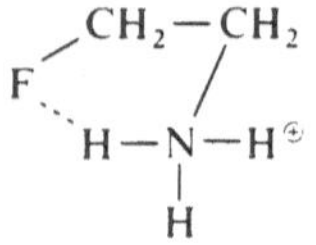

$$
\begin{array}{c}
\text{CH}_2\text{---CH}_2 \\
\text{F} \cdots \diagup \\
\text{H---N---H}^{\oplus} \\
| \\
\text{H}
\end{array}
$$

4

Table 9

Polar substituent effects on gas-phase base strengths

$$\text{NH}_{4(g)}{}^{+} + \text{B}_{(g)} + \text{BH}_{(g)}{}^{+} + \text{NH}_{3(g)}$$

1. Fluoro-substituted ethyl amines

B	$-\delta_R\Delta G^0_{i(g)}$/kcal mol^{-1} [a]		$-\delta_R\Delta G^0_{i(aq)}$/kcal mol^{-1}	
CF$_3$CH$_2$NH$_2$	−1.40	(−13.20)[b]	−4.96	(−6.92)[b]
HCF$_2$CH$_2$NH$_2$	4.00	(−7.80)	−2.94	(−4.90)
FCH$_2$CH$_2$NH$_2$	8.00	(−3.80)	−0.62	(−2.58)
CH$_3$CH$_2$NH$_2$	11.80	(0.00)	1.96	(0.00)

2. Trifluoromethyl substituent effect

B	$-\delta_R\Delta G^0_{i(g)}$/kcal mol^{-1} [a]		$-\delta_R\Delta G^0_{i(aq)}$/kcal mol^{-1} [c,d]	
CF$_3$CH$_2$NH$_2$	−1.40		−4.96	
CH$_3$CH$_2$NH$_2$	11.80	(−13.20)[e]	1.96	(−6.92)[e]
CF$_3$(CH$_2$)$_2$NH$_2$	6.70		−0.82	
CH$_3$(CH$_2$)$_2$NH$_2$	13.0	(−6.30)	1.81	(−2.63)
CF$_3$(CH$_2$)$_3$NH$_2$	10.10		0.64	
CH$_3$(CH$_2$)$_3$NH$_2$	13.60	(−3.50)	1.90	(−1.26)

3. α-Substituted trimethyl amines

B	$-\delta_R\Delta G^0_{i(g)}$/kcal mol^{-1} [a]		σ_I [f]
NC—CH$_2$NMe$_2$	7.10	(12.90)[g]	0.56
F$_3$C—CH$_2$NMe$_2$	10.90	(9.10)	0.45
H—CH$_2$NMe$_2$	20.00	(0.00)	0.00
H$_3$C—CH$_2$NMe$_2$	22.40	(2.40)	−.05

4. Substituted pyridines

Reaction: $X-\langle\bigcirc\rangle NH^+ + \langle\bigcirc\rangle N \rightleftharpoons \langle\bigcirc\rangle NH^+ + X-\langle\bigcirc\rangle N$

X	$\delta_R \Delta G^0_{(g)obs}$ [h]	$\delta_R \Delta G^0_{(g)calc}$ [i]	$\delta_R \Delta G^0_{(aq)obs}$ [j]	$\delta_R \Delta G^0_{(aq)calc}$ [i]
OMe	6.70	6.90	1.87	1.84
Me	4.00	3.30	1.12	1.20
H	(0.00)	(0.00)	(0.00)	(0.00)
F	-3.50	-3.50[k] $(-1.3$[l]$)$	—	-1.43[l]
Cl	-2.80	-3.00	-1.88	-1.91
CF$_3$	-7.30	-7.50	-3.52	-3.46
CN	-9.8	-9.70	-4.57	-4.41
NO$_2$	-11.10	-11.20	-5.21	-5.56
		$-\rho_I = 14.80$		$-\rho_I = 7.02$
		$-\rho_R = 10.70$		$-\rho_R = 3.67$
		$\rho_R/\rho_I = 0.71$		$\rho_R/\rho_I = 0.52$

[a] Reference [86].
[b] Values relative to ethyl amine.
[c] Reference [8].
[d] M. C. Raasch, *J. Org. Chem.*, **27**, 1406 (1962); we are indebted to Dr. W. A. Sheppard for providing these compounds.
[e] Incremental values.
[f] Reference [92].
[g] Values relative to trimethylamine.
[h] Reference [87].
[i] Calculated by the DSP equation using the ρ_I and ρ_R values given and values of σ_I and σ_R^+ listed in Reference [92].
[j] A. Fischer, W. J. Galloway, and J. Vaughn, *J. Chem. Soc.*, **3**, 591 (1964).
[k] Calculated using $\sigma_I = 0.65$.
[l] Calculated using the usual $\sigma_I = 0.50$ (Reference [92]).

4 is involved [8, 88, 89] at least with β-fluoroethyl and β,β-difluoroethyl ammonium ions in the gas phase. In aqueous solution, the availability of basic water molecules dictates that

$$
\begin{array}{c}
H\cdots OH_2 \\
| \\
FCH_2CH_2\overset{}{N}-H^+\cdots OH_2 \\
| \\
H\cdots OH_2
\end{array}
$$

5

solvation in **5** destroys the above chelation. In the gas phase chelation of this kind is apparently a common means for stabilizing cations [8, 88, 89], giving rise to very high base strengths of molecules such as ethylene diamine in gaseous proton-transfer equilibria [89]. The results for trifluoroethylamine, as well as for trifluoropropylamines and trifluorobutylamines (series 2 of Table 9), are affected by the corresponding chelation in the gas phase to an uncertain extent. Clearly, the much weaker base strength of a trifluoromethyl compared with a monofluoromethyl tends to reduce the importance of the chelation in gaseous conjugate acids of these bases. The loss of entropy accompanying the chelation, together with the feeble fluorine base strength, probably tends to eliminate any appreciable contributions to the gas-phase base strengths. This conclusion is supported by the comparison of gas and aqueous phase results for series 2 of Table 9. The attenuation factors, $\delta_R \Delta G^0_{i(g)}/\delta_R \Delta G^0_{i(aq)}$, for aqueous solvent in the effects of the CF_3 group are as follows: $CF_3CH_2NH_2$, 1.91; $CF_3(CH_2)_2NH_2$, 2.40; $CF_3(CH_2)_3NH_2$, 2.78. The increasing values perhaps may be understood either in terms of coulombic forces acting through increasing effective dielectric constants [90, 91], or substituent effects on cation hydrations which have smaller fall-off factors than do the gas-phase substituent effects. On the other hand, if chelation in the gas phase were the predominant effect, the attenuation factor would be expected to decrease rather than increase in the above sequence. This conclusion follows from the expected increase of chelate stability with increased ring size [8, 88, 89].

The relative gas-phase base strengths of the amines in series 3, i.e. α-substituted trimethylamines, are correlated by the inductive substituent parameter [92] σ_I; that is, relative to Me_3N, $-\delta_R \Delta G^0_{i(g)} \simeq (-23)\sigma_I$. While this result indicates promise for the application of substituent-parameter treatment to gas-phase proton-transfer equilibria, the present data are too limited to provide critical evidence on this issue. Further work is in progress.

The effects of 4-substituents on the gas-phase base strengths of pyridine have been more extensively studied [87]. The results are shown for series 4 of Table 9. The gas-phase substituent effects $\delta_R \Delta G^0_{(g)}$ are roughly linear with the corresponding aqueous-phase effects $\delta_R \Delta G^0_{(aq)}$, as shown in Figure 5. However, the points for $-R$ (π-electron donor) substituents all lie above the relatively precise correlation line for H and $+R$ substituents (CF_3, CN, and NO_2). Data for both phases are described with precision (cf. Table 9) by the dual-substituent parameter treatment [92]:

$$\delta_R \Delta G^0 = \rho_I \sigma_I + \rho_R \rho_R{}^+$$

using the ρ_I and ρ_R values given in Table 9 and the σ_I and $\sigma_R{}^+$ values of Ehrenson, Brownlee and Taft [92]. This treatment indicates a

greater attenuation factor of water solvent for the resonance or
π-delocalization effects in the pyridinium proton-transfer equilibria

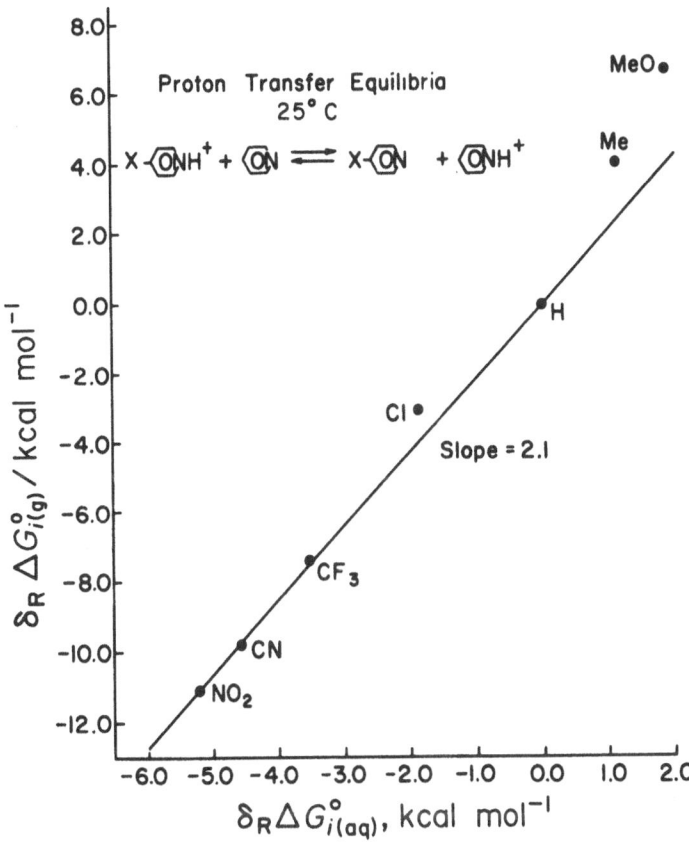

Figure 5. Comparison of base strengths of 4-substituted pyridines at 298 K in the gas phase and in solution.

$$\text{ordinate} = \delta_R \Delta G_{i(g)}^0, \text{kcal mol}^{-1}$$
$$\text{abscissa} = \sigma_R \Delta G_{i(aq)}^0, \text{kcal mol}^{-1}$$

$(10.7/3.67 = 2.92)$ than for polar effects $(14.8/7.02 = 2.11)$. An
alternative statement is that the gas-phase results involve a higher
blend (*i.e.*, ρ_R/ρ_I ratio) of resonance to polar effects than do the
aqueous results: $\rho_R/\rho_{I(g)} = 0.71 > \rho_R/\rho_{I(aq)} = 0.52$.

Yamdagni, McMahon and Kebarle have measured the relative
acidities of *p*-substituted benzoic acids in the gas phase at 600 K
by high pressure mass-spectrometry [47]. They report results indi-
cating that $\rho_R/\rho_{I(g)} < \rho_R/\rho_{I(aq)}$, which is a result opposite to that

obtained for the 4-substituted pyridinium ions. However, these results are indeed those expected on the basis that hydration of the pyridinium ion centre would reduce the relative importance of the interaction form, e.g., 6 [87], whereas hydration of the carboxyl group would increase the relative importance of the form 7 [47]

The 'attenuation factors' of water have been ascribed [87] to a combination of two effects, solvation by hydrogen bonding [1, 16, 41, 42] and effective dielectric constant considerations [90, 91] but the relative contributions of these have yet to be evaluated. For example, on either or both grounds, the observed results [47, 87] $\rho_{I(g)}/\rho_{I(aq)} \simeq 2$ for the pyridines and the larger ratio $\rho_{I(g)}/\rho_{I(aq)} \cong 10$ for the benzoic acids are expected.

The use of σ_I and σ_R^+ parameters based upon solution measurements is shown by the present results to apply to gas-phase equilibria with at least moderate success for those substituents which have little tendency to hydrate (or solvate). However, the relatively low gas-phase base strength of 4-fluoropyridinium is much better described (as shown in Table 9) by the use of $\sigma_I = 0.65$ rather than the usual value of 0.50. This result is under further study.

2.3.6 Comparison with theoretical calculations

The observed standard free energy changes $\delta_R \Delta G^0_{i(g)}$ for proton-transfer equilibria in the gas phase serve as an especially interesting and critical test of theoretical energy calculations. Professor W. J. Hehre has carried out molecular orbital calculations (at the STO-3G level of approximation [94]) of the energy of proton transfer for a number of nitrogen bases [95]. Table 10 summarizes the experimental and theoretical energetics for the indicated proton-transfer equilibria. For the purposes of this comparison the standard free energy changes have been converted to standard enthalpy changes; it is assumed that a small correction applies [96] for changes in the ratio of rotational symmetry numbers for the various B and BH^+.

Table 10

Comparison of experimental and theoretical (STO-3G) energies
of proton transfer in the gas phase[a]
Experimental values of $\delta_R \Delta H^0_{i(g)}$ (second column); calculated
values of $\delta_R \Delta E^0_{(g)}$ (third column) kcal mol^{-1}

$$NH_{4(g)}{}^+ + B_{(g)} = BH_{(g)}{}^+ + NH_{3(g)} \qquad (1)$$

B	$-\delta_R \Delta H^0_{i(g)}$/kcal mol^{-1}	$-\delta_R \Delta E^0_{(g)}$/kcal mol^{-1}
NH_3	(0.0)	(0.0)
$C_6H_5NH_2$	6.5	6.4
$MeNH_2$	8.9	9.0
C_5H_5N	15.2	18.4
Me_3N	19.2	19.8

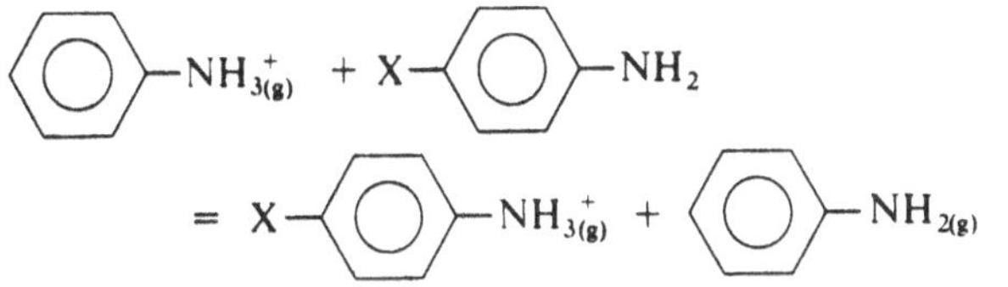

$$\text{(2)}$$

X	$-\delta_R \Delta H^0_{i(g)}$/kcal mol^{-1}	$-\delta_R \Delta E^0_{(g)}$/kcal mol^{-1}
MeO	3.1	3.6
Me	2.5	2.4
H	(0.0)	(0.0)
F	-2.0	-1.2

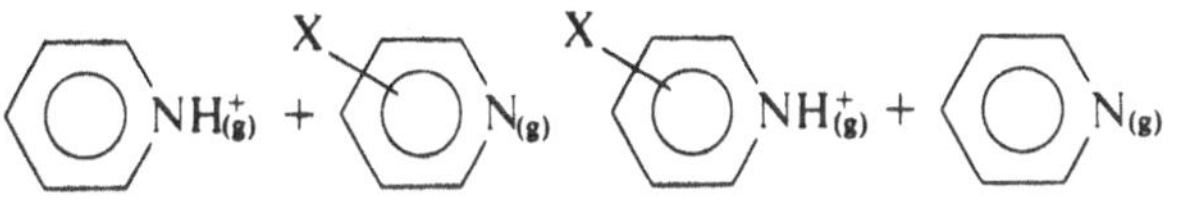

$$\text{(3)}$$

3-position

X	$-\delta \Delta H^0_{i(g)}$/kcal mol^{-1}	$-\delta_R \Delta E^0_{(g)}$/kcal mol^{-1}
Me	2.5	2.5
H	(0.0)	(0.0)
F	-6.4	-6.3
CN	-10.5	-12.4

4-position

X	$-\delta_R \Delta H^0_{i(g)}$/kcal mol^{-1}	$-\delta_R \Delta E^0_{(g)}$/kcal mol^{-1}
MeO	6.7	8.5
Me	4.0	5.0
H	(0.0)	(0.0)
F	-3.5	-0.2
CN	-9.8	-11.4

[a] A small correction for rotational symmetry number ratios has
been applied to observed $-\delta_R \Delta G^0_{i(g)}$ values to obtain the $-\delta_R \Delta H^0_{i(g)}$
value.

The general accord between the experimental values for 'live' molecules and the theoretical values for 'dead' molecules is indeed gratifying. Reasonably, the largest discrepancies appear for the more severe structural changes, e.g., proton transfer from NH_4^+ to pyridine, or from pyridinium ion to 4-fluoropyridine. For 4-substituted pyridines, all differences are consistent with some over-estimation of π-electron delocalization effects in the theoretical calculations. For 4-substituted anilines, observed substituent effects are of the order of one-half as large as for corresponding substituted pyridines. The accord between experimental and theoretical values is accordingly improved for the former series. It is also of interest that the theoretical calculations show an increase in the $C-N$ bond length in aniline of 0.067 Å on protonation. π-electron delocalization in the free base (Section 2.3.3), i.e. partial π bonding, decreases the bond length.

2.4 Hydrogen-atom transfer equilibria between cation radicals and saturated cations

The results summarized in Table 11 reveal markedly different structural effects on the heats of homolytic bond dissociation energies (for the hydrogen-transfer reaction 6) and heterolytic bond dissociation energies (for the proton-transfer reaction 2) of ammonium [18] and phosphonium ions [50]. Relative to ammonium ions, nitrogen cation ion radicals (cf. series 1) are stabilized by methyl groups, for example, by roughly twice as much as the ammonium ions are

Table 11

Comparison of heats of hydrogen atom transfer and proton transfer in the gas phase

$$NH_{4(g)}^+ + \cdot B_{(g)}^+ = BH_{(g)}^+ + \cdot NH_{3(g)}^+ \tag{6}$$

$$NH_{4(g)}^+ + :B_{(g)} = BH_{(g)}^+ + :NH_{3(g)} \tag{2}$$

	1. Methyl amines		
B	$\delta_R \Delta H^0_{(6)}$/kcal mol^{-1}	$-\delta_R \Delta H^0_{(2)}$/kcal mol^{-1}	Ref.
$MeNH_2$	19.3	8.9	[18]
Me_2NH	29.9	15.1	[18]
Me_3N	35.5	19.2	[18]

Table 11 continued

2. Ethyl amines

B	$\delta_R \Delta H^0_{(6)}$/kcal mol^{-1}	$-\delta_R \Delta H^0_{(2)}$/kcal mol^{-1}	Ref.
$EtNH_2$	19.1	11.6	[18]
Et_2NH	28.6	19.8	[100]
Et_3N	38.5	25.9	[100]

3. α-Me substituted amines

B	$\delta_R \Delta H^0_{(6)}$/kcal mol^{-1}	$-\delta_R \Delta H^0_{(2)}$/kcal mol^{-1}	Ref.
$MeNH_2$	19.3	8.9	[18]
$EtNH_2$	19.1	11.6	[18]
$i\text{-}PrNH_2$	20.0	13.9	[18]
$t\text{-}BuNH_2$	19.9	15.9	[18]

4. Fluorine substituted amine

B	$\delta_R \Delta H^0_{(6)}$/kcal mol^{-1}	$-\delta_R \Delta H^0_{(2)}$/kcal mol^{-1}	Ref.
$CF_3CH_2NH_2$	13.9	1.6	[100]
$MeCH_2NH_2$	19.1	11.6	[18]

5. Tertiary amines

B	$\delta_R \Delta H^0_{(6)}$/kcal mol^{-1}	$-\delta_R \Delta H^0_{(2)}$/kcal mol^{-1}	Ref.
	32.2	25.6	[51]
Et_3N	38.5	25.9	[51]
	45.6	22.8	[51]

6. Phosphines

$$PH^+_{4(g)} + \cdot B^+_{(g)} = BH^+_{(g)} + \cdot PH^+_{3(g)} \tag{6}$$

$$PH^+_{4(g)} + :B_{(g)} = BH^+_{(g)} + :PH_{3(g)} \tag{2}$$

B	$\delta_R \Delta H^0_{(6)}$/kcal mol^{-1}	$-\delta_R \Delta H^0_{(2)}$/kcal mol^{-1}	Ref.
PH_3	(0.0)	(0.0)	[50]
$MePH_2$	5.6	13.9	[50]
Me_2PH	8.9	25.5	[50]
Me_3P	10.6	34.4	[50]

stabilized relative to the neutral bases (actually the former show even larger saturation effects than those discussed in section 2.3.2 for the latter). On the other hand, Staley and Beauchamp find that, relative to phosphonium ions, phosphorus cation ion radicals are stabilized by methyl groups by roughly only 35 percent as much as the phosphonium ions are stabilized relative to their neutral bases (series 6). It is also to be noted that, relative to the neutral bases, methyl groups stabilize phosphonium ions by about 1.7 times as much as ammonium ions [50]. Finally, as noted previously, methyl, ethyl, i-Pr, and t-butyl groups stabilize ammonium ions relative to free base by successive increments of 2.7, 2.3, and 2.0 kcal mol^{-1}. There is, however, very little difference between these substituents in their stabilizing effects on nitrogen cation radicals ($\delta_R \Delta H^0_{(6)}$ values of series 3) [18].

The large effects of methyl (as well as other alkyl) groups which stabilize nitrogen cation radicals relative to their ammonium ions has been attributed to delocalization of charge and spin density into these substituents, for instance:

$$H-\underset{\underset{H}{|}}{\overset{\overset{H}{|}}{C}}-N\overset{.\ +}{\diagdown} \leftrightarrow H^+ \cdot \underset{\underset{H}{|}}{\overset{\overset{H}{|}}{C}}-N\overset{..}{\diagdown} \leftrightarrow H\cdot \underset{\underset{H}{|}}{\overset{\overset{H}{|}}{C}}=N\overset{+}{\diagdown} \leftrightarrow H\cdot \underset{\underset{H}{|}}{\overset{\overset{H}{|}}{C}}{\overset{-}{\diagdown}N}\overset{+}{\diagdown}$$

Since this type of delocalization is hindered by poor orbital overlap of phosphorus, methyl groups stabilize phosphorus cation radicals

In nitrogen cation radicals, this type of delocalization to hydrogen and alkyl groups is hindered by the CF$_3$ group. Consequently, the $\delta_R \Delta H^0_{(6)}$ value is appreciably smaller for CF$_3$CH$_2$NH$_2\cdot^+$ than for CH$_3$CH$_2$NH$_2\cdot^+$ (series 4) [75].

The approximately 1.65-fold greater effects of methyl groups in increasing the base strengths of phosphines compared to amines is probably to be ascribed largely to hybridization effects at phosphorus compared to nitrogen [28, 50, 98] although there may be some hyperconjugative stabilization of phosphonium ions [50]. Phosphines have been shown to possess a low degree of s character in their bonding orbitals [97], whereas amines approach sp^3 hybridization in bonding [98]. Therefore, protonation of phosphorus is accompanied by a substantially larger change in electron deficiency (due to the higher increment in s character) in its bonding orbitals than is the protonation of nitrogen (both phosphonium and ammonium ions being sp^3 in character). Consequently, on protonation methyl groups respond to the greater change in electron deficiency at phosphorus with larger effects. Also, any hyperconjugative stabilization possible in the phosphonium ions through use of empty d-π orbitals [28, 50], presumably does not contribute in the case of the ammonium ions:

$$
\begin{array}{ccc}
\text{H} & & \text{H} \\
| & & | \\
\text{H}-\text{C}-\text{P} & \leftrightarrow & \text{H}^+\text{C}=\text{P} \\
| & & | \\
\text{H} & & \text{H}
\end{array}
$$

8

Staley and Beauchamp [51] have also found the delocalization energy [101, 102] of the diazobicyclooctane cation radical, **9**

9

to be about $14\,\text{kcal mol}^{-1}$ greater than for the quinuclidine cation radical **10**

10

(series 5 of Table 11). The fact that the $\delta_R \Delta H^0_{(6)}$ value for the latter is lower than for the cation radical of triethylamine is attributed to the greater constraint from coplanarity about $\text{N} \cdot {}^+$.

2.5 Acid strengths

McIver and Eyler [15] have studied the following specific proton-transfer equilibrium in the gas phase:

$$\text{SH}^- + \text{HCN} \rightleftharpoons \text{CN}^- + \text{H}_2\text{S}$$

A moderate medium effect of water is found: $K_{(g)} = 10$; $K_{(aq)} = 6.6 \times 10^{-3}$; $K_{(aq)}/K_{(g)} = 6.6 \times 10^{-4}$. It will be of interest to evaluate this medium effect further, by obtaining the relative thermodynamic properties for solution of the two gaseous neutral acids and hence of the two gaseous anions.

The gas-phase acidities of acetylene and hydrogen fluoride relative to methanol have been quantitatively determined at 298 K [39]. The value of $-\delta_R \Delta G^0_i/\text{kcal mol}^{-1}$ is 4.9 for acetylene (with statistical correction applied) and 6.6 for HF (the more positive value indicates greater acid strength). Kebarle [45–47] lists gaseous acidities at

600 K, increasing in the order: $H_2O < HF < H_2S <$ phenol $<$ formic acid $<$ HCl $<$ HBr $<$ trifluoroacetic acid $<$ HI. Here again, the orders of acidity among the acid functions with the simplest substituents are subject to such large general substituent effects [27, 39, 45–47], that this order is not generally applicable.

Table 12

Comparison of the effects of alkyl groups on acidity of alcohols and the basicity of amines

$$ROH_{(g)} + {}^-OMe_{(g)} \rightleftharpoons RO^-_{(g)} + MeOH_{(g)} \qquad (4)$$

$$RNH_{2(g)} + {}^+H_3NMe_{(g)} \rightleftharpoons RNH^+_{3(g)} + MeNH_{2(g)} \qquad (2)$$

| R | $-\delta_R \Delta G_i^0 / \text{kcal mol}^{-1}$ 298 K | |
	Reaction (4)[a]	Reaction (2)[b]
Me	0.0	0.0
Et	2.9	2.7
i-Pr	4.8	5.0
t-Bu	5.6	7.0
neo-Pentyl	6.9	5.8
(t-Bu)$_2$CH	8.6 ($\pm$0.3)	—
(t-Bu)$_3$C	11.6 ($\pm$1.0)	—

[a] *Cf.* reference [40]; [b] *cf.* Table 3.

The acidity of alcohols in the gas phase increases with increasing carbon substitution [39, 40]; this order is the opposite of the acidity order in solution [40]. The free-energy data are shown in Table 12; they confirm the qualitative gas-phase results first reported by Brauman and Blair [27]. The effects of successive substitution of methyl or t-butyl for hydrogen on the α-C atom are not additive but show a regular 'saturation' effect which is even greater than the one discussed for base strengths in Section 3.2 for the corresponding substitution in methylamine; the methyl and t-butyl substituent effects are in the ratio 1.00:1.60:2.00, compared with 1.00:1.85:2.60.

Brauman and Blair have argued that the polarization stabilization effect of alkyl groups should be essentially the same for negative as for corresponding positive ions [27]. Qualitative results for acidities of amines and alcohols were compared with corresponding results for basicities of amines in support of their argument.

Table 12 compares the quantitative data for the effects of methyl substitution at the α-carbon atom on the acidities of alcohols and on the basicities of corresponding amines. The substituent effects are

66

indeed parallel, but the substitutional patterns differ significantly in quantitative details. Thus, for the tertiary butyl group, reaction (2) gives a substantially larger effect (by 1.4 kcal-mol^{-1}) than does reaction (4). On the other hand for the neopentyl group, the converse holds (reaction (4) gives the larger effect by 1.1 kcal-mol^{-1}). The explanation may involve the internal inductive effect arising from the greater electronegativity of hydrogen than of sp^3 carbon [103] which is superimposed upon the larger but not necessarily equal corresponding polarization effects [27] for reactions (2) and (4). However additional comparisons of other (more polar) substituents for corresponding reactions (2) and (4) are required for critical tests of the hypothesis.

Table 13

Effects of alkyl substituents on the gas-phase acidity of acetylenes

$$RC\equiv CH_{(g)} + ^-\!:C\equiv C-H_{(g)} \rightleftharpoons RC\equiv C\!:^-_{(g)} + HC\equiv CH_{(g)} \quad (4)$$

R	$-\delta_R\Delta G^0_{i(g)}$/kcal mol^{-1} [b]
H	(0.0)[a]
t-Bu	-1.3
i-Pr	-2.9
Me	-4.7

[a] Statistical correction has been applied.
[b] Reference [39] and private communication from R. T. McIver, Jr.

Table 13 gives results for the effects of alkyl substituents upon the gas phase acidities of acetylenes [39]. The results are interpreted [39] as due to destabilization of negative charge by the methyl substituent's hybridization-inductive and hyperconjugative effects, which substantially predominate over the polarization effect. With t-butyl substitution, the stabilizing polarization effect most strongly opposes the destabilizing hybridization-inductive and hyperconjugative effects. Similar alkyl substituent effects appear in the gas-phase acidities of phenol [38] (Table 14). In the ortho position both methyl and t-butyl substituents increase acidity (the polarization effect being predominant), but in the more distant para position the methyl substituent decreases acidity (hybridization-polar and hyperconjugative effects predominant), while the t-butyl substituent increases acidity (polarization remains predominant). In the aqueous acidities of the phenols, which are also given in Table 14, the polarization effects of the alkyl group become minor or disappear, so that all alkyl groups decrease

Table 14

Substituent effects on the acidities of ortho, meta, and para substituted phenols

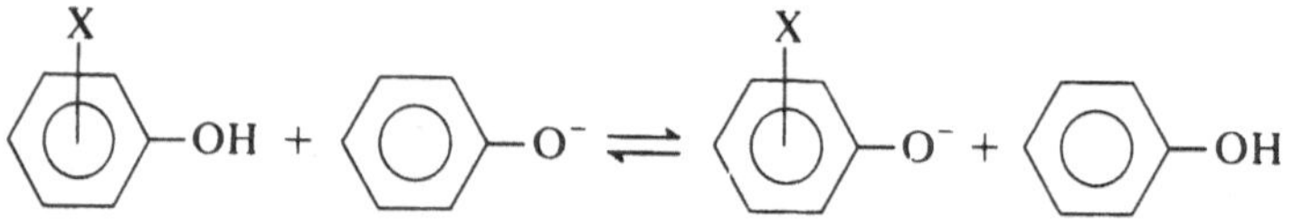

	$-\delta_R \Delta G_i^0$/kcal mol^{-1}					
	ortho		meta		para	
X	(g)	(aq)	(g)	(aq)	(g)	(aq)
Me	0.30	−0.40	−0.50	0.12	−1.20	−0.35
i-Pr	2.00				−0.20	
t-Bu	3.40	−1.80	0.50		0.60	−0.35
H	(0.00)	(0.00)	(0.00)	(0.00)	(0.00)	(0.00)
F	2.80	1.76	4.80	1.06	2.10	0.12
Cl	4.60	2.01	6.10	1.23	2.90	0.79

the acidity at all three ring positions. Steric inhibition of solvation of the phenoxide is probably an additional factor in the acidity of the ortho t-butyl phenol [104]. In this connection, evaluation of the thermodynamic properties of solution of the gaseous phenoxide ions would be of interest.

McIver has shown that a plot of the values of $-\delta_R \Delta G_{i(g)}^0$ against the corresponding values of $-\delta_R \Delta G_{i(aq)}^0$ in Table 14 for *m*- and *p*-CH$_3$, F, and Cl substituents gives rise to a roughly linear relationship of slope about 4.8 [38]. This attenuation factor for water solvent is appreciably larger than that, about 2, for the 4-substituted pyridines (Section 2.3.5). The results can be accounted for either by the model that assumes ion hydration by hydrogen bonding or by the model based on the effective dielectric constant. The observed gas-phase acidities of 2-fluoro- and 2-chloro-phenols are appreciably smaller (about 6 and 5 kcal mol^{-1} respectively) than expected from the aqueous acidities and the attenuation factor 4.8. This behaviour undoubtedly arises through internal hydrogen-bonding [105] stabilizing the molecular acids in the gas phase **11**:

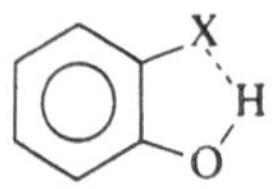

11

Internal hydrogen-bonding chelation in molecular acids may also play a very significant role in the gaseous acidities of halo (and other) substituted carboxylic acids.

Kebarle and his associates [45–47] have reported the gas-phase acidities of a number of carboxylic acids at 600 K. The results show general trends similar to those of aqueous acidities, but the correlation of the two is relatively poor. In addition to chelation effects, several other factors may be involved [45–47].

The gas-phase acidity of a molecule HX is related to the homolytic bond dissociation energy D_{H-X}, and the electron affinity (EA) of the variable X radical. With the gas-phase acidity and the D_{H-X} energy known, a method is provided for evaluating the electron affinity. Several interesting results have been obtained in this manner [39, 45–47], but a discussion of electron affinities is beyond the scope of this chapter.

2.6 Ion-molecule attachment equilibria

Kebarle has given an excellent summary [9] of his important studies on the energetics of attachment of 'solvent' molecules to ions in the gas phase. These studies include successive hydrations of alkali ions, halide ions and hydrogen ion; competitive solvations of hydrogen ion by water, methanol, and ammonia; and competitive solvations of chloride ion by various molecules, such as benzene, water, alcohols, aniline, acetonitrile, and phenol. In this section a very brief discussion is given of the results of preliminary studies of equilibria which appear to be of the types represented by equations 7, 8, 9, and 10 below.

McIver, Scott, and Riveros [37] have reported equilibrium constants at 298 K, for example, of each of reaction types 7, and 9:

$$CH_3O^-\cdots HOCH_{3(g)} + C_2H_5OH_{(g)}$$
$$= C_2H_5O^-\cdots HOCH_{3(g)} + CH_3OH_{(g)}; \qquad (7)$$
$$-\Delta G^0 = 1.2 \pm 0.2 \, \text{kcal mol}^{-1}$$

$$C_2H_5O^-\cdots HOCH_{3(g)} + C_2H_5OH_{(g)}$$
$$= C_2H_5O^-\cdots HOC_2H_{5(g)} + CH_3OH_{(g)}; \qquad (9)$$
$$-\Delta G^0 = 0.4 \pm 0.2 \, \text{kcal mol}^{-1}$$

This reaction (and subsequent ones) is written so as to differentiate ions and molecules and there is no intention to imply structures for the ion-molecule complexes. Since the simple proton-transfer equilibrium for unsolvated ions, equation 4, gives $-\Delta G^0 = 1.9 \, \text{kcal mol}^{-1}$,

$$CH_3O_{(g)}^- + C_2H_5OH_{(g)} = C_2H_5O_{(g)}^- + CH_3OH_{(g)} \qquad (4)$$

McIver, Scott, and Riveros interpret the result for reaction 7 to indicate that the attachment of a single molecule has started to bring about the inversion in acid strengths of the two alcohols which occurs between the gas phase and solution. The mechanism by which the 'solvent' molecule performs this action is of great interest and is the subject of continuing investigations both with anions and cations. The result for reaction 9 corresponds to the formation of stronger hydrogen bonds by the stronger proton donor [35]. However, it is likely that this generalization holds only for hydrogen-bonds having the same framework, i.e. $X \cdots H \cdots Y$ with the same X and Y atoms [23].

Equilibrium constants for reactions of the type 8 and 10 have been obtained by three groups. For the reaction:

$$Me_3NH^+ \cdots NHMe_{2(g)} + Me_3N$$

$$\rightleftharpoons Me_3NH^+ \cdots NMe_{3(g)} + NHMe_{2(g)} \quad (10a)$$

Wolf, McIver, and Taft [20] have obtained $-\Delta G^0 = 0.5\,\text{kcal mol}^{-1}$ at 298 K, and Yamdagni and Kebarle [19] report $-\Delta G^0 = 0.1\,\text{kcal mol}^{-1}$ at 550 K. Since reactions of type 10 may have significant values of ΔC_p, the values of ΔH^0 and ΔS^0, as well as ΔG^0 values, may be too temperature-dependent to be reliably compared over such a wide temperature interval. In this light, the agreement appears satisfactory. The result indicates stronger attachment to a given cationic acid by the more basic 'solvent' molecule, at least when the same hydrogen-bond framework is involved.

For the reaction:

$$Me_3NH^+ \cdots NMe_{3(g)} + Et_2NH_{(g)}$$

$$\rightleftharpoons Me_3NH^+ \cdots NHEt_{2(g)} + Me_3N_{(g)} \quad (10b)$$

Bowers, Aue, and Webb [21] have obtained $-\Delta G^0 = 1.2\,\text{kcal mol}^{-1}$, and Wolf, McIver, and Taft [20] have found $-\Delta G^0 = 0.9\,\text{kcal mol}^{-1}$, both at 298 K. The reaction is spontaneous in the direction of attachment of the amine molecule of greater base strength, but the value of $-\Delta G^0$ is greater than the base-strength difference (Table 3) of only $0.2\,\text{kcal mol}^{-1}$. Evidently, the ion-molecule complexes having different base moieties (hetero-dimers) rather than the same (homo-dimers) have the greater stability, provided that the acid and base strengths involved are sufficiently similar and steric effects are unimportant. The favourable formation of hetero-dimer (by factors greater than symmetry number effects [21]) is clearly illustrated by the results of

Wolf, McIver, and Taft [20] for two additional equilibria of type 10:

$$Me_3NH^+ \cdots NMe_{3(g)} + C_5H_5N_{(g)}$$
$$\rightleftharpoons Me_3NH^+ \cdots NC_5H_{5(g)} + NMe_{3(g)}; \qquad (10c)$$
$$-\Delta G^0 = 1.0 \, kcal \, mol^{-1}$$

$$Et_2NH^+ \cdots NHEt_{2(g)} + C_5H_5N_{(g)}$$
$$\rightleftharpoons Et_2NH^+ \cdots NC_5H_{5(g)} + NHEt_{2(g)}; \qquad (10d)$$
$$-\Delta G^0 = 1.3 \, kcal \, mol^{-1}$$

Both reactions are substantially spontaneous, even though the gaseous base-strength of pyridine is 4.0 kcal mol^{-1} *less* than that of trimethylamine and 4.2 kcal mol^{-1} less than that of the diethylamine (Table 3). Besides hetero-dimer formation, reaction 10d probably involves also, an unfavourable steric effect for the diethylamine homo-dimer [21].

The limited preliminary studies of reactions of type 8 include the following specific example of interest [20]:

$$C_5H_5NH^+ \cdots NC_5H_{5(g)} + Me_3N_{(g)}$$
$$\rightleftharpoons Me_3NH^+ \cdots NC_5H_{5(g)} + C_5H_5N_{(g)}; \qquad (8)$$
$$-\Delta G^0 = 2.3 \, kcal \, mol^{-1}$$

This result apparently does not correspond to that of McIver for reaction 7. While the attachment of the pyridine 'solvent' molecule does reduce the magnitude of the structural effect on the gaseous proton transfer ($-\Delta G^0 = 4.0 \, kcal \, mol^{-1}$), this is in the opposite direction to the larger 'structural effect' observed in solution (Table 3), *i.e.* $-\Delta G^0_{(2)(aq)} = 6.25 \, kcal \, mol^{-1}$ and $-\Delta G^0_{(2)(MeCN)} = 6.58 \, kcal$ mol^{-1}. The latter values are more favourable than that for the gas-phase proton-transfer reaction because there is a polarization effect favouring $C_5H_5NH^+_{(g)}$ over $Me_3NH^+_{(g)}$ which opposes the predominant hybridization-inductive effect (Section 2.3.3). Clearly the pyridine 'solvent' molecule attached to the acidic cations in reaction 8 does not here play the same kind of role (Section 2.3.3) as the bulk solvent.

2.7 Summary

The effects of molecular structure on acid and base strengths in proton-transfer equilibria have been found to exhibit gross differences in the gas phase compared to the corresponding reactions in solvents.

The following are examples from among many which are given in Table 3 which illustrate and dramatize these differences. Aniline is 4.6 powers of ten less basic than ammonia (reaction 2) in water, whereas the former is 4.9 powers of ten *stronger* as a base in the gas phase. Similarly, pyridine is 4.0 powers of ten less basic than ammonia in water, but is 11.7 powers of ten *stronger* in the gas phase. While trimethylamine is only 0.55 powers of ten stronger base than ammonia in water, it is 14.6 powers of ten stronger in the gas phase. Trimethylamine is more basic than trimethylphosphine in water by 1.1 powers of ten, but in the gas phase the former is *weaker* by 0.9 power of ten. Trimethylamine is 6.7 powers of ten more basic than tripropargyl amine in water, yet in the gas phase the former is only 3.7 powers of ten stronger than the latter. Finally, tetramethylethylene diamine has about the same base strength as ammonia in water, yet in the gas phase it is 22.2 powers of ten stronger.

In the light results such as these, one is immediately faced with the possible need to abandon theories which have evolved from, or utilized for support, measurements of the effects of molecular structure on acid and base strengths in solution. However, more careful scrutiny at this time indicates instead that the gas-phase results offer the most compelling evidence yet achieved in support of many of these theories. It is the major alterations in the relative importances of the various effects of molecular structure in the gas phase as compared to solution which lead to the observations which at first sight appear so drastically unrelated to prior experience. In this and the companion chapter by Arnett we have presented a current assessment of how the observed effects of molecular structure on both gas-phase and solution proton-transfer equilibria may be accounted for in terms of existing theories.

Resonance (or π-electron delocalization) and hybridization effects of hydrocarbon substituents have been identified and evaluated in gas-phase proton-transfer equilibria involving nitrogen bases. Neither of these effects is strongly solvent-dependent. A similar situation may apply for internal inductive effects, but more critical experiments are required to establish this point. Polarization effects of hydrocarbon substituents are a major structural effect in gas-phase base and acid strengths, and depend strongly on the number of carbon atoms in proximity to the cationic centre. Although this effect is frequently the predominant one in the gas phase, it is not easily recognized in proton-transfer equilibria in solution. The means by which large polarization effects on the standard free energy change in the gas phase are reduced to near zero in aqueous solution appears to involve dispersal of the cationic charge by hydrogen bonding. Hydrogen bonding chelation is a second interaction of importance in gas-phase proton-transfer equilibria which is strongly modified or erased by bulk (protonic) solvent.

Dipolar substituents give rise to much larger polar effects in gas-phase proton-transfer equilibria than for the corresponding equilibria in polar solvents. The magnitude of the aqueous solvent 'attenuation' factor has been found to increase with the length of the chain between the substituent and the charge centre, as well as with increasing hydrogen-bonding solvation of the charge centre.

The relationship of solvent effects on proton-transfer equilibria to the number and the acidity of acid protons in cationic conjugate acids (BH^+) is considered in detail in Chapter 3. A similar analysis of solvent effects on the acidities of hydroxylic compounds has not been achieved but can be expected for the near future.

The absence of steric effects in gas-phase proton-transfer equilibria has been shown to aid greatly the evaluation of steric effects on solvation and in the formation of Lewis acid-base complexes.

A general level of agreement has been established between experimental results and theoretical calculations of the energetics of proton transfer in the gas phase. The results indeed appear encouraging for the use of theory to obtain needed molecular structures of organic cations and anions.

Marked contrasts in the effects of molecular structure on hydrogen-atom transfer reactions (equation 6) and the corresponding proton-transfer reactions (equation 2) have been found. The former reaction involves no change in charge (all species are univalent cations). Consequently polarization effects of hydrocarbon substituents are minimal, but large effects of delocalization of charge and spin densities in the cation radicals are observed.

The attachment of single molecules to cations and anions in the gas phase offers potential insights into the manner of action of bulk solvent molecules. Some results from preliminary studies of two types of attachment exchange reactions for cations and anions (equations 7, 8 and 9, 10) have been presented.

Acknowledgements

It is a pleasure to acknowledge the assistance of all of my collaborators in this work. Particular thanks are due to Professors J. L. Beauchamp and R. T. McIver, Jr. for their instruction with the ICR measurements, as well as for valuable comments, criticisms, and permission to quote unpublished results. Professor E. M. Arnett has contributed greatly to the collaboration on thermodynamic analyses of amine basicities in solution and provided many helpful comments. I am indebted to Professors D. H. Aue and M. T. Bowers for sharing unpublished results and for helpful discussions. Professors D. K. Bohme and P. Kebarle have generously provided unpublished results. Professor W. J. Hehre has kindly made available the results of his STO-3G calculations. I am indebted to him for this and for numerous helpful discussions. Thanks are due for partial support of this work from grants received from the National Science Foundation and the Public Health Service. Finally, I acknowledge the use and inspiration of ideas of many others; in particular I have drawn heavily from the authors of references 1–7.

REFERENCES

[1] R. P. Bell, *The Proton in Chemistry*, Cornell University Press, Ithaca, N.Y. (1973).

[2] L. P. Hammett, *Physical Organic Chemistry*, McGraw-Hill Book Co., N.Y. (1970).

[3] G. E. K. Branch and M. Calvin, *The Theory of Organic Chemistry*, Prentice-Hall, N.Y. (1941).

[4] C. K. Ingold, *Structure and Mechanism in Organic Chemistry*, Cornell University Press, Ithaca, N.Y. (1953).

[5] G. W. Wheland, *Resonance in Organic Chemistry*, John Wiley and Sons, N.Y. (1955).

[6] C. A. Coulson, *Valence*, Oxford University Press, Oxford (1956).

[7] H. C. Brown, D. H. McDaniel, and O. Häfliger, *Determination of Organic Structures by Physical Methods*, E. A. Braude and F. C. Nachod, Editors, Academic Press, N.Y., 1, 634–643 (1955).

[8] P. Love, R. B. Cohen and R. W. Taft, *J. Amer. Chem. Soc.*, **90**, 2455 (1968).

[9] P. Kebarle, *Ions and Ion Pairs in Organic Reactions*, M. Szwarc, Ed., Wiley-Interscience, N.Y., (1972) Chapter 2.

[10] D. K. Bohme and L. B. Young, *J. Amer. Chem. Soc.*, **92**, 3301 (1970).

[11] R. T. McIver, Jr., *Rev. Sci. Instrum.*, **41**, 555 (1970).

[12] T. B. McMahon and J. L. Beauchamp, *ibid.*, **43**, 509 (1972).

[13] R. T. McIver, Jr. and R. C. Dunbar, *Int. J. Mass Spectrom. Ion. Phys.*, **7**, 471 (1971).

[14] M. T. Bowers, D. H. Aue, H. M. Webb, and R. T. McIver, Jr., *J. Amer. Chem. Soc.*, **93**, 4313 (1971).

[15] R. T. McIver, Jr. and J. R. Eyler, *ibid.*, **93**, 6334 (1971).

[16] D. H. Aue, H. M. Webb, and M. T. Bowers, *ibid.*, **94**, 4726 (1972).

[17] J. L. Beauchamp, *Ann. Rev. Phys. Chem.*, **22**, 527 (1971).

[18] W. G. Henderson, M. Taagepera, D. Holtz, R. T. McIver, Jr., J. L. Beauchamp, and R. W. Taft, *J. Amer. Chem. Soc.*, **94**, 4728 (1972).

[19] R. Yamdagni and P. Kebarle, *J. Amer. Chem. Soc.*, **95**, 3504 (1973).

[20] J. F. Wolf, R. T. McIver, Jr., and R. W. Taft, unpublished results.

[21] D. H. Aue, H. M. Webb, and M. T. Bowers, *Abstracts, 21st Ann. Conf. on Mass Spectrometry and Applied Topics*, San Francisco, May 20, 1973, p. 183.

[22] H. Hiraoke, E. P. Grimarud, and P. Kebarle, *J. Amer. Chem. Soc.*, **96**, 3359 (1974).

[23] R. W. Taft, D. Gurka, L. Joris, P. von R. Schleyer, and J. W. Rakshys, *J. Amer. Chem. Soc.*, **91**, 4801 (1969).

[24] L. Joris, J. Mitsky, and R. W. Taft, *ibid.*, **94**, 3438 (1972).

[25] J. Mitsky, L. Joris, and R. W. Taft, *ibid.*, **94**, 3442 (1972).

[26] M. S. B. Munson, *ibid.*, **87**, 2332 (1965).

[27] J. I. Brauman and L. K. Blair, *ibid.*, **90**, 5636, 6501 (1968); **91**, 2126 (1969); **92**, 5986 (1970); **93**, 3911, 4315 (1971).

[28] D. H. McDaniel, N. B. Coffman, and J. M. Strong, *ibid.*, **92**, 6697 (1970).

[29] M. A. Haney and J. L. Franklin, *J. Phys. Chem.*, **73**, 4328 (1969).

[30] D. K. Bohme, E. Lee-Ruff, and L. B. Young, *J. Amer. Chem. Soc.*, **94**, 5153 (1972).

[31] M. S. Foster and J. L. Beauchamp, *ibid.*, **94**, 2425 (1972).

[31a] D. Holtz, J. L. Beauchamp, W. G. Henderson, and R. W. Taft, *Inorg. Chem.*, **10**, 201 (1971).

[32] P. C. Isolani, J. M. Riveros, and P. W. Tiedeman, *J. Chem. Soc., Faraday Transactions II*, **69**, 1023 (1973).

[33] J. Long and B. Munson, *J. Amer. Chem. Soc.*, **95**, 2427 (1973).

[34] J. P. Briggs, R. Yamdagni, and P. Kebarle, *J. Amer. Chem. Soc.*, **94**, 5128 (1972).

[35] R. Yamdagni and P. Kebarle, *ibid.*, **95**, 3504 (1973).

[36] D. H. Aue, H. M. Webb, and M. T. Bowers, *J. Amer. Chem. Soc.*, **95**, 2699 (1973).

[37] R. T. McIver, Jr., J. A. Scott, J. M. Riveros, *J. Amer. Chem. Soc.*, **95**, 2706 (1973).

[38] R. T. McIver, Jr., and J. H. Silvers, *ibid.*, **95**, 8462 (1973).

[39] R. T. McIver, Jr. and J. S. Miller, *ibid.*, **96**, 4323 (1974).

[40] E. M. Arnett, L. E. Small, R. T. McIver, Jr., and J. S. Miller, *ibid.*, **96**, 5638 (1974). Subsequent revision (McIver, private communication) of these results has been made.

[41] R. W. Taft, M. Taagepera, K. D. Summerhays, and J. Mitsky, *J. Amer. Chem. Soc.*, **95**, 3811 (1973).

[42] E. M. Arnett, F. M. Jones III, M. Taagepera, W. G. Henderson, J. L. Beauchamp, D. Holtz, and R. W. Taft, *ibid.*, **94**, 4724 (1972). Values reported in this paper have subsequently been revised slightly.

[43] J. Echols, M. Taagepera, K. D. Summerhays, and R. W. Taft, unpublished results.

[44] D. H. Aue, H. M. Webb, and M. T. Bowers, private communication.

[45] R. Yamdagni and P. Kebarle, *J. Amer. Chem. Soc.*, **95**, 4050 (1973).

[46] K. Hiraoka, R. Yamdagni, and P. Kebarle, *ibid.*, **95**, 6833 (1973).

[47] R. Yamdagni, T. B. McMahon, and P. Kebarle, *J. Amer. Chem. Soc.*, **96**, 4035 (1974).

[48] D. K. Bohme, E. Lee-Ruff, and L. B. Young, *J. Amer. Chem. Soc.*, **93**, 4608 (1971).

[49] D. K. Bohme, private communication.

[50] R. H. Staley and J. L. Beauchamp, *J. Amer. Chem. Soc.*, **96**, 6252 (1974).

[51] R. H. Staley and J. L. Beauchamp, *ibid.*, **96**, 1604 (1974).

[52] R. T. McIver, Jr. and P. W. Tiedeman, private communication.

[53] K. M. A. Refaey and W. A. Chupka, *J. Chem. Phys.*, **48**, 5205 (1968).

[54] J. Echols, K. D. Summerhays, and R. W. Taft, unpublished results.

[55] J. Echols and R. W. Taft, unpublished results.

[56] W. G. Henderson, M. Taagepera, D. Holtz, J. L. Beauchamp, and R. W. Taft, unpublished results.

[57] M. Taagepera, K. D. Summerhays, T. B. McMahon, J. L. Beauchamp, and R. W. Taft, unpublished results.

[58] M. Taagepera and R. W. Taft, unpublished results.

[59] K. D. Summerhays and R. W. Taft, unpublished results.

[60] M. Taagepera, J. Mitsky, B. M. Wepster, and R. W. Taft, unpublished results.

[61] K. D. Summerhays, F. E. Condon, and R. W. Taft, unpublished results.

[62] J. Mitsky and R. W. Taft, unpublished results.

[63] E. M. Arnett, private communication, based upon critical compilation of literature values.

[64] Taken from a critical review of literature values, F. M. Jones III, Ph.D. Thesis, University of Pittsburgh, 1970; *cf.* also F. M. Jones III and E. M. Arnett, *Progr. Phys. Org. Chem.*, **11**, 263 (1974).

[65] D. Northcolt and R. E. Robertson, *J. Phys. Chem.*, **73**, 1559 (1969).

[66] L. Spialter and J. A. Pappalardo, *The Acyclic Aliphatic Tertiary Amines*, The Macmillan Company, N.Y. (1965).

[67] C. L. Liotta, E. M. Perdue, and H. P. Hopkins, Jr., *J. Amer. Chem. Soc.*, **95**, 2439 (1973).

[68] N. F. Hall and M. R. Sprinkle, *ibid.*, **54**, 3469 (1932).

[69] H. E. Folkers and O. Runquist, *J. Org. Chem.*, **29**, 830 (1964).

[70] B. M. Wepster, *Progr. in Stereochemistry*, Vol. 2, W. Klyne and P. B. de la Mare, Editors, Butterworths, London, (1958).

[70a] H. C. Brown, S. H. Brewster, and H. Schechter, *J. Amer. Chem. Soc.*, **76**, 467 (1954); J. W. Eastes, M. H. Aldridge, R. R. Minesinger, and M. J. Kamlet, *J. Org. Chem.*, **36**, 3847 (1971), A. J. Kresge, P. H. Fitzgerald and Y. Chiang, *J. Amer. Chem. Soc.*, **96**, 4698 (1974).

[71] B. M. Wepster, reference [70], gives pK_A values in 50 per cent aq. EtOH.

[72] P. Paoletti, J. H. Stern, and A. Vacca, *J. Phys. Chem.*, **69**, 3759 (1965).

[73] R. L. Hinman, *J. Org. Chem.*, **23**, 1587 (1958).

[74] D. D. Perrin, *Dissociation Constants of Organic Bases in Aqueous Solution*, Butterworths, London (1965).

[75] W. A. Henderson and C. A. Streuli, *J. Amer. Chem. Soc.*, **82**, 5791 (1960).

[76] L. Sacconi, P. Paoletti, and M. Ciampolini, *ibid.*, **82**, 3831 (1960).

[77] D. Landini, G. Modena, G. Scorrano, and F. Taddei, *J. Amer. Chem. Soc.*, **91**, 6703 (1969).

[78] J. F. Coetzee, *Progr. Phys. Org. Chem.*, **4**, 76 (1967).

[79] D. Gurka, R. W. Taft, L. Joris, and P. von R. Schleyer, *J. Amer. Chem. Soc.*, **89**, 5957 (1967).

[80] J. I. Brauman and L. K. Blair, *J. Amer. Chem. Soc.*, **92**, 5986 (1970).

[81] B. M. Wepster, *Recl. Trav. Chim. Pays-Bas*, **71**, 1171 (1952).

[82] F. E. Condon, *J. Amer. Chem. Soc.*, **87**, 4485 (1965).

[83] W. J. Hehre, private communication.

[84] J. D. Cox and G. Pilcher, *Thermochemistry of Organic and Organo-metallic Compounds*, Academic Press (1970), p. 557, 570.

[85] H. C. Brown *et al.*, *J. Amer. Chem. Soc.*, **75**, 1 (1953).

[86] W. G. Henderson, M. Taagepera, J. L. Beauchamp, and R. W. Taft, unpublished results.

[87] R. W. Taft and M. Taagepera, Abstracts of the 167th ACS National Meeting, Los Angeles, April, 1974, Orgn-61; an earlier report based upon less accurate methods has appeared: M. Taagepera, W. G. Henderson, R. T. C. Brownlee, J. L. Beauchamp, D. Holtz, and R. W. Taft, *J. Amer. Chem. Soc.*, **95**, 1369 (1972).

[88] T. H. Morton and J. L. Beauchamp, *J. Amer. Chem. Soc.*, **94**, 3671 (1972).

[89] D. H. Aue, H. M. Webb, and M. T. Bowers, *ibid.*, **95**, 2699 (1973).

[90] J. G. Kirkwood and F. H. Westheimer, *J. Chem. Phys.*, **6**, 506 (1938).

[91] F. H. Westheimer and J. G. Kirkwood, *ibid.*, **6**, 513 (1938).

[92] S. Ehrenson, R. T. C. Brownlee, and R. W. Taft, *Progr. Phys. Org. Chem.*, **10**, 1 (1973).

[93] R. Yamdagni, T. B. McMahon, and P. Kebarle, unpublished results.

[94] W. J. Hehre, R. F. Stewart, and J. A. Pople, *J. Chem. Phys.*, **51**, 2657 (1969).

[95] W. J. Hehre, unpublished results.

[96] S. W. Benson, *J. Amer. Chem. Soc.*, **80**, 5151 (1958).

[97] M. F. Guest, I. H. Hillier, and V. R. Saunders, *J. Chem. Soc.*, **68**, 867 (1972).

[98] J. R. Weaver and R. W. Parry, *J. Inorg. Chem.*, **5**, 718 (1966).

[99] J. H. Gibbs, *J. Chem. Phys.*, **22**, 1460 (1954).

[100] R. H. Staley and J. L. Beauchamp, private communication.

[101] R. Hoffmann, A. Imamura, and W. J. Hehre, *ibid.*, **90**, 1499 (1968).

[102] R. Hoffmann, *Accts. Chem. Res.*, **4**, 1 (1971).

[103] W. Moffitt, *Proc. Roy. Soc.* (*London*) A202, 548 (1950).

[104] C. H. Rochester, *J. Chem. Soc.*, London, 4603 (1965).

[105] A. W. Baker, H. O. Kerlinger, and A. T. Shulgin, *Spectrochim. Acta.*, **20**, 1467 (1964).

[106] J. F. Wolf, I. Koppel, R. H. Staley, R. T. McIver, Jr., J. L. Beauchamp, and R. W. Taft, unpublished results.

[107] J. F. Wolf, P. G. Harsch, and R. W. Taft, *J. Amer. Chem. Soc.*, in press.

[108] J. F. Wolf, P. G. Harsch, R. W. Taft, and W. J. Hehre, *ibid.*, in press.

[109] N. C. Deno, P. T. Groves, and G. Saines, *ibid.*, **81**, 5790 (1959).

[110] H. C. Brown and B. Kanner, *ibid.*, **88**, 986 (1966).

3

Edward M. Arnett

PROTON-TRANSFER AND THE SOLVATION OF AMMONIUM IONS

3.1 Introduction

In the following article* we will compare the relative protonation energies of a number of amines at 25°C in three media; the gas phase, fluorosulfuric acid (HSO_3F), and water. By means of a simple thermodynamic cycle we will analyse the liquid-phase data to derive the relative solvation energies of the ammonium ions from the gas phase to HSO_3F and to water. We will discuss these results in terms of the interplay between the internal and external factors which stabilize ions by charge distribution.

In several recent articles we have treated closely-related aspects of this problem, namely the general approach to gas-phase proton-transfer study and its significance for solvation, the thermodynamics of ionization and solution of amines in water [3], the ionization of bases and solvation of their ions [4]. Although they are published separately these four articles are intended to be complementary. We will use some of the data tabulated in the other articles, but the focus of the present one is quite different and we have tried to reduce overlap to a minimum.

Amines will be considered here almost exclusively, and only those amines for which there are sufficient data to allow a complete analysis of proton transfer thermodynamics in solution and the gas phase.

* This chapter is a progress report to June 1, 1974 on a collaborative study with Professor Robert Taft, the author of the previous chapter [1]. Although his contribution might have been written without reference to ours, we are entirely dependent on his work and that of Professors J. L. Beauchamp (California Institute of Technology) and D. H. Aue (University of California at Santa Barbara) and their students who have shared freely the gas phase proton-transfer data which are essential to this study.

The amines have been chosen for this purpose since they are the only series of neutral bases for which the thermodynamics of ionization and solution can be completely and unequivocally determined in aqueous solution at this time. By means of appropriate substitution it is possible to produce considerable variation in their thermodynamic properties for proton transfer. In some cases substituent changes can be related convincingly to inductive, resonance and steric effects [1]. Furthermore, some of the important structural requirements for solvation may be specified.

It is appropriate here to emphasize the contributions of Professor R. P. Bell and his students to this field. Much of the pioneering experimental work about solvent effects on acid-base equilibria was done in his laboratory. His book [5] has been a primary resource and inspiration to students of proton transfer. Of particular importance to the discussion below is his work with Bascombe [6] and Trotman–Dickenson [7], which used the concepts of specific hydrogen bonding from ammonia and ammonium ions to water to interpret acid-base phenomena in aqueous media. Twenty-five years later we can see that their basic formulation is essentially correct.

Experimental approach. Before presenting the results of our investigations, a brief description of the experimental methods used here may be in order. The key to calculating solvation energies of ammonium ions is provided by several new mass-spectrometric methods which allow the determination of equilibrium constants for proton exchange reactions such as

$$NH_4^+ + B \rightleftharpoons BH^+ + NH_3 \qquad (1)$$

where B is a Brønsted base, in the gas phase [1, 2]. These methods use various means of approaching a Boltzmann distribution of energies at 25°C through allowing an adequate number of ion-molecule collisions for the dissipation of excitation energy. Ion cyclotron resonance [1] achieves this by holding the ions for a comparatively long time in a circular orbit in the presence of other molecules and ions. Methods for studying proton-transfer reactions in aqueous solution are described and referenced extensively elsewhere [2, 3, 4].

Whenever discrepancies arise, as they commonly do, between gas-phase and condensed-phase proton-transfer thermodynamics, the source may be assigned unequivocally to solvent effects. These factors may in turn be analysed into contributions from the solvation of the neutral basic species (NH_3 and B) and the corresponding ions (NH_4^+ and BH^+), by means of the simple thermodynamic cycle below. Here P stands for any thermodynamic property (G, H, S) and the other symbols mean: g, gas; l, liquid; i, ionization; s, solvation

$$\Delta P_i^0(g) \quad NH_4^+(g) \;+\; B(g) \;\rightarrow\; NH_3(g) \;+\; BH^+(g)$$

$$\left\downarrow{\scriptstyle \Delta P_s(NH_4^+)} \qquad \left\downarrow{\scriptstyle \Delta P_s(B)} \qquad \left\downarrow{\scriptstyle \Delta P_s(NH_3)} \qquad \left\downarrow{\scriptstyle \Delta P_s(BH^+)}$$

$$\Delta P_i^0(l) \quad NH_4^+(l) \;+\; B(l) \;\rightarrow\; NH_3(l) \;+\; BH^+(l)$$

Figure 1

from gas to liquid phase. We will generally express the values for other amines (B) and their ions (BH$^+$) relative to those for NH_3 or NH_4^+ by using the symbol δ [8]. Thus $\delta\Delta G_s^0(BH^+)$ would mean: the standard free energy of solvation of the ammonium ion, BH$^+$, relative to that for NH_4^+. The use of relative energies removes the question of standard states from discussion here, although it is referred to later for comparison with other published data.

The necessary thermodynamic properties for solution of the neutral amines are tabulated elsewhere [3]. Values of ΔH_s^0 may be determined directly by calorimetry or from temperature coefficients of Henry's Law constants. Values of ΔS_s^0 are obtained directly from ΔH_s^0 and ΔG_s^0 through the Gibbs–Helmholtz equation. Finally, it follows from the above cycle, Figure 1, that $\delta\Delta P_s(BH^+) = \delta\Delta P_i(l) - \delta\Delta P_i(g) + \delta\Delta P_s(B)$.

All three properties $(\Delta G^0, \Delta H^0, \Delta S^0)$ are available for all four processes shown in the cycle for quite a few amines when water is the liquid phase. In favourable cases it is therefore possible to work through a complete analysis in terms of all twelve thermodynamic properties. This was done for twelve amines in our initial paper on this topic [9, 10].

A complete enthalpy analysis is also possible for ionization and solvation in HSO_3F. This super-acid [11] is a particularly good medium for the study of 'onium ions [4, 11, 12] for a number of reasons, including its low basicity, its probably high dielectric constant [13], and its fluidity down to $-90°C$ [14]. Virtually all amines (and most other Brønsted bases) are converted so completely to their conjugate acids in this medium that it is impossible to study their equilibrium properties for proton transfer or solution. Therefore ΔG_i^0, ΔG_s^0 and the corresponding entropy terms cannot be estimated at this time.

However, no such problems apply to determining the heats of solution $\Delta H_s(B)$ of amines or other bases in HSO_3F [4, 15]. The only difficulty is that B not only dissolves in the acid; it is converted instantly to BH^+. In order to correct for this, the heat of solution of B is also measured in an 'inert' solvent such as carbon tetrachloride or o-dichlorobenzene and it is assumed that these ΔH_s values will be proportional to those which would be found in HSO_3F if protonation did not occur. The pros and cons of this assumption are considered elsewhere [4, 15] and it is unlikely that it would contribute more than a two per cent error to most of our estimates of $\delta\Delta H_i(BH^+)$ in HSO_3F.

3.2 Proton-transfer energies in the gas phase

Values for $\delta\Delta G_i^0$ in the first column of Table 1 are the best currently available for gas-phase proton-transfer energies of the species shown and are quoted directly from Taft [1]. In this section there is some repetition of points developed by him in much greater detail. Our discussion is designed to give the reader a general orientation for comparing gas-phase basicities to those in solution. In most cases the relative values are probably accurate to ± 0.4 kcal mol^{-1} and the standard errors are in the neighbourhood of ± 0.2 kcal mol^{-1}. Most of the measurements are based on a series of stepwise equilibrations of compounds whose basicities are similar enough to permit observation of both neutral molecules and their ions as a function of the respective partial pressures of the neutral molecules.

In most cases the position of protonation is unequivocal. However, in others there are several alternative basic sites (e.g. the amino function and the benzene ring in aniline), and there is no direct way of proving which group is the actual proton acceptor when such bases undergo reaction[1] in the gas phase.

In using these data later we will differentiate between the free energy term [$\delta\Delta G_i^0$(gas)] and the enthalpy value [$\delta\Delta H_i^0$(gas)]. There is a small formal entropy difference between these properties, depending on the number of acidic protons on the nitrogen of the ammonium ion [9]. This difference amounts to at most ± 0.5 kcal mol^{-1} in comparing primary with tertiary ions, and is small compared to accumulated errors from other sources. The experimental entropy terms are not known exactly because they are at present too small to measure, a fact which qualitatively supports the small formal difference. One may expect that whenever comparisons are made between bases whose molecular frameworks are of very different size and rigidity small additional entropy terms will be found.

The effect of molecular structure on ΔG_i^0 (gas) shows that in the absence of solvent, charge delocalization within the base is extremely sensitive to substitution. For the first time it is possible to consider unequivocally such 'inherent' factors as inductive, resonance and hybridization effects in relation to experimental data. Such quantitative gas-phase protolysis data often show inverted basicity orders compared to the familiar condensed-phase ones [9, 16, 17].

Comparison of the simple primary, secondary, and tertiary alkylamines displays the classical inductive effect in its clearest experimental terms. Alkyl groups are obviously quite polarizable both for stabilizing positive or negative [16, 17] charges in the gas phase. Closely analogous patterns are also found for the protonation of other basic heteroatoms in the alcohols, ethers, mercaptans, sulfides and phosphines [1, 2, 18, 19]. As Libit and Hoffmann have demonstrated recently [20] the ability of methyl groups to stabilize charges is by no means simple in ultimate theoretical terms. For our purposes, and probably those of most experimentalists, the inductive effect as revealed by gas-phase proton-transfer reactions is thought of most conveniently as dispersion of positive or negative charge through polarization of the various bonds in the molecule. In these terms the notion that alkyl groups reduce the stability of negative ions by 'pushing' electrons is obsolete. That reasonable but erroneous conclusion was founded previously on the demonstrated ability of alkyl groups to stabilize positive charges. Now we can see that this fact does not require that they destabilize negative ones.

Comparison of alkylamines, especially those of varying substitution on the nitrogen atom, shows that the ability of alkyl groups to stabilize the ammonium ion is sharply dependent on the proximity of the alkyl groups to the seat of the charge. Thus the following isomeric $C_4H_9NH^+$ ions fall in order of decreasing stability relative to their neutral molecules: $Me_2EtNH^+ > Et_2NH_2^+ > $ t-butyl $NH_3^+ \approx$ i-BuNH$_3^+ > $ n-BuNH$_3^+$. Again the charge-stabilizing effect of a terminal methyl group decreases rapidly in the series $MeNH_3^+ > EtNH_3^+ > $ n-PrNH$_3^+ > $ n-BuNH$_3^+$, and the charge destabilizing effect of fluorines is subject to similar rules [1]. Other electronegative groups such as NH_2, CN, allyl and propargyl are also strongly base weakening. Comparison of the difference in $\delta\Delta G_i^0$ between n-propylamine and propargylamine (6.3 kcal mol^{-1}) with that between tri-n-propylamine and tri-propargylamine (13.7 kcal mol^{-1}) shows that the effect may fall far short of additivity.

Comparison of the cyclic amines, both secondary and tertiary, with their acylic analogues shows the same feeble response to increasing carbon number after the first two or three methylenes. At this point

Table 1

Relative proton-transfer energies in kcal mol^{-1} in gas phase, HSO_3F and H_2O at 25°C

$$NH_4^+ + B \rightleftharpoons NH_3 + BH^+$$

Base	$-\delta\Delta G_i^0$ gas	$-\delta\Delta H_i^0$ HSO_3F	$-\delta\Delta G_i^0$ H_2O	$-\delta\Delta H_i^0$ H_2O
Primary RNH$_2$				
MeNH$_2$	9.1	3.00	1.92	0.70
EtNH$_2$	11.8	3.50	1.96	1.23
n-Pr—NH$_2$	13.0	2.90	1.80	1.36
n-Bu—NH$_2$	13.6	2.90	1.90	1.49
i-Pr—NH$_2$	14.1	6.0	1.94	1.48
i-Bu—NH$_2$	16.1	—	—	—
t-Bu—NH$_2$	16.1	5.5	1.96	1.87
F$_3$C—CH$_2$—NH$_2$	−1.4	−3.9	−4.96	—
F$_2$CH—CH$_2$—NH$_2$	4.0	—	−2.94	—
F CH$_2$CH$_2$—NH$_2$	8.0	—	−0.62	—
CF$_3$CH$_2$CH$_2$NH$_2$	6.7	—	−0.82	—
F$_3$C(CH$_2$)$_3$NH$_2$	10.1	—	0.64	—
NC(CH$_2$CH$_2$)NH$_2$	3.0	—	−2.04	—
HC≡C—CH$_2$NH$_2$	6.7	—	−1.50	—
H$_2$C=CH—CH$_2$NH$_2$	11.3	—	0.38	0.58
Me$_2$N—NH$_2$	15.2	—	−2.90	—
H$_2$N—NH$_2$	3.8	—	−1.77	−2.76
cyclo-C$_6$H$_{11}$NH$_2$	16.3	—	1.34	1.82
Anilines				
m-Cl	4.2	−9.1	−7.8	−6.00
p-F	4.7	−6.1	−6.30	−4.70
p-Cl	5.1	−7.7	−7.20	−5.80
H	6.7	−9.3	−6.30	−5.10
p-Me	9.2	−6.4	−5.70	−4.60
p-MeO	9.8	—	—	—
Secondary amines				
Me$_2$NH	15.5	4.5	2.09	−0.45
Me Et NH	17.9	—	—	—
Et$_2$ NH	20.2	4.4	2.42	0.25
n-Pr$_2$ NH	(22.2)	—	2.39	0.68
n-Bu$_2$ NH	(23.1)	—	2.74	1.18
i-Bu$_2$ NH	(23.6)	—	1.71	0.89
i-Pr$_2$ NH	(23.9)	—	2.67	1.07
s-Bu$_2$ NH	(25.8)	—	2.27	1.54
▷NH	11.2	—	−1.79	−3.43
◇NH	18.0	—	2.79	0.10
⬠NH	19.8	—	2.81	0.54
⬡NH	21.1	—	2.56	0.28

Table 1—Continued

Base	$-\delta\Delta G^0_i$ gas	$-\delta\Delta H^0_i$ HSO_3F	$-\delta\Delta G^0_i$ H_2O	$-\delta\Delta H^0_i$ H_2O
Tertiary				
Me_3N	20.0	4.2	0.76	−3.67
$Me_2Et\,N$	22.4	—	—	—
$Et_2Me\,N$	24.6	—	—	—
Et_3N	26.7	5.9	2.01	—
$(n\text{-}Pr)_3\,N$	28.7	4.8	1.93	—
N—Me (pyrrolidine)	24.3	—	1.66	—
N—Me (piperidine)	25.7	—	1.14	—
quinuclidine	27.1	—	2.60	−1.34
Dabco	23.5	—	−0.90	−5.19
$NC{-}CH_2N\,Me_2$	7.1	—	−6.82	—
$(HC{\equiv}C{-}CH_2)_3N$	15.0	—	−8.43	—
$(H_2C{=}CH{-}CH_2)_3\,N$	24.7	—	−1.32	−3.66
Pyridines				
4-CF$_3$	8.7	−7.2	−9.02	−10.27
4-CN	6.2	−3.8	−10.07	−11.64
H	16.0	−4.7	−5.50	−7.69
4-Me	20.0	−4.2	−4.34	−6.47
4-MeO	22.7	−0.2	−3.63	−4.64
2,6-Dimethyl	22.8	−2.6	−3.34	−5.25
2,6-Di-t-butyl	28.0	−5.1	−7.74	—
2,4,6-Trimethyl	—	−0.6	−2.26	—
Pentafluoro	—	−31.6	—	—
3-CF$_3$	8.4	—	−9.30	—
2-Cl	10.2	−11.6	−11.70	—
2-Br	10.2	−13.1	−11.40	—
4-NO$_2$	4.9	—	−10.70	—
3-CN	5.5	—	−10.80	—
2-F	6.7	—	−13.13	—
4-Cl	13.2	—	−7.38	—

All values in kcal mol^{-1} at 25°C, relative to ammonia, for which we take a gas-phase proton affinity of 207 kcal mol^{-1}. Values in parenthesis are from reference 10 and are normalized to Taft's data by subtracting 2.4 kcal mol^{-1}. Methylamine has a different reference point because Aue *et al.* [10] did not compare it with ammonia.

we are in no position to separate the various conformational factors which might be involved from 'simple' electrostatic polarization, in either the cyclic or the acyclic series. Likewise, we are still in no position to sort out how much of the latter operate through space or through bonds. To the author it seems likely that many of these conceptual distinctions are based on models which are too crude to be completely interpretable at this time in quantum mechanical terms [20, 21], but are too fine to be resolved clearly by experiment. The single factor which definitely has been separated from the structure-reactivity data in the gas-phase column of Table 1 is solvation, and comparison with the other three columns shows that this can play a tremendously important role.

Examination of the gas-phase basicities ($\delta \Delta G_i^0$) for substituted anilines and pyridines falls closely in line with the familiar patterns found for the response of these systems to electron-demanding reactions in solution. It is comforting to find that the enormous collection of data which has been stored in the various aromatic substituent parameters [22, 23, 24] applies to solvent-free systems and this gives verisimilitude to their interpretation in quantum-mechanical terms such as resonance or bond hybridization.

This did not need to be so. Quantum-mechanical explanations have sometimes been applied erroneously to systems where solvation is the dominating influence, rather than internal electron delocalization. As a case in point, consider the classical examples of the low basicities $[\delta \Delta G_i^0(H_2O)]$ of aniline and pyridines compared to ammonia in aqueous systems. For a generation students have been led to understand that the two aromatic bases are weaker than ammonia because of resonance and hybridization factors. Initial examination of Table 1 leads to the refreshing conclusion that the facts are otherwise; in the gas phase both aniline and pyridine are much more basic than ammonia, therefore the elementary interpretation of the basicity order *for these three compounds* is false. However, further examination of the data shows that both aniline and pyridine really are weakly basic if proper non-aromatic models are used. In view of the enormous additive ability of C—C and C—H bonds to stabilize positive charges it is appropriate to compare aniline and pyridine with their alicyclic cognates of similar carbon number, instead of with ammonia. When this is done we see that the difference in ΔG_i^0 between cyclohexylamine and aniline is 9.6 in the gas phase and 8.2 in water; between N-methylpyrrolidine and pyridine it is 8.3 in the gas phase and 7.2 in water. These results isolate clearly the internal-energy terms which are to be interpreted if the correct model is used, and furthermore we learn that in appropriate cases the energy differences may be affected scarcely at all by solvent.

The role of hybridization and basicity in the gas phase has been discussed further by Aue [25], who draws attention to the low basicity of aziridine as against dimethylamine when compared with the other pairs of cyclic and acyclic compounds, and points out that acetonitrile is about 34 kcal mol^{-1} *lower* in proton affinity than is ethylamine. Beauchamp [26] has considered the importance of hybridization in comparing the effects of alkyl substitution on phosphines compared to amines.

It has been an article of reasonable faith that the steric requirements of the unsolvated proton are nil. Therefore, the very low pK_a of 2,6-di-t-butyl-pyridine compared to other pyridines in water has long been cited as a classic argument for steric inhibition to solvation [27, 28], assuming that the bulky t-butyl groups are acting to exclude *something* from what should otherwise be a rather basic nitrogen atom. This hypothesis is confirmed in the first instance by the enormous proton affinity of this base in the gas phase compared to the other pyridines in general and to 2,6-dimethylpyridine in particular. Although the energy change for transferring a proton to 2,6-di-t-butylpyridine is thermodynamically very favourable, the gas-phase equilibrium was unusually difficult to achieve because of the steric obstacles to forming a transition state which would permit proton-transfer from a more acidic ammonium ion. We shall consider the exact magnitude of the solvation factor in a later section.

3.3 Proton transfer in solution

A swift perusal of Table 1 is all that is needed to convince the reader that no general correlation exists between $\delta\Delta G_i^0$ (gas) and $\delta\Delta G_i^0$ (H$_2$O). A plot of all such data from Table 1 results in a random scatter diagram (see Figure 4 in Chapter 2), and the same may be said for the comparison of $\delta\Delta G_i^0$ (gas) with $\delta\Delta H_i^0$ (HSO$_3$F). However, Figure 5 in Chapter II of this book demonstrates that for the pyridines a good linear correlation is found. This is especially interesting in view of Taft's recent demonstration of the close correlation between polar substituent effects on the pK_a's of 4-substituted pyridines and 4-substituted quinuclidines [29]. In view of the great difference in bonding within these two types of cyclic systems, this result argues in favour of the transmission of charge through the field effect rather than through electron-pair bonds. It will be interesting to see whether the same type of correlation will hold when quinuclidines are studied in the gas phase. Taft's results lead to the prediction that it will.

Returning to Table 1, we have noted previously a rather good general correlation between $\delta\Delta H_i^0$ (HSO_3F) and $\delta\Delta G_i^0$ (H_2O) for a large range of primary amines, and to a lesser extent many other bases [15]. Excluding 4-cyanopyridine, which in HSO_3F undergoes hydrogen bonding to the cyano group, we find for all the amines in Table 1 a correlation coefficient of 0.978 and a slope of 1.22. This is strong evidence that these two different properties (ΔG_i^0 in H_2O and ΔH_i^0 in HSO_3F) are measuring the same property – the potential energy of protonation.

Probably, the most significant general result which comes from comparing the last three columns of Table 1 with each other and with the first column is that in general (but not always) substituent effects for protonating amines are much larger in the gas phase than they are in solution. Most dramatically, the extreme range for aliphatic and alicyclic amines is nearly 29 kcal mol^{-1} in the gas phase, while in HSO_3F it is about 6 kcal mol^{-1} and in water it is scarcely 3 kcal mol^{-1}. Thus in general these solvents are able to distribute and stabilize charges to such an extent that substituent effects in solution are severely levelled; solvent molecules can interact externally with ammonium ions to cancel the internal effects of charge-stabilizing substituents.

Bell has documented, from a number of sources, the rough tendency for basicity-acidity orders to be preserved for series of similar compounds in different solvents [5]. This has been generally found for amines [3], even in low-dielectric solvents, although small inversions occasionally arise. Apparently, if an ammonium ion cannot stabilize itself by hydrogen bonding or ion-dipolar interactions to a good solvent it will do so by ion-pairing. We may thus suggest that the comparisons presented in Table 1 between gas and condensed phases are rather general and not limited strictly to water and HSO_3F.

3.4 Solvation of ammonium ions

The data in Table 2 for solvation energies of ammonium ions have been calculated from the ionization energies in the gas phase and solution in combination with the solvation energies of the neutral bases, using equation 1. The solvation energies of many of the bases are presented in reference 3 and also the method for deriving them from Henry's Law constants or heats of vaporization and dilution. It must be borne constantly in mind when interpreting fine differences in Table 2 that each figure there has been derived by combining three or four independent measurements, so that their cumulative experimental errors are in the range 0.5–1 kcal mol^{-1}. However, the trends in such errors for a related series of compounds may be sufficiently similar that some of the regular trends in the derived data are significant.

Several fascinating facts emerge from Table 2. Perhaps the most important of these is the tendency of heats of solvation in water $[\delta\Delta H_s^0(BH^+)H_2O]$ to fall into three major groups, depending on the number of acidic hydrogens on the ammonium ion. Thus, regardless of their size or shape, the primary ammonium ions have values that fall largely within a range of ± 1 kcal mol^{-1} from an average figure of 6.8 kcal mol^{-1}, while those for the secondary ions are mostly grouped within ± 1 kcal mol^{-1} of 13 kcal mol^{-1} and those for the tertiary ions, including pyridine, are close to 19 kcal mol^{-1}. The obvious interpretation of these results is that a major factor in differentiating the solvation energies of these ions is the number of hydrogen bonds which the ammonium ion can present to the solvent or counterions in the series [3, 7, 9, 32].

$$NH_4^+ > RNH_3^+ > R_2NH_2^+ > R_3NH^+$$

The increments of 6–7 kcal mol^{-1} are of the correct magnitude to fit this interpretation. We shall return to this point later. A second and surprising fact is the close correspondence between the heats of solvation of BH^+ in HSO_3F and those in water. The reader is reminded that the former values $[\delta\Delta H_s(BH^+)HSO_3F]$ were derived from estimated heats of ionization in HSO_3F. These in turn have an uncertainty of about ± 0.5 kcal mol^{-1}, since they were derived by means of the assumption that the reference solvent is inert. One may say therefore that the relative solvation energies of ammonium ions are within experimental error the same as those in water. Since SO_3F^- is about 23 powers of ten less basic to protons than neutral water ($\Delta H^0 \approx -16$ to 17), and HSO_3F must be even less basic than SO_3F^-, it is surprising that the two media are so comparable in solvating ability. We have shown previously [33, 34] that there is no direct relation between Brønsted basicity and hydrogen-bonding, and there is reason to believe that SO_3F^- might be a rather good hydrogen-bond acceptor, since SO_4H^- apparently is [35]. It is therefore *not* necessary to dismiss the importance of hydrogen bonding to the solvent as an important differentiating factor between ammonium ions of different degrees just because their behaviour in HSO_3F and H_2O is so similar.

Table 2

Standard heats and free energies of solution of ammonium ions from gas phase to HSO_3F and H_2O at 25°C

Ammonium ion	$\delta\Delta H_S(BH^+)$ HSO_3F	$\delta\Delta H_S(BH^+)$ H_2O	$\delta\Delta G_S^0(BH^+)$ H_2O
Ammonium			
1. NH_4^+	0.0	0.0	0.0
Primary amines			
2. $MeNH_3^+$	4.9	5.8	6.9
3. $EtNH_3^+$	6.2	5.9	9.6
4. $n\text{-}PrNH_3^+$	6.7	6.3	11.1
5. $n\text{-}BuNH_3^+$	6.3	6.3	11.7
6. $i\text{-}PrNH_3^+$	7.9	7.6	12.1
7. $i\text{-}BuNH_3^+$	—	7.1	—
8. $t\text{-}BuNH_3^+$	8.8	8.4	14.1
9. $cyclo\text{-}C_6H_{11}NH_3^+$	—	7.8	—
10. $C_6H_5NH_3^+$	—	7.2	—
Secondary amines			
11. $Me_2NH_2^+$	8.9	10.9	13.4
12. $Et_2NH_2^+$	11.9	12.8	18.0
13. $n\text{-}Pr_2NH_2^+$	13.1	(12.5)	—
14. $n\text{-}Bu_2NH_2^+$	11.7	(10.9)	—
15. $i\text{-}Bu_2NH_2^+$	—	(15.1)	—
16. $i\text{-}Pr_2NH_2^+$	—	(13.9)	—
17. $sec\text{-}Bu_2NH_2^+$	—	(14.4)	—
18. ▷$NH_2^{\oplus}$	—	10.1	—
19. ◇$NH_2^{\oplus}$	—	11.8	16.0
20. ⬠$NH_2^{\oplus}$	—	12.3	18.2
21. ⬡$NH_2^{\oplus}$	—	13.3	20.1
22. $Me_2\overset{H}{\overset{|}{N^+}}{-}NH_2$	—	13.4	—
Tertiary amines			
23. Me_3NH^+	13.4	18.2	20.3
24. Et_3NH^+	20.3	19.9	26.0
25. ⬠$\overset{\oplus}{N}{-}Me$ with H	—	20.7	25.3

Table 2—Continued

Ammonium ion	$\delta\Delta H_s(BH^+)$ HSO_3F	$\delta\Delta H_s(BH^+)$ H_2O	$\delta\Delta G_s^0(BH^+)$ H_2O
26.	—	—	—
Pyridines			
27. 4-CF$_3$-Pyr H$^+$	14.9	15.0	—
28. 4-CN-Pyr H$^+$	9.7	15.5	—
29. Pyr H$^+$	20.0	19.5	23.6
30. 4-Me-Pyr H$^+$	21.8	21.0	26.2
31. 4-MeO-Pyr H$^+$	18.9	21.0	—
32. 3-Me-Pyr H$^+$	—	20.9	25.3
33. 2-Me-Pyr H$^+$	—	20.4	26.0
34. 2,6-Me$_2$-Pyr H$^+$	23.2	21.0	28.3
35. 2,6-di-t-butyl-Pyr H$^+$	29.2	—	

All values are in kcal mole^{-1} at 25°C, and are taken relative to NH$_4$$^+$, for which $\Delta H_s(BH^+)$ in water is -83.8 kcal mol^{-1}. Values in parentheses are from reference 10 and are normalized to the same scale as those calculated from Taft's data by subtracting 0.5 kcal mol^{-1}. See reference 3 for tabulations of solvation energies of free amines.

Another property which may contribute to the close agreement between solvation energies in these two drastically different media is their high dielectric constant. Undoubtedly, dielectric solvation is by far the largest term contributing to the solvation energy of the ammonium ions in water. This amounts to an overall heat of solvation of -83.8 kcal mol^{-1} for NH$_4$$^+$ and a standard free energy change of -77.1 kcal mol^{-1} for transfer of a mole of the ion at one atmosphere in the gas phase to a hypothetical molal aqueous solution [3]. However, the data in Table 2 are *relative* values describing differentiation between the ions and ammonia (so that formal contributions from difference standard states cancel out). Now the important question arises as to how much of this differentiation is due to the number and strength of hydrogen-bonds to the solvent and how much of it is due purely to factors determined by the effective ionic radius. Conceptually, the notion of solvation through hydrogen bonding lies at one extreme of the dichotomy between 'specific' and 'continuum' models for solvation, and electrostatic solvation lies at the other. However, the exact quantitative separation of these terms is impossible to apply to the present case in any thoroughly convincing way, for several reasons, since neither factor can be accounted for in unequivocal terms. The Born equation 2 [36] for the loss in standard electrostatic free energy for

$$\Delta G^0 = -\left(1 - \frac{1}{D}\right)\frac{e^2}{2r} \qquad (2)$$

transfer of a sphere of charge e and radius r from a vacuum to a medium of dielectric constant D cannot be applied rigorously to the systems in question. The ions are not spherical and their interactions with water or HSO_3F are of precisely the kind which would be expected to elicit a maximum degree of discontinuous character from the solvent as it adjusts itself to the dispersed charge in the hydrocarbon part of the ion and the more concentrated charge on nitrogen. The concept of an ionic radius is hard to apply even to the smaller ammonium ions where the charge is most symmetrical around nitrogen and which should be most amenable to the Born treatment. In fact the very idea of a single radius in solution for such an entity as NH_4^+ is probably naive, since the lengths of the hydrogen bonds doubtless vary according to their strength [37]. This in turn depends on the nature of the ion or solvent molecule which is accepting the bond. Nightingale [38] has demonstrated convincingly the frequently large discrepancies between crystal radii of ions and the radii which might be applied to solvated ions. For small ions such as the alkali cations and NH_4^+, errors of 100 per cent (e.g. 1.48 compared with 3.31) are common.

A second related difficulty in accounting for the relative contributions of electrostatic and hydrogen-bonding interactions is that there is no useful theory of the hydrogen bond which could be used for such a separation in systems as complex as these. We have discussed elsewhere [33] the conceptual difficulties which ultimately limit most attempts to separate hydrogen-bonding cleanly from other types of intermolecular or interionic forces. Although there is no generally applicable quantum theory of hydrogen-bonding available for comparing systems as complex as these, most of the presently-accepted models endow the hydrogen-bond with considerable electrostatic character. This coulombic component to the wave equation for $-N^+-H--OH_2$ should be very considerable for hydration of an ammonium ion. We see no possibility of separating this from other types of electrostatic solvation of such systems, and hence consider that the distinction is almost meaningless for the ammonium ion in water.

Close examination of Table 2 shows that within each series of ions (primary, secondary, tertiary) there is a consistent tendency for the smallest ions to have the most exothermic solvation energies (i.e., *least* endothermic relative to NH_4^+). There is also a slight but distinct trend for the larger ions in each series to have lower solvation energies than the smaller ions, both in water and in HSO_3F. This trend is in qualitative agreement with the predictions of a Born-type electrostatic treatment in which solvation energy is inversely related to ionic volume. It is also consistent with the notion of specific hydrogen-bonding solvation. We consider that the strength of an

$$R-\overset{\displaystyle |}{\underset{\displaystyle |}{N}}{}^+-H\cdots O$$ bond to specific solvent molecules is weakened by

delocalizing positive charge within the ion, or to solvent dipoles in the neighbourhood of its alkyl groups. Weakening the hydrogen-bond would lower the solvation energy. It is therefore possible to interpret the trends for aliphatic ions of the same series by qualitative arguments for both continuum and specific solvation.

A more refined test of this question is provided by the work of Miss L. J. Bell in our laboratory. In the pyridinium ions listed at the end of Table 2 it is possible to vary the electron density in the $-N^+-H$ bond by changing substituents at sterically remote sites while having little effect on the cavity size or shape. The trend of solvation enthalpies in water $[\Delta H_s(BH^+)H_2O]$ is clearly in the direction of increasingly strong hydrogen bonding to solvent as the positive charge is concentrated on nitrogen in going from 4-methoxypyridine to 4-trifluoromethylpyridine. We ascribe the failure of 4-methoxypyridine and 4-cyanopyridine to follow this trend proportionately in HSO_3F to the strong interactions of these two substituents with this solvent.

The reader will note the extremely low solvation energy for 2,6-di-t-butylpyridinium ion in HSO_3F; this is in complete accordance with the previously stated picture of steric hindrance to solvation or ion-pairing at the ionic site. We have not yet been able to determine the heat of solution for the free base in water because of its extreme insolubility. This is the single fact which is lacking to complete the cycle for calculation of the solvation of this important ion in water.

The mode of solvation of ammonium ions is a substantial problem. Proton transfer is the most general and important reaction in chemistry and, as we have seen, the energetics of protolysis are often dominated completely by solvation of the ions. The solvation energies of the ammonium ions will probably be more accurately known than those for any other series of organic ions for some time to come. It is therefore important to draw what rules we can for their behaviour, even though its complete dissection into quantum mechanical terms may not be possible at present.

Despite the complexities to which we have referred above when dealing with the sizes of hydrated ions, it is possible to compare the effective hydrated volumes of many ammonium ions in water in terms of their partial molal volumes, $\bar{V}_2^0$. Millero's comprehensive review [39] provides such data for NH_4^+, a number of quaternary ions, and some primary ones. In Figure 2 these have been plotted against

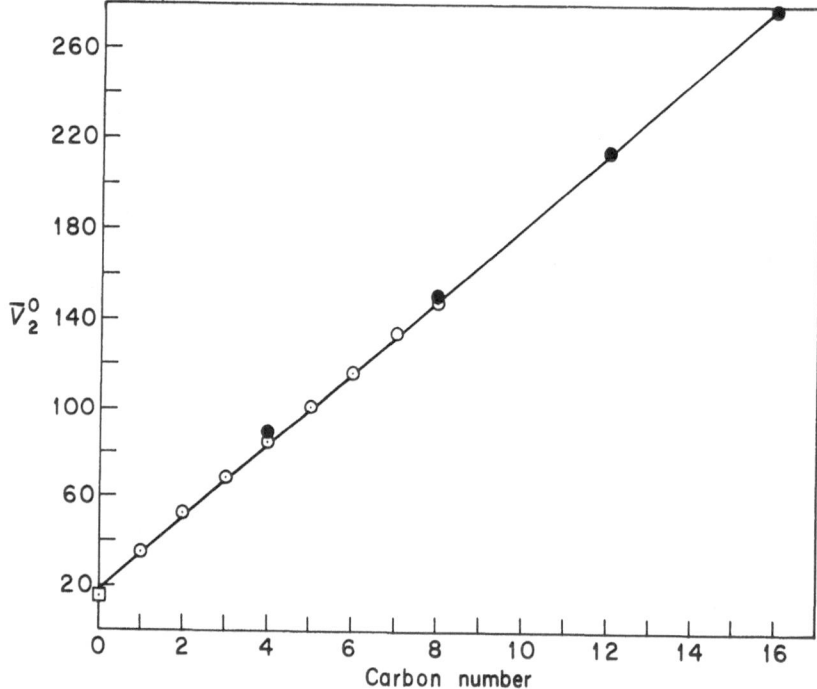

Figure 2. Plot of partial molal volumes ($\bar{V}_2^0$) of ammonium ions in water at 25° against carbon number. Solid points (●) refer to R_4N^+; crosses (x) refer to RNH_3^+; intercept ⊡ refers to NH_4^+. Data are taken from Millero's review [39].

the carbon number of the ammonium ion; the plot indicates a remarkable degree of additivity. Unfortunately, all of the alkyl groups on the R_4N^+ and RNH_3^+ ions used to generate Figure 2 are normal straight chains. However, it is both surprising and helpful to find that the two groups of ammonium ions which differ most in shape and symmetry behave so simply and extrapolate so perfectly to NH_4^+. On the strength of this we have used Figure 2 to interpolate values of $\bar{V}_2^0$ for all the other ammonium ions, including secondary and tertiary ones, whose hydration energies appear in Table 2. If we make the crude assumption that the effective electrostatic radius of the ion may be approximated by the cube root of $\overline{V}_2^0$, the Born equation may be tested by plotting $(\overline{V}_2^0)^{-\frac{1}{3}}$ against $\delta\Delta G_s^0(BH^+)$. Those ions for which complete data are available are displayed in Figure 3. Through the overall random pattern of points, three lines may be drawn, converging on the prototype ion of greatest solvation energy, *i.e.* NH_4^+. We note that the lines are fitted nearly as well by lop-sided primary ions as by the more symmetrical secondary and tertiary ions. Obviously, there is partial adherence to electrostatic behaviour, but it is differentiated by class for primary, secondary and tertiary ions, and the differentiation is greater as the ionic size increases. We interpret this to mean that the separation into three lines is due to varying degrees of hydrogen-bonding to the solvent determined by the number of acidic hydrogens on nitrogen, and that for large ions this is the chief factor. Smaller ions seem to show a slightly greater tendency to be grouped on the basis of radius.*

There is probably no solvent in which correlations between free energy changes and corresponding enthalpy changes are more 'badly behaved' than water. The source of these discrepancies is identified (but in no sense explained) as compensation effects related to changes of water structure [30]. From what is generally known about heats of solution of ions and electrolytes in water we should not expect to find parallel behaviour between $\delta\Delta G_s^0(BH^+)$ as a function of $(\overline{V}_2^0)^{-\frac{1}{3}}$ and that for $\delta\Delta H_s(BH^+)$. Figure 4 shows such a plot and clearly it is quite different in pattern from Figure 3. The Born charging enthalpy (equation 3) obtained from the temperature-derivative of

$$\Delta H = -\frac{e^2}{2r}\left(1 - \frac{1}{D}\right)\left[1 - \frac{T\,\delta D/\delta T}{D(D-1)} + \frac{T\,\delta r}{r\,\delta T}\right] \qquad (3)$$

equation 2 reduces practically to the Born equation in water at 25°C, since the last two terms in brackets are small. Thus on purely electrostatic grounds Figures 3 and 4 should look alike, and their failure to do so argues strongly against a major electrostatic component in the differences in solvation energy between these ions. In Figure 4 we see a much more consistent differentiation between primary, secondary and tertiary ions than in Figure 3. Again the best adherence to each family line is found for the most symmetrical ions. The increasingly *exothermic* solvation of the ions as the hydrocarbon portion is

---

* The reader is warned again that NH_4^+ is the only symmetrical ion on the plot, and that after appropriate corrections for charge distributions all of these ions might be brought into line with classical electrostatics. This seems most unlikely to the author.

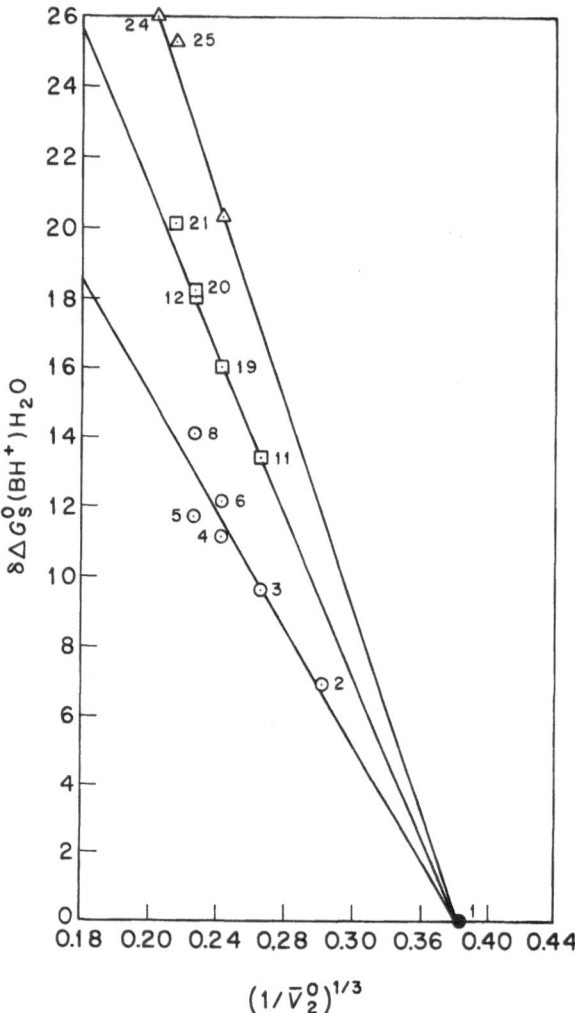

Figure 3. Attempted Born plot of relative standard free energies of ammonium ions against estimated reciprocal radius (see text).

increased is typical of non-polar entities for water [30], and is exactly counter to the usual pattern for the free energy of solution, which is dominated by unfavourable entropy terms.

Finally, we note that for purely electrostatic processes in water the standard entropy change is related to the corresponding free-energy term by the relation [40] (from the middle term on the right-hand side of equation 3)

$$\delta \Delta S^0 = \delta \Delta G^0 \frac{d \ln D}{d T}$$

96

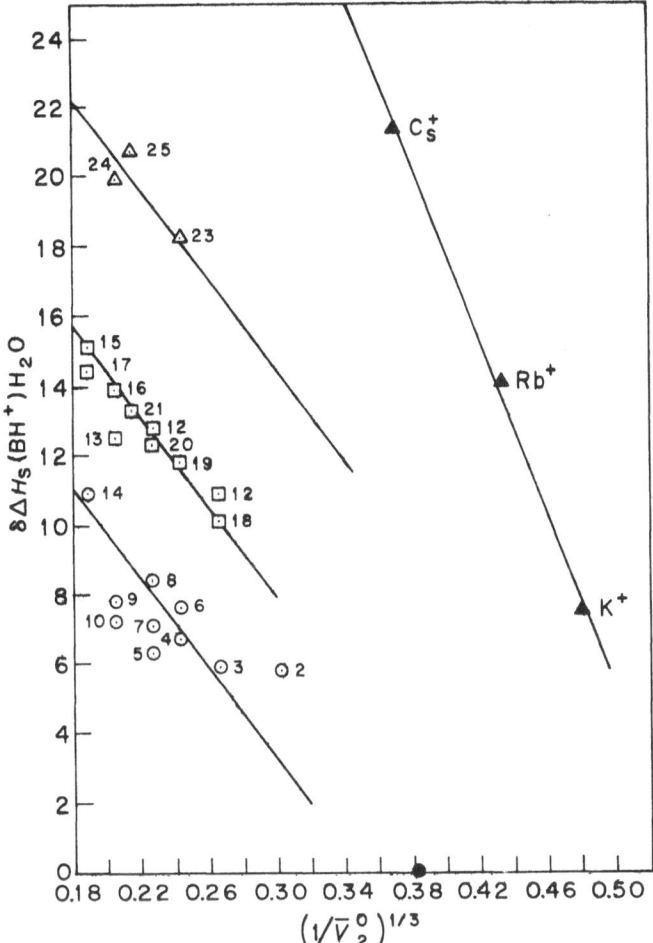

Figure 4. Attempted Born plot of relative heats of hydration of ammonium ions in water against estimated reciprocal radius.

where $d \ln D/dT = -0.005$ for water. If this is applied to the solvation data for aliphatic ammonium ions [9], the estimated entropies are grossly in error both in magnitude and trend (see Table 3).

Thus, although Born charging energy is of major significance in solvating all of the ions under discussion here, we have shown that it cannot be applied exclusively either for interpretative or predictive purposes. On the contrary, the facts cited above support the notion that hydrogen-bonding is the chief differentiating factor in ammonium ion solvation. As will be shown below, our picture not only has broad interpretative value but can also lead to predictions for manipulating the behaviour of ions in solution.

Table 3

Comparison of experimentally-determined solvation entropies with those calculated from pure electrostatic terms

All data are in units of cal $mol^{-1} K^{-1}$

	$-\delta\Delta S^0$	$-\delta\Delta G_S^0(BH^+) \times 0.005$
$MeNH_3^+$	4.36	36.5
$Me_2NH_2^+$	6.72	68.5
Me_3NH^+	7.05	103.5
$EtNH_2^+$	12.75	49.0
$Et_2NH_2^+$	17.45	92.0
Et_3NH^+	20.47	132.0

The solvation energies for a considerable variety of ammonium ions can be ranked in terms of the number of acidic hydrogens on the nitrogen atom, rather than their sizes or shapes. Secondly, the average values for solvation energies of tertiary, secondary, primary and NH_4^+ ions are separated from each other by 6–7 kcal mol^{-1}, a very reasonable increment for strong N^+H--OH_2 bonds. Of significance to this argument is the fact that Pearson and Vogelsong [41] have observed exactly comparable effects on the ion-pairing behaviour of ammonium ions in non-polar media of low dielectric constant. From their data and several estimates of solvation energies *including a modified Born treatment* they predicted almost exactly the observed increment between solvation energies for different classes of ammonium ions and ascribed it to hydrogen bonding. Likewise, we have observed protonation energies for amines with HSO_3F in CH_2Cl_2 as solvent very similar to those in pure HSO_3F. The relative invariance of these solvation or ion-pairing energies with respect to the dielectric constant of the medium constitutes behaviour much closer to hydrogen-bonding [33] than to the Born-charging model (equation 2).

A dramatic application of the hydrogen-bonding interpretation which gives strong support to its use for prediction is found in the work of Dr. James Wolf in our laboratory. On the basis of scattered and inconclusive evidence that hydrogen bonds to solvent from phosphonium ($-P^+-H\cdots S$) and sulphonium ions ($S^+-H\cdots S$) are weak, Wolf measured the heats of protonation of PH_3, cyclohexylphosphine, di-cyclohexylphosphine and trimethylphosphine [42]. Likewise H_2S, MeSH and Me_2S were compared. It was reasoned that if the ions obtained on protonating these basic molecules could not stabilize themselves by hydrogen-bonding to the medium in the way that ammonium ions can, then they would place greater electron-releasing demands on their substituent groups. This should lead to the observation of larger substituent effects for the phosphonium and sulphonium series than had previously be found for the ammonium ions. In gratifying accordance with expectations, Wolf did in fact find that the substituent effects comparable to those reported for the gas-phase protonation of these compounds could be realized. To the best of our knowledge, his observation of a difference of 14 kcal mol^{-1} between the heats of ionization of H_2S and MeSH, and 16 kcal mol^{-1} between PH_3 and cyclohexylphosphine, are by far the largest recorded substituent effects for simple alkyl groups for a thermodynamic process in solution. These bases and their ions are of similar size and shape to the ammonium ions discussed above in this chapter. The drastic difference in their behaviour cannot therefore be reasonably explained by Born charging and must be due to some other factor. Whether or not the hydrogen-bonding theory on which this experiment was based is ultimately correct the results show that by using it we have been able to demonstrate in solution substituent effects which are as large as those in the gas phase.

We look forward to rapid and often dramatic development of the field presented here in the next few years. Probably, some of the data in Tables 1 and 2 will prove to be incorrect. However, we have every reason to believe that the main features of the results and viewpoint in this article are sound.

Addendum

This work was supported by N.S.F. Grant G.P.-31565-X for which the author is most grateful. He also appreciates close collaboration with Professor Robert Taft. Helpful comments were provided by Professors Henry Frank and J. F. Coetzee. Dr. D. Oancea provided calculations and plots for Figures 3 and 4.

REFERENCES

[1] R. W. Taft, chapter 2.
[2] E. M. Arnett, *Accounts Chem. Research*, **6**, 404 (1973).
[3] F. M. Jones, III and E. M. Arnett, *Progr. Phys. Org. Chem.*, **11**, 263 (1974).
[4] E. M. Arnett and G. Scorrano, *Adv. Phys. Org. Chem.*, in press.
[5] R. P. Bell, *The Proton in Chemistry*, second edition, Chapman and Hall, London (1973).
[6] R. P. Bell and K. N. Bascombe, *Disc. Faraday Soc.*, **24**, 158 (1957).

[7] R. P. Bell and A. F. Trotman-Dickenson, *J. Chem. Soc.*, 1293 (1949); A. F. Trotman-Dickenson, *ibid.*, 1293.

[8] J. E. Leffler and E. Grunwald, *Rates and Equilibria of Organic Reactions*, J. Wiley and Sons, New York (1963).

[9] E. M. Arnett, F. M. Jones III, M. Taagepera, W. G. Henderson, J. L. Beauchamp, D. Holtz., and R. W. Taft, *J. Amer. Chem. Soc.*, **94**, 4724 (1972).

[10] D. H. Aue, H. M. Webb, and M. T. Bowers, *ibid.*, 4726.

[11] R. J. Gillespie, chapter 1.

[12] R. J. Gillespie, *Accounts Chem. Research*, **1**, 202 (1968).

[13] R. J. Gillespie, personal communication (June 1974).

[14] G. A. Olah, A. M. White and D. H. O'Brien, *Chem. Rev.*, **70**, 561 (1970).

[15] E. M. Arnett, R. P. Quirk, and J. J. Burke, *J. Amer. Chem. Soc.*, **92**, 1260 (1970).

[16] J. I. Brauman and L. K. Blair, *ibid.*, **92**, 5986 (1970); **91**, 2126 (1969); **90**, 5636 (1968).

[17] E. M. Arnett, L. E. Small, R. T. McIver, Jr., and J. Scott Miller, *J. Amer. Chem. Soc.*, **96**, 5638 (1974).

[18] J. L. Beauchamp, *Ann. Rev. Phys. Chem.*, **22**, 527 (1971).

[19] D. Holtz, J. L. Beauchamp, and J. R. Eyler, *J. Amer. Chem. Soc.*, **92**, 7045 (1970).

[20] L. Libit and R. Hoffmann, *ibid.*, 1370.

[21] J. A. Pople, *Accounts Chem. Research*, **3**, 217 (1970).

[22] *Advances in Linear Free Energy Relationships*, ed. N. B. Chapman and J. Shorter, Plenum Press, New York (1972).

[23] C. D. Ritchie and W. F. Sager, *Prog. Phys. Org. Chem.*, **2**, 323 (1964).

[24] S. Ehrenson, *ibid.*, p. 195.

[25] D. A. Aue, *Structure-Reactivity Conference*, San Juan, (January, 1974).

[26] R. H. Staley and J. L. Beauchamp, *J. Amer. Chem. Soc.*, **96**, 6252 (1974).

[27] F. E. Condon, *J. Amer. Chem. Soc.*, **87**, 4494 (1965).

[28] H. C. Brown and B. Kanner, *ibid.*, **88**, 986 (1966).

[29] R. W. Taft and C. A. Grob, *ibid.*, **96**, 1236 (1974).

[30] E. M. Arnett and D. R. McKelvey in *Solute-Solvent Interactions*, ed. J. Coetzee and C. D. Ritchie, M. Dekker, New York (1969).

[31] E. M. Arnett, J. J. Burke, J. V. Carter, and C. F. Douty, *J. Amer. Chem. Soc.*, **94**, 7837 (1972).

[32] R. W. Taft, M. Taagepera, K. D. Summerhays, and J. Mitsky, *ibid.*, **95**, 3811 (1973).

[33] E. M. Arnett, E. J. Mitchell, and T. S. S. R. Murty, *ibid.*, **96**, 3875 (1974); **93**, 4052 (1971).

[34] R. W. Taft, D. Gurka, L. Joris, P. von R. Schleyer, and J. W. Rakshys, *ibid.*, **91**, 4801 (1969).

[35] R. P. Taylor and I. D. Kuntz, Jr., *ibid.*, **94**, 7963 (1972).

[36] R. A. Robinson and R. H. Stokes, *Electrolyte Solutions*, 2nd ed. Butterworths, London (1959).

[37] G. C. Pimentel and A. L. McClellan, *The Hydrogen Bond*, W. H. Freeman, San Francisco (1960).

[38] E. Nightingale, Jr., *J. Phys. Chem.*, **63**, 1381 (1959).

[39] F. Millero, *Chem. Rev.*, **71**, 147 (1971).

[40] L. P. Hammett, *Physical Organic Chemistry*, McGraw-Hill, New York (1940).

[41] R. G. Pearson and D. C. Vogelsong, *J. Amer. Chem. Soc.*, **80**, 1038 (1958).

[42] E. M. Arnett and J. F. Wolf, *J. Amer. Chem. Soc.*, **95**, 978 (1973).

4

Ernest Grunwald and Daniel Eustace

PARTICIPATION OF HYDROXYLIC SOLVENT MOLECULES

4.1 Introduction

Proton transfer reactions are often depicted as simple bimolecular processes in which a proton is transferred directly from the acid to the base. However, with the development of proton magnetic resonance techniques for measuring the rates of fast reactions, it has become clear that many proton transfer reactions in hydroxylic solvents actually proceed with participation of one or two solvent molecules. The solvent molecule then acts as a bifunctional catalyst, that is, both as a proton acceptor and a proton donor. Thus, termolecular proton transfer from an acid HA to a base B with participation of a water molecule is shown in equation 1.

$$A-H + O-H + B \rightarrow A^- + H-O + HB^+ \qquad (1)$$
$$\qquad\quad |\qquad\qquad\qquad\quad |$$
$$\qquad\quad H\qquad\qquad\qquad\quad H$$

Participation of solvent molecules has been observed particularly with oxygen and nitrogen acids and bases, and for reactions that are very fast. Indeed, in many cases, the second order rate constants approach the diffusion-controlled limit of 10^9–10^{10} dm^3 mol^{-1} s^{-1}. Because of the high speed of these reactions, it is unlikely that the transition state is formed from three truly independent molecules. It is far more likely that the participating solvent molecule is already found in the inner solvation shell of one of the reactants. If this be granted, then proton transfer reactions with solvent participation become of special interest because they enable us to probe the fine-structure of the solvation shell at the hydrogen bonding site of the acid or base.

The idea that hydroxylic solvent molecules can participate in acid-base reactions, acting as bifunctional catalysts, is one of the classical concepts of solution chemistry. In essence (though in archaic language) the idea was stated as early as 1806 by C. J. D. von Grotthuss, in an attempt to account for the electrical conductivity of aqueous solutions [1]. While Grotthuss envisioned polymolecular reactant chains, the theory that termolecular reactions should be especially favourable energy-wise was proposed by Lowry and Faulkner [2] and generalized by Swain [3]. However, for bifunctional proton transfer to be favoured, conformations in which the double proton transfer is possible must be of low energy [4].

That water can participate in very fast proton-transfer reactions was demonstrated by Grunwald, Lowenstein, and Meiboom [5] using proton magnetic resonance (p.m.r.) techniques. Some years later, Luz and Meiboom [6] extended the technique, by using O^{17} labelled water, to measure the actual number of water molecules that participate.

4.2 Measurement of solvent participation by means of proton exchange

Suppose that proton transfer from the acid HA to the base B proceeds with participation of n solvent molecules SOH. Suppose further that we can measure the proton exchange rates d[HA]/dt in the acid and d[SOH]/dt in the solvent. The mole ratio n is then given in equation 2.

$$n = \frac{d[SOH]/dt}{d[HA]/dt} \tag{2}$$

In many cases, the actual proton exchange is the result of two or more parallel reactions. To evaluate n for a specific reaction, it is necessary, in applying equation 2, to take the ratio of corresponding kinetic terms [7].

For proton exchange with half-lives that are longer than about 10 seconds, reaction rates can be measured by isotopic labelling [8, 9]. For reactions with shorter half-lives, the p.m.r. method is especially useful [10]. Briefly, when a proton leaves its original site and enters a new site at which its resonance frequency is different, the p.m.r. spectrum indicates this fact by a characteristic broadening or coalescence of lines. The line shapes can then be interpreted quantitatively to yield accurate rates of exchange. Since p.m.r. lines can be assigned unambiguously to specific sites, the reaction producing the exchange is clearly identified. The measurements are made at dynamic equilibrium. As the exchange may involve protons only, the rates need not be complicated by kinetic isotope effects.

Apart from giving information about solvent participation, the nuclear magnetic resonance (n.m.r.) spectrum can reveal other features of the reaction mechanism as well. The reason for this is that the spectrum often consists of several resonances the independent analysis of which gives different and complementary kinetic information. Thus, for asymmetrically substituted amines, it is possible to measure the rate of Walden inversion as well as the rate of proton exchange [11, 12]. In other cases it is possible to compare the rate of proton exchange (as measured by n.m.r.) with that of proton *transfer* (as measured by relaxation spectrometry), thus evaluating statistical factors in proton exchange [13].

4.3 When will solvent molecules participate?

Our working hypothesis is that solvent molecules participate in proton transfer if the following conditions are met:

(1) Either the acid or the base, or both, are so strongly solvated at the reaction site that the solvation complex is best regarded as a single kinetic unit. Thus, even though the proton transfer formally involves three or more molecules, it actually involves only two kinetic units. The activation entropy ($\Delta S^{\ddagger}$) is therefore typical of a bimolecular reaction and high rate constants are possible.

(2) The conformation of the solvent molecule at the reaction site is favourable for proton transfer.

In agreement with this hypothesis, proton transfer with solvent participation for sulphur and phosphorus acids and bases is slow, compared both to simple bimolecular proton transfer and to the corresponding reactions of oxygen and nitrogen analogues [8, 14–17]. These facts are consistent with the relative weakness of hydrogen bonds formed by sulphur and phosphorus acids and bases. For carbon acids and bases, rates of proton transfer tend to be too slow for solvent participation to be directly measurable. However, for the reaction of carbanion bases with sufficiently strong oxygen acids in aqueous solution, the rate constants approach the theoretical diffusion-controlled limits and the possibility of solvent participation cannot be dismissed [18].

For oxygen and nitrogen acids and bases, proton transfer with solvent participation is very common. Typical results for symmetrical proton transfer (from an acid to its conjugate base) in water and methanol are listed in Tables 1 and 2. Further results, including reactions in t-butyl alcohol-water mixtures, anhydrous t-butyl alcohol, and anhydrous acetic acid, will be described later.

Table 1

Rate constants for symmetrical proton transfer with participation by water molecules in aqueous solutions

Acid	$k/dm^3\,mol^{-1}\,s^{-1}$	Temp. (°C)	Reference
$(C_2H_5)_3NH^+$	1.8×10^8	30	[19]
$(C_6H_5 \cdot CH_2)_2\overset{+}{N}H \cdot CH_3$	1.5×10^7	30	[20]
PuH^a	1.0×10^8	20	[21]
LuH^{+b}	1.05×10^8	25	[22]
LuH^{+c}	0.26×10^8	25	[22]
ImH^{+d}	1.07×10^8	25	[23]
C_6H_5OH	5.7×10^8	25	[24]
$CH_3 \cdot \overset{+}{N}H_2 \cdot CH_2 \cdot CO_2^-$	1.5×10^8	21	[25]
$(HO \cdot CH_2 \cdot CH_2)_3NH^+$	3.6×10^6	30	[26]

[a] PuH = purine.
[b] LuH^+ = 2,4-Lutidinium.
[c] Rate constant for proton exchange in HOD $-$ D_2O (5 atom % H).
[d] ImH^+ = Imidizolium.

Table 2

Rate constants for symmetrical proton transfer with participation of methanol molecules in methanol

Acid	$k/dm^3\,mol^{-1}\,s^{-1}$	Temp. (°C)	Reference
$p\text{-}CH_3 \cdot C_6H_4 \cdot NH_3^+$	0.81×10^8	25.00	[15]
$C_6H_5 \cdot CO_2H$	1.2×10^8	24.80	[7]
$m\text{-}O_2N \cdot C_6H_4 \cdot CO_2H$	2.5×10^8	24.80	[27]
$p\text{-}O_2N \cdot C_6H_4 \cdot CO_2H$	2.7×10^8	24.80	[27]
$o\text{-}O_2N \cdot C_6H_4 \cdot CO_2H$	1.8×10^8	24.80	[27]
$Ar\overset{+}{N}H(C_2H_5)_2{}^a$	4.2×10^5	30.00	[28]

[a] $Ar\overset{+}{N}H(C_2H_5)_2$ = N,N-dimethyl-m-toluidinium.

As shown in Table 1, participation of water molecules has been observed in proton transfer reactions of aliphatic and heterocyclic amines, amino acids, carboxylic acids, phenols, and a variety of inorganic acids. As shown in Table 2, participation of methanol molecules has been observed in proton transfer reactions of amines and anilines, carboxylic acids, and phenols.

Proton transfer with solvent participation always competes with direct bimolecular proton transfer, at least in principle, and comparison of the two rates is of theoretical interest. Unfortunately, data are available only in rare instances. The most complete series is that for the methylamines in water; second-order rate constants for reaction with (equation 3) and without (equation 4) water participation are compared in Table 3 (p. 111)

$$B + \underset{\underset{H}{|}}{HO} + HB^+ \xrightarrow{k_{2w}} BH^+ + \underset{\underset{H}{|}}{OH} + B \tag{3}$$

$$B + HB^+ \xrightarrow{k_2} BH^+ + B \tag{4}$$

According to the data in Table 3, the patterns of reactivity in 3 and 4 are very different. While k_{2w} varies roughly as the base strength (K_B) of the amine, k_2 shows a marked decrease with methyl substitution. However, k_2 can be roughly correlated with the rate constant (k_H) for breaking the hydrogen bond between the amine and an adjacent water molecule (equation 5).

$$HOH(aq) + B \cdot HOH \xrightarrow{k_H} B \cdot HOH + HOH(aq) \tag{5}$$

In interpreting these facts, we note that in the bimolecular process, 4, water molecules must be displaced from the hydrogen-bonding sites in the acid and base before proton transfer can take place. If this requires an activation energy, the rate of the bimolecular reaction will be correspondingly low. For ionic species, there is a long chain of evidence that the removal of water molecules from the inner solvation shell requires substantial amounts of work. Among the more dramatic manifestations of this effect are the great enhancements of the nucleophilic reactivity of small anions (such as Cl^- or acetate$^-$) and of the basicity of hydroxide salts in aprotic solvents [29–31].

Thus, when bimolecular proton transfer is relatively slow, we surmise that the displacement of water molecules from the solvation shells requires an activation energy and is the rate-determining step for reaction. The fact that proton transfer *with* solvent participation remains fast when k_2 becomes small, suggests that the proton-transfer step is inherently fast; and the fact that k_2 is small also for the reaction of $(CH_3)_3N$ with NH_4^+ suggests that steric hindrance is not the primary factor [32].

By the same token, reaction with participation by *one* water molecule will be relatively slow when *both* reactive sites are firmly hydrated. In that case, we may observe fast reaction with participation of *two* water molecules. However, in no case is reaction with solvent participation expected to be fast unless the conformation of the firmly bound solvent molecule is favourable for proton transfer. Thus, the relatively small values of the rate constants for reaction of *tris-β-hydroxyethylamine* compared to triethylamine in water (Table 1) and of N,N-diethyl-*m*-toluidine compared to *p*-toluidine in methanol (Table 2) suggest that polar substituents or centres of van der Waals attraction can modify the hydrogen-bonded structure in the solvation shell.

4.4 Higher-than-termolecular proton transfer reactions

In the great majority of cases of proton-transfer in which the participation of solvent molecules has been determined, the reaction is termolecular – that is, reaction involves one bifunctional solvent molecule in addition to the proton donor and acceptor. However, a few examples have been established in which the proton transfer involves more than one solvent molecule. Thus, for symmetrical proton transfer from a nitrogen acid to its conjugate base in aqueous solution, the known examples and the average number (n) of participating water molecules are listed in equations 6–8 [22, 33, 34].

$$n = 1.4 \text{ in HOH}$$
$$n = 1.0 \text{ in t-BuOH--HOH} \tag{6}$$

$$n = 1.74 \text{ in HOH}$$
$$n = 1.8 \text{ in HOD--D}_2\text{O} \tag{7}$$

$$n = 1.7 \text{ in HOH}$$
$$n = 2.0 \text{ in HOD--D}_2\text{O} \tag{8}$$

For other proton transfer reactions in aqueous solution, the rate constant (and hence the encounter diameter σ of the reactants in the reactive complex) is so great as to require the intervention of two or

108

more water molecules. According to the theory of Debye [35] and Eigen, [36] the second-order rate constant k_e (in units of $dm^3 mol^{-1} s^{-1}$) for the formation of an encounter complex from two molecular species A and B, is given by equation 9 if one species is electrically neutral, and by equation 10 if both species are ions. In these equations, D_A and D_B denote the phenomenological diffusion coefficients

$$k_e = 0.004\pi N\sigma(D_A + D_B) \tag{9}$$

$$k_e = 0.008\pi Nq(D_A + D_B)/(\exp[2q/\sigma] - 1) \tag{10}$$

and q the Bjerrum radius, i.e. $q = Z_A Z_B e^2/2\varepsilon kT$. Equations 9 and 10 may overestimate k_e because specific effects, due to the partial desolvation required in forming the encounter complex, are neglected. In estimating σ by this approach, one uses the observed reaction-rate constant k in place of k_e and solves for σ. Since $k \leqslant k_e$, [37] this estimate will be a lower limit.

A few examples, in which the estimate of σ suggests the participation of two or more water molecules, are listed in equations 11–13 [38–40], where 1 pm = 0.01 Å.

$$H_3O^+ + OH^- \; ; k = 1.4 \times 10^{11} \, s^{-1} \, dm^3 \, mol^{-1} \quad \text{at } 25°C$$

$$\sigma > 800 \, pm \tag{11}$$

$$H_2PO_4^- + HPO_4^{2-} \; ; k = 1.4 \times 10^9 \, s^{-1} \, dm^3 \, mol^{-1} \quad \text{at } 25°C$$

$$\sigma > 550 \, pm \tag{12}$$

$$Al(OH_2)_6^{3+} + HOAl(OH_2)_5^{2+} \; ; k = 9 \times 10^8 \, s^{-1} \, dm^3 \, mol^{-1}$$

$$\text{at } 30°C$$

$$\sigma > 1160 \, pm \tag{13}$$

In interpreting these data, we compare the σ-values with the following mean 0–0 distances in liquid water: nearest neighbours, 290 pm; next-nearest neighbours, $\sim$450 pm; third-nearest neighbours, $\sim$600 pm. A higher-than-termolecular mechanism is clearly indicated in reactions 11 and 13 and probable also in case 12.

4.5 London dispersion and hydrophobic forces

In examples 6–8, the increase in n above the value of unity associated with a termolecular reaction mechanism was ascribed to quatermolecular reaction in which the reactive complex has the formula $A \cdot W \cdot W \cdot B$, where W represents a water molecule and each dot represents a hydrogen bond. However, this interpretation is not unique.

An alternative mechanism leading to $n > 1$ involves a termolecular reactive complex, $A \cdot W \cdot B$, in which the water molecule is loosely held and exchanges readily with water molecules from the bulk solvent equation 14.

$$A \cdot \underline{W} \cdot B + W(aq) \rightarrow A \cdot W \cdot B + \underline{W}(aq) \qquad (14)$$

Thus, during the lifetime of an encounter between A and B, more than one water molecule comes into a position in which it can receive a new proton. This mechanism requires that proton transfer be fast compared to the separation of A and B.

The alternative mechanism is especially attractive for acids and bases with bulky and polarizable substituents, because then the complex of A and B can be held largely by London dispersion forces, while the water molecule simply slips into the space between the reactive sites, as in 15. That water molecules from bulk water are in

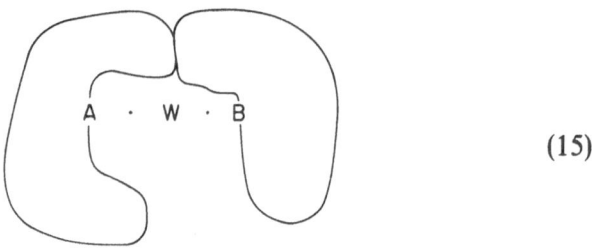

(15)

dynamic equilibrium with water molecules trapped on suitable sites in larger structures, such as proteins or micelles, is well known [42, 43]. However, the alternative mechanism must be ruled out at least in cases 7 and 8, because n is practically the same (within the experimental error) in H_2O and in $HOD-D_2O$, while the rate constant shows a substantial solvent isotope effect. If the alternative mechanism were operating, one would have expected n to be smaller in $HOD-D_2O$. Thus these reactions are thought to be truly quatermolecular [22].

There is other evidence that the exchange of water molecules between the site in $A \cdot W \cdot B$ and bulk solvent is slow compared to both proton transfer within the complex and the separation of A from B. Regarding the former, the fact that so many proton-transfer reactions in which ΔG^0 is negative are diffusion-controlled proves that the proton-transfer step is fast compared to the dissociation of the complex. Regarding the latter, if the departure of a water molecule from $A \cdot W \cdot B$ were fast compared to the dissociation of the complex, one would expect to find more examples of rapid direct bimolecular proton transfer without solvent participation.

Returning to the model of A · W · B shown in 15, the attraction of the hydrophobic groups in A and B by London dispersion forces is well known and must be expected to slow down the separation of A from B. Similarly, it is becoming increasingly clear that London dispersion forces are helpful also in binding the water molecule to the hydrogen-bonding site. For instance, the introduction of alkyl groups into ammonia greatly reduces the rate constant k_H for breaking the $R_3N \cdot HOH$ hydrogen bond (Table 3). The observed substituent

Table 3
Rate constants for proton exchange according to equations 3 to 5 at 30°C[a]

B	$10^{-8}k_2/$ dm³ mol⁻¹ s⁻¹	$10^{-8}k_{2w}/$ dm³ mol⁻¹ s⁻¹	$10^{-10}k_H/$ s⁻¹	$10^4 K_B/$ mol dm⁻³
NH_3	12.7 ± 0.5	0.50 ± 0·04	22.0	0.183
CH_3NH_2	4.4 ± 0.4	5.8 ± 0.4	6.2	4.20
$(CH_3)_2NH$	0.5 ± 0.4	9.9 ± 0.3	—	6.30
$(CH_3)_3N$	0.0 ± 0.3	3.7 ± 0.3	1.0	0.73
$(CH_3)_3N + NH_4^+$	0.4 ± 0.5	7.6 ± 0.4	1.0[b]	0.73[b]

[a] Based on data of Grunwald and Ku [32] and Grunwald and Ralph [44].
[b] For $(CH_3)_3N$.

effects on the free energy of activation $(-RT\delta_R \ln k_H)$ are in this instance well correlated with the predicted changes in the London dispersion interaction energy [20].

In mixed solvents such as t-butyl alcohol-water mixtures, the study of solvent participation in proton transfer reactions can elucidate differences in solvation forces due to the two solvent components. In an alcohol-water mixture, it is sometimes possible to measure separate rate constants for proton transfer with participation of water and of alcohol [23]. In the kinds of systems in which this was done, it was shown that at least the reaction with water is termolecular. Results obtained in 88.5 mole per cent water–11.5 mole per cent t-butyl alcohol are listed in Table 4. Here k_{2w} has the same significance as in equation 3, and k_{2ROH} is the analogous second-order rate constant for reaction with alcohol participation.

If certain reasonable assumptions are made, the results in Table 4 provide an index of relative solvation. In brief, the ratio k_{2w}/k_{2ROH} is believed to measure the mole ratio of the corresponding solvation complexes of the ionic reactant [26]. Thus,

$$[ImH^+ \cdot OH_2]/[ImH^+ \cdot (OH)R] \approx 13;$$

$$[o\text{-}BrC_6H_4O^- \cdot HOH]/[o\text{-}BrC_6H_4O^- \cdot HOR] \approx 14;$$

$$[m\text{-}O_2NC_6H_4O^- \cdot HOH]/[m\text{-}O_2NC_6H_4O^- \cdot HOR] \approx 32.$$

Table 4

Kinetic results for solvent participation in 88.5 mole per cent water—11.5 mole per cent t-butyl alcohol at 15.7° [26]

A	B	k_{2w}/dm^3 $mol^{-1} s^{-1}$	k_{2ROH}/dm^3 $mol^{-1} s^{-1}$	$\dfrac{k_{2w}}{k_{2ROH}}$
ImH^+	Im^a	2.00×10^8	1.6×10^7	13 ± 2
$o\text{-}BrC_6H_4OH$	$o\text{-}BrC_6H_4O^-$	1.25×10^8	9.1×10^6	14 ± 3
$m\text{-}O_2NC_6H_4OH$	$m\text{-}O_2NC_6H_4O^-$	5.40×10^8	1.7×10^7	32 ± 8
Mole ratio in solvent				7.44

[a] Im = imidazole, at 25°C.

Since the mole ratio [HOH]/[ROH] in the bulk solvent is 7.44, we have in each case an enrichment of water at the hydrogen-bonding site in the ionic solvation shell. However, the striking difference between $m\text{-}O_2NC_6H_4O^-$ and $o\text{-}BrC_6H_4O^-$ could very well be due to van der Waals-London dispersion forces between the *ortho*-bromo substituent and the t-butyl group.

4.6 Acid and base dissociation

The acid dissociation of a Brønsted acid HA to give the conjugate base B and the lyonium ion may be a bimolecular process equation 16 or it may proceed with solvent participation equation 17.

$$A-H + SOH \underset{k_{-a}}{\overset{k_a}{\rightleftharpoons}} B + \underset{H}{SOH^+} \tag{16}$$

$$A-H + \underset{S}{OH} + SOH \underset{k'_{-a}}{\overset{k'_a}{\rightleftharpoons}} B + \underset{S}{HO} + SOH_2^+ \tag{17}$$

For acid dissociation in water, the two mechanisms are readily distinguished by means of $HOD-D_2O$ solvent isotope effects *if the reverse reaction* (rate constant k_{-a} or k'_{-a}) *is diffusion controlled*. For bimolecular proton transfer according to equation 18 the resulting lyonium ion is HOD_2^+, while for proton transfer with water participation equation 19 it is D_3O^+.

$$A-H + OD_2 \rightleftharpoons B + HOD_2^+ \tag{18}$$

$$A-H + OD_2 + OD_2 \rightleftharpoons B + HOD + DOD_2^+ \tag{19}$$

Since HOD_2^+ is thermodynamically a more stable species than D_3O^+

(see Chapter 9), it can be calculated that the kinetic isotope effect, $k_a^{HOD-D_2O}/k_a^{H_2O}$, should be 0.37 for the bimolecular reaction 18 and 0.25 for the higher-than-bimolecular reaction 19. For the acid dissociation of $Al(OH_2)_6^{3+}$ in water, $k_a^{HOD-D_2O}/k_a^{H_2O}$ is found to be 0.34 ± 0.02, indicating equation 18, while for the acid dissociation of $Pt(NH_3)_6^{4+}$ it is found to be 0.23 ± 0.02, indicating equation 19 [13, 40].

A special version of the acid-dissociation mechanism, 17, leads to the formation of a conjugate base solvate, as shown in equation 20.

$$A-\underset{\underset{S}{|}}{H} + \underset{\underset{S}{|}}{OH} + OH \underset{k_{-a}}{\overset{k_a}{\rightleftharpoons}} B \cdot HOS + SOH_2{}^+ \qquad (20)$$

This mechanism can be recognized by its characteristic rate law. Note that in $B \cdot HOS$, the acidic proton of $A-H$ remains connected to its original site by means of a hydrogen bond. Thus, simple reversal of acid dissociation will restore the original $A-H$ molecule, without proton exchange. In order to complete the proton exchange, the $B \cdot HOS$ hydrogen bond must break, as shown formally in equation 21.

$$B \cdot HOS + HOS(\text{solvent}) \overset{k_H}{\rightarrow} B \cdot HOS + HOS(\text{solvent}) \qquad (21)$$

Subsequent reversal of acid dissociation then leads to proton exchange.

$$B \cdot HOS + SOH_2{}^+ \overset{k_{-a}}{\longrightarrow} AH + \underset{\underset{S}{|}}{OH} + SOH \qquad (22)$$

The overall mechanism, 20 to 22, leads to the rate law, 23.

$$\text{Rate of proton exchange} = \frac{k_a k_H [HA]}{k_H + k_{-a}[SOH_2{}^+]} \qquad (23)$$

Equation (23) has been found to fit rates of proton exchange in the acid dissociation of aliphatic ammonium ions in water, methanol, and t-butyl alcohol–water mixtures. When the kinetic data are combined with values obtained for the acid dissociation constant $K_a (= k_a/k_{-a})$, all three rate constants in equation 23 can be evaluated. [44]. Results obtained for k_a and k_{-a} by this approach are in good agreement with relaxation spectrometry; representative values of k_H are included in Table 3.

The only available direct measurement of solvent participation in base dissociation is for the dissociation of p-bromo-phenoxide ion in methanol, which is termolecular (equation 24) [45].

$$p\text{-BrC}_6\text{H}_4\text{O}^- + \underset{\overset{|}{\text{CH}_3}}{\text{HO}} + \underset{\overset{|}{\text{CH}_3}}{\text{HO}} \rightarrow p\text{-BrC}_6\text{H}_4\text{OH} + \underset{\overset{|}{\text{CH}_3}}{\text{OH}} + \underset{\overset{|}{\text{CH}_3}}{\text{O}^-} \qquad (24)$$

4.7 Acid and base ionization

It is often useful to dissect the overall process of acid or base dissociation into separate steps of *ionization* to form an ionized intermediate
equation 25, and *dissociation* of that intermediate equation 26.

$$A\text{—H} + \underset{\overset{|}{S}}{\text{O—H}} \overset{K_i}{\rightleftharpoons} A^- \cdot \underset{\overset{|}{S}}{\text{H—O—H}^+} \qquad (25)$$

$$A^- \cdot \underset{\overset{|}{S}}{\text{H—O—H}^+} \overset{K_d}{\rightleftharpoons} A^- + \text{SOH}_2{}^+ \qquad (26)$$

If this mechanism is correct, that is, if the ionized intermediate has a
finite lifetime, its presence should be detectable by kinetic studies of
proton exchange.

For carboxylic acids in water and methanol, there is good evidence
that such an intermediate exists. The rate constant for the acid dissociation of acetic acid in water has been measured by relaxation
spectrometry and found to be $8 \times 10^5 \text{ s}^{-1}$ at 25°C [46, 47]. On the
other hand, the first order rate constant for proton exchange between
acetic acid and water under the same conditions is $1.0 \times 10^8 \text{ s}^{-1}$,
more than one hundred times greater [48]. Similarly, for the acid
dissociation of benzoic acid in methanol, at 25°C the rate constant
for acid dissociation is predicted to be $<30 \text{ s}^{-1}$, while the first order
rate constant for proton exchange is $7 \times 10^4 \text{ s}^{-1}$, more than two

114

thousand times greater [7]. To explain the greater rates of proton exchange, it is proposed that acid dissociation involves an initial ionization step in which the ionized intermediate has a cyclical structure, as shown in equation 27. For benzoic acid in methanol, it was found that this intermediate contains two methanol molecules, as indicated.

If proton transfer from A to B took place in a time that is short compared to the lifetime of this intermediate, then reversal of ionization would lead to proton exchange between the carboxyl group and methanol without formation of the dissociated ions, RCO_2^- + $CH_3OH_2^+$. Because the process, 27, entails no net chemical reaction, its existence cannot be detected by relaxation spectrometry.

To find out whether the cyclic intermediate involved in proton exchange is indeed an ion pair, both the strength of the acid and the solvent were varied. For a series of substituted benzoic acids in methanol it [27] turned out that the rate constant (k) for proton exchange increases monotonically with K_a, approximately according to $k \propto K_a^{\frac{1}{2}}$. Changing the solvent from methanol to water similarly raises the rate. On the basis of these facts it seems probable that the cyclic intermediate is a hydrogen-bonded ion pair, as shown in equation 27. The dissociation of the ion pair into free ions is slow compared to the re-formation of the unionized acid.

A closely related reaction is proton exchange of neopentyl alcohol in dry acetic acid in the presence of various acid catalysts. In this case the facts again suggest that reaction proceeds via an ionized intermediate, but this time the intermediate contains only one molecule of alcohol [49].

Because of the symmetrical structure of the carboxylate ion, the probability that reversible ionization leads to proton exchange with solvent is high, approaching 0.5 if the proton-transfer step, 27, is fast enough. In the absence of two equivalent sites, proton transfer in an ionized intermediate would require the breaking of a hydrogen bond, and hence be more difficult. A possible mechanism is rotation of the lyonium ion, as shown in equation 28.

$$B \cdot H{-}O{-}H^+ \;\rightarrow\; B \cdot H{-}O{-}H^+ \qquad (28)$$
$$\underset{S}{\mid} \qquad\qquad\qquad \underset{S}{\mid}$$

As a matter of fact, proton exchange of imidazolium ion in aqueous solutions is significantly faster than acid dissociation. Since the activation energies for proton exchange and acid dissociation are equal, it is probable that the two processes have the same rate-determining step, formation of an ionized intermediate [23].

A more direct approach is to study ionization in solvents of low dielectric constant, such as acetic acid ($\varepsilon = 6.27$ at 25°). Then, if the product of ionization is an ion pair, dissociation into free ions is stoichiometrically insignificant. Equilibrium constants for, and proton exchange resulting from ionization have been measured for a series of substituted anilines in acetic acid [50–53]. The results are listed in Table 5. Because the anilines are fairly weak bases, the actual reactant in the kinetic studies is introduced as a p-toluenesulphonate salt, which is in dynamic equilibrium with the acetate salt.

Table 5

Ionization and proton exchange of substituted anilines in acetic acid at 30°C

Substituent	K_i [50]	k_e/s^{-1}	k_H/s^{-1}	Reference
p-OCH$_3$	32.8[a]	1.00×10^8	3.00×10^9	[53]
p-CH$_3$	19.20[a]	1.04×10^8	2.00×10^9	[52]
m-CH$_3$	9.56[a]	3.50×10^8	3.30×10^9	[53]
p-F	4.66[a]	5.60×10^8	2.60×10^9	[53]
m-OCH$_3$	4.71[a]	3.80×10^8	1.80×10^9	[53]
N-(CH$_3$)$_2$	12.70[b]	1.01×10^8	1.30×10^9	[51]
N-(C$_2$H$_5$)$_2$	109.00[b]	5.50×10^6	0.60×10^9	[51]
N-(n-C$_3$H$_7$)$_2$	30.50[b]	8.80×10^6	0.27×10^9	[51]

[a] Measurement by Ceska and Grunwald [50].
[b] Data at 26°C.

The measured proton exchange is that of the N—H proton in the anilinium acetate salt, BH^+OAc^-, with the carboxyl proton of the solvent. This rate is relevant to the present discussion because BH^+OAc^- is the ionized form of the given aniline in acetic acid. It turns out that the rate constant k_e for proton exchange is correlated with the basicity of the aniline, decreasing as the basicity increases. This suggests a two-step reaction mechanism, *i.e.* equations 29 and 30.

$$\text{BH}^+ \cdot \text{OAc}^- \underset{k_r}{\overset{k_f}{\rightleftharpoons}} \text{B} \cdot \text{HOAc} \qquad (29)$$

$$\text{B} \cdot \text{HOAc} + \text{HOAc(solvent)} \overset{k_H}{\longrightarrow} \text{B} \cdot \text{HOAc} + \text{HOAc(solvent)} \qquad (30)$$

If proton transfer in the first step is rate-determining, the experimental rate constant k_e is equal to k_f. If the first step is fast and reversible, and the second step is rate-determining, we have $k_e = k_f k_H/(k_H + k_r) \approx k_H k_f/k_r = k_H/K_i$, where K_i denotes the base ionization constant, *i.e.* the equilibrium constant for the reverse of 29. A mechanism in which the first step is rate-determining seems improbable, because the values

k_r evaluated on this basis do not follow the expected Brønsted relationship with base strength. On the other hand, the values of k_H calculated on the assumption that the second step is rate-determining, show an analogous dependence on N-alkyl substitution as the values of k_H for aliphatic amine hydrates listed in Table 3. Assuming that this is the correct mechanism, we have listed values of k_H in Table 5. For the *meta* and *para*-substituted anilines, these values show the same lack of dependence on base strength as do the values of k_H for aliphatic amines in water. Unfortunately, the data do not disclose the number of acetic acid molecules that participate in the ionization process.

4.8 Acid-base reactions of ion pairs

Ion pairs of the type BH^+X^-, formed from the interaction of an acidic cation with a basic anion, are often bound by a hydrogen bond in addition to the bond due to the opposite electrical charges [54]. In order to determine the effect of an adjacent anion on the reactivity of an acidic cation in proton transfer reactions, Cocivera and Grunwald [55] measured the rates of proton exchange between methyl-substituted ammonium salts and their conjugate amines in t-butyl alcohol solution. This reaction proceeds with solvent participation. Owing to the fairly low dielectric constant ($\varepsilon = 12.47$ at $25°C$), the salts exist largely in the form of ion pairs at the given concentration, and the fact that the rates are proportional to the salt concentration indicates that ion pairs, rather than free ions, are reactants. Thus the reaction is represented appropriately by equation 31, in which R = H or CH_3 and Bu = t-butyl.

$$R_3\overset{\oplus}{\underset{\overset{\cdot\cdot}{\underset{X^\ominus}{}}}{N}H} + \overset{\overset{Bu}{|}}{O}H + NR_3 \overset{k_2'}{\rightarrow} R_3\underline{N} + H\overset{\overset{Bu}{|}}{O} + H\overset{\oplus}{\underset{\overset{\cdot\cdot}{\underset{X^\ominus}{}}}{N}R_3} \quad (31)$$

Kinetic results are listed in Table 6. For the di- and monomethyl-ammonium salts, the second-order rate constants (k_2') are nearly independent of the nature of the anion and nearly equal to each other. For the trimethylammonium salts, the rate constants are highly sensitive to the nature of the anion and much smaller.

These results are represented well by the theory that in the ion pair, one of the NH protons is hydrogen-bonded to the anion; any others are hydrogen-bonded to a solvent molecule. Further, the NH proton that is hydrogen-bonded to an anion is considerably less reactive than

Table 6

Values of k'_2 for methyl-substituted ammonium salts in
t-butyl alcohol at 35°C [49]

Cation	Anion	$10^5 k'_2/dm^3\,mol^{-1}\,s^{-1}$
$(CH_3)_3NH^+$	Cl^-	1.1 ± 0.12
	OTs^{-a}	7.0 ± 0.70
	TFA^{-b}	1.8 ± 0.40
$(CH_3)_2NH_2^+$	Cl^-	270 ± 40
	OTs^{-a}	310 ± 30
	TFA^{-b}	220 ± 20
$CH_3NH_3^+$	Cl^-	260 ± 30
	OTs^{-a}	240 ± 40

[a] *p*-toluenesulphonate.
[b] trifluoroacetate.

one bound to solvent molecules. Thus, in monomethylammonium and dimethylammonium salts, there are, respectively, two protons and one proton that remain available for hydrogen bonding to the t-BuOH molecules. Proton transfer with solvent participation is relatively easy, requiring only a minor motion of the anion towards the new cationic centre. In trimethylammonium salts, proton exchange can take place only if the hydrogen bond to the anion is first broken or loosened. This process seems to require an activation energy.

By an extension of this argument, proton transfer to or from a hydrogen-bonded ion pair will proceed with a high, diffusion-controlled rate constant only if the reaction mechanism does not require the breaking of the hydrogen bond joining the cation and anion. For example, rate constants for proton transfer from *p*-toluenesulfonic acid (HOTs) to substituted anilinium acetate ion pairs in acetic acid have been found to be greater than $10^{10}\,dm^3\,mol^{-1}\,s^{-1}$ [51, 52]. The reaction is therefore thought to result in a solvent-separated ion pair, as in equation 32.

$$BH^+ \cdot OAc^- + HOTs \rightarrow BH^+ \cdot AcOH \cdot OTs^- \qquad (32)$$

4.9 Concluding remarks

We hope to have shown in this chapter that the study of solvent participation in proton transfer reactions is rewarding, not only because it leads to a sharper definition of the reaction mechanisms, but also because it elucidates the nature of solvation shells and the rates for the breaking of hydrogen bonds. Although much has already been learned, many problems remain. For instance, our knowledge of deuterium isotope effects and of their mechanistic implications in termolecular and quatermolecular proton transfer is far from satisfactory. So is our knowledge of bifunctional catalysis, particularly of its geometrical and stereo-electronic requirements. So is our knowledge of the interpenetration of solvation shells, both for separate molecules and for polar groups in the same molecule. These and other problems concerning the interaction of acids, bases, and hydroxylic solvent molecules will, it is hoped, yield to attack by methods such as those described in this chapter.

REFERENCES

[1a] C. J. D. von Grotthuss, *Ann. Chem.*, **58**, 54 (1806).

[1b] W. J. Moore, *Physical Chemistry*, 4th edn, Prentice-Hall, Inc., Englewood Cliffs, N.J., pp. 428, 435 (1972).

[2] T. M. Lowry and I. J. Faulkner, *J. Chem. Soc.*, **127**, 2883 (1925).

[3] C. G. Swain, *J. Amer. Chem. Soc.*, **70**, 1119 (1948).

[4] C. G. Swain and J. F. Brown, *J. Amer. Chem. Soc.*, **74**, 2538 (1952).

[5] E. Grunwald, A. Lowenstein, and S. Meiboom, *J. Chem. Phys.*, **27**, 630 (1957).

[6] Z. Luz and S. Meiboom, *J. Chem. Phys.*, **39**, 366 (1963).

[7] E. Grunwald, C. F. Jumper, and S. Meiboom, *J. Amer. Chem. Soc.*, **85**, 522 (1963).

[8] A. I. Brodskii, *J. Gen. Chem.* (*U.S.S.R.*) (Eng. Trans.), **24**, 421 (1954).

[9] C. G. Swain and M. M. Labes, *J. Amer. Chem. Soc.*, **79**, 1084 (1957).

[10] F. A. Cotton and L. M. Jackman, Eds. *Dynamic Nuclear Magnetic Resonance*, Academic Press, New York (1974).

[11] M. Saunders and F. Yamada, *J. Amer. Chem. Soc.*, **85**, 1882 (1963).

[12] D. E. Leyden and R. E. Channell, *J. Phys. Chem.*, **77**, 1562 (1973).

[13] E. Grunwald and D.-W. Fong, *J. Amer. Chem. Soc.*, **94**, 7371 (1972).

[14] B. Silver and Z. Luz, *J. Amer. Chem. Soc.*, **85**, 786 (1961).

[15] M. Cocivera, E. Grunwald, and C. F. Jumper, *J. Phys. Chem.*, **68**, 3234 (1964).

[16] M. M. Kreevoy, D. S. Sappenfield, and W. Schwabacher, *J. Phys. Chem.*, **69**, 2287 (1965).

[17] J. F. Whidby and D. E. Leyden, *J. Phys. Chem.*, **74**, 202 (1970).

[18] F. Hibbert, F. A. Long, and E. A. Walters, *J. Amer. Chem. Soc.*, **93**, 2829 (1971).

[19] E. K. Ralph, III and E. Grunwald, *J. Amer. Chem. Soc.*, **89**, 2963 (1967).

[20] E. Grunwald and E. K. Ralph, III, *J. Amer. Chem. Soc.*, **89**, 4405 (1967).

[21] T. H. Marshall and E. Grunwald, *J. Amer. Chem. Soc.*, **91**, 4541 (1969).

[22] D. Rosenthal and E. Grunwald, *J. Amer. Chem. Soc.*, **94**, 5956 (1972).

[23] E. K. Ralph, III and E. Grunwald, *J. Amer. Chem. Soc.*, **91**, 2422, 2429 (1969).

[24] E. Grunwald and M. S. Puar, *J. Phys. Chem.*, **71**, 1842 (1967).

[25] M. Scheinblatt, *J. Chem. Phys.*, **39**, 2005 (1963).

[26] E. Grunwald, D.-W. Fong, and E. K. Ralph, III, *Bull. Israel Chem. Soc.*, **9**, 287 (1971).

[27] E. Grunwald and S. Meiboom, *J. Amer. Chem. Soc.*, **85**, 2047 (1963).

[28] E. Grunwald, R. L. Lipnick, and E. K. Ralph, III, *J. Amer. Chem. Soc.*, **91**, 4333 (1968).

[29] A. J. Parker, *Chem. Rev.*, **69**, 1 (1969).

[30] C. H. Langford and R. L. Burwell, *J. Amer. Chem. Soc.*, **82**, 1503 (1960).

[31] D. Dolman and R. Stewart, *Can. J. Chem.*, **45**, 911 (1967).

[32] E. Grunwald and A. Y. Ku, *J. Amer. Chem. Soc.*, **90**, 29 (1968).

[33] E. K. Ralph, III and E. Grunwald, *J. Amer. Chem. Soc.*, **90**, 517 (1968).

[34] E. K. Ralph, III, personal communication, January 21, 1974.

[35] P. Debye, *Trans. Am. Electrochem. Soc.*, **82**, 265 (1942).

[36] M. Eigen, *Z. Phys. Chem.* (*Frankfurt*), **1**, 176 (1954).

[37] M. V. Smoluchowski, *Physik. Z.*, **17**, 557, 585 (1916); *Z. phys. Chem.*, **113**, 35 (1924).

[38] G. Ertl and H. Gerischer, *Z. Elektrochem.*, **66**, 560 (1962).

[39] Z. Luz and S. Meiboom, *J. Amer. Chem. Soc.*, **86**, 4764 (1964).

[40] D.-W. Fong and E. Grunwald, *J. Amer. Chem. Soc.*, **91**, 2413 (1969).

[41] G. W. Brady and W. J. Romanow, *J. Chem. Phys.*, **32**, 306 (1960).

[42] F. H. C. Crick and J. C. Kendrew, *Adv. Protein Chem.*, **12**, 133 (1957).

[43] F. M. Menger, *J. Amer. Chem. Soc.*, **88**, 3081 (1966).

[44] E. Grunwald and E. K. Ralph, III, *Accounts Chem. Res.*, **4**, 107 (1971).

[45] E. Grunwald, C. F. Jumper, and M. S. Puar, *J. Phys. Chem.*, **71**, 492 (1967).

[46] M. Eigen and J. Schoen, *Z. Elektrochem.*, **59**, 483 (1955).

[47] M. Eigen and E. M. Eyring, *J. Amer. Chem. Soc.*, **84**, 3254 (1962).

[48] Z. Luz and S. Meiboom, *J. Amer. Chem. Soc.*, **85**, 3923 (1963).

[49] M. Cocivera and E. Grunwald, *J. Amer. Chem. Soc.*, **87**, 2070 (1965).

[50] G. W. Ceska and E. Grunwald, *J. Amer. Chem. Soc.*, **89**, 1371 (1967).

[51] E. Grunwald and M. S. Puar, *J. Amer. Chem. Soc.*, **89**, 6842 (1967).

[52] M. R. Crampton and E. Grunwald, *J. Amer. Chem. Soc.*, **93**, 2987 (1971).

[53] F. M. Jones, III, D. Eustace, and E. Grunwald, *J. Amer. Chem. Soc.*, **94**, 8941 (1972).

[54] M. M. Davis, *Acid-Base Behavior in Aprotic Organic Solvents*, National Bureau of Standards Monograph 105, Washington, D.C. (1968).

[55] M. Cocivera and E. Grunwald, *J. Amer. Chem. Soc.*, **87**, 2070 (1965).

5

Brian H. Robinson

HYDROGEN-BONDING AND PROTON-TRANSFER REACTIONS IN APROTIC SOLVENTS

5.1 Introduction

In this chapter, the thermodynamics and kinetics of the steps involved in proton-transfer reactions in aprotic solvents of low dielectric constant will be considered, and special attention will be paid to dissecting the overall reaction 1 into a sequence of elementary processes, each of which will be discussed and related to the measured kinetic parameters k_f and k_b.

$$AH + B \underset{k_b}{\overset{k_f}{\rightleftharpoons}} A^- + HB^+ \tag{1}$$

To date few detailed kinetic studies on ion-pair formation in aprotic solvents have been carried out, perhaps because fast reaction techniques, *e.g.* stopped-flow and temperature-jump, are generally required. However, these methods are now well established and rate constants can be obtained to the same order of accuracy as by conventional techniques.

The general features of proton-transfer reactions in hydroxylic solvents are now becoming more clearly understood. Details of the

121

mechanism, in particular the nature of the transition state, can be discussed through the application of the free-energy relationships of Brønsted and of Marcus, and through the analysis of primary and secondary isotope effects. The role played by the solvent has been increasingly stressed, and revealed most clearly in hydroxylic solvents through solvent isotope effects and nmr studies.

In aqueous media, proton transfers are complicated by specific solute-solvent interactions. Oxygen and nitrogen acids and bases can form strong hydrogen bonds to the solvent, which because of its amphoteric nature, can allow fast proton transfer between the 'solvated' reactants by a Grotthuss-chain type mechanism. For carbon acids, hydrogen bonding is weak and proton transfer is generally slower and can take place between the acid and base in direct contact.

In aprotic solvents of low dielectric constant (ε), specific solute-solvent interactions are much reduced (but not eliminated) and proton transfer is expected to be direct. Thus the mechanism is much simpler, and this consideration, coupled with the lack of solvent structure, suggests that proton transfers will be more easily analysed and interpreted in such media. In water, coulombic interactions (e.g. ion–ion and ion–dipole forces) are relatively unimportant, but they are likely to dominate proton-transfer considerations in solvents of low ε.

However, there may well be complications involving the so-called dipolar aprotic solvents with values of ε between 20 and 50, due primarily to the uncertainty in the nature of the proton-transferred species. Consequently, the thermodynamic analysis of acid–base interactions in these solvents is generally unsatisfactory.

The thermodynamic and analytical aspects of acid-base reactions in aprotic solvents are surveyed in reviews by Davis [1, 2]. The correlation of acid-base strength in water and aprotic solvents is of major importance. Early kinetic work by Bell and co-workers on the acid catalysis of (i) the ethyldiazoacetate–phenol interaction [3] (ii) the rearrangement of N-bromoacetanilide [4] and (iii) the inversion of *l*-menthone [5] established an order of acid strengths in aprotic media and the importance of intra-molecular hydrogen bonds (*e.g* in picric acid). A thermodynamic method using reference acids and bases is more direct, and Bell and Bayles [6] employed the indicator acid Bromophenol Blue to obtain a basicity order for weak amine bases. Kinetic measurements on these systems have recently been made, and are considered in detail in Section 7.

5.2 Thermodynamics of hydrogen-bond formation in aprotic solvents

A proton-transfer reaction can only occur through the initial formation of a hydrogen-bonded complex and this step, although rapid, may significantly affect the kinetic analysis of a proton-transfer reaction, since a factor involving the thermodynamics of hydrogen-bond formation will be incorporated in the rate expression.

The thermodynamics of hydrogen-bond formation and acid-base behaviour in aprotic solvents in general have been comprehensively reviewed by Davis [1, 2] in the two indispensable monographs already mentioned. The book by Vinogradov and Linnell [7] brings up to date the classic reference work of Pimentel and McClellan [8], and, in addition, papers by Schleyer et al. [9], have discussed in detail the ability of a wide range of groups (often in so-called inert aprotic solvents, e.g. $CHCl_3$, CH_3CN) to act as proton donors or acceptors.

5.2.1 Experimental methods

Several techniques provide fundamental thermodynamic information about hydrogen-bond formation. The main methods are (i) infra-red spectroscopy, (ii) nmr spectroscopy, (iii) uv-visible spectrophotometry, (iv) calorimetry, and (v) dipole-moment measurement.

In the infra-red method, it is usual to observe the frequency shift (Δv) of the A—H stretching band to longer wavelengths on hydrogen-bond formation. The absorption is broadened and shifts of 30 to 500 cm^{-1} are typical. Analysis is based on the Beer-Lambert law and ΔG^0, ΔH^0 and ΔS^0 are readily determined. There is usually a reasonable correlation between Δv and the enthalpy of hydrogen-bond formation ($-\Delta H^0$) for bonds of intermediate strength (3–6 kcal mol^{-1}). In the nmr technique, the chemical shift of the A—H proton is similarly analysed, and from line-broadening studies, kinetic information can also be obtained. The refinements of signal-averaging and Fourier-transform methods enable lower concentrations of reagents to be used, so that hydrogen-bond formation between different acids and bases can be studied, free from the effects of self-association.

Of particular interest is the use of uv-visible spectrophotometry, since there is the added possibility of following the kinetics of simple hydrogen-bond formation by means of the temperature-jump relaxation technique with a conventional optical detection system. The method is suitable for acid or base species which contain a suitable chromophore, for example 2,4-dinitrophenol (2,4-DNP), tetrabromophenolsulphonephthalein (Bromophenol Blue; **1**) and tetrabromophenolphthalein ethyl ester (Magenta E; **2**). Both these substituted phenols have been extensively used as reference acids to determine relative base strengths in aprotic solvents. Suitable reference bases are tolyldipropyl Nile Blue (**3**) and pyridine-2-azo-dimethyl-aniline (PADA; **4**).

A feature of the chromophores in the above compounds is that they can detect and distinguish between hydrogen bonds and ion-pairs. Hydrogen-bond formation is accompanied (as in the infra-red method) by a small (10–30 nm) bathochromic shift of the main absorption peak, with little change in extinction coefficient. However, for strong acids and bases, ion-pairs will be formed, indicated by bathochromic shifts of 100–150 nm, coupled with an extinction coefficient enhancement by a factor of two. Typical examples of systems that have been studied are (i) phenol and dioxan in isooctane [10] and (ii) 2,4-DNP and pyridine in toluene [11].

5.2.2 Hydrogen bond strengths

In principle it is not difficult to distinguish thermodynamically between hydrogen-bond and ion-pair formation, since for the former reaction

$$AH + B \overset{K}{\rightleftharpoons} AH\cdots B \qquad (2)$$

it is found that $K \leqslant 10^2 \, dm^3 \, mol^{-1}$ at 298 K and $-\Delta H^0 \leqslant +7 \, kcal \, mol^{-1}$. If higher values are obtained, the formation of ion pairs is indicated (Table 1). Dipole-moment measurements provide confirmatory evidence.

However, thermodynamic data obtained by different methods are not always consistent, and published results are of relatively low accuracy (ΔH^0 to ± 1 kcal mol^{-1}). It is gratifying, therefore, that a careful study by infra-red and calorimetry [13] on hydrogen-bond formation between p-fluorophenol and pyridine in carbon tetra-chloride gives very good agreement between the two methods (which is further confirmed [14] by ^{19}F nmr). Results, which are representative for hydrogen bonds between oxygen and nitrogen, are $\Delta H^0 = -7.1 \pm 0.1$ kcal mol^{-1}, $\Delta S^0 = -15.2(\pm 0.3)$ cal K^{-1} mol^{-1}, and $\Delta G^0 = -2.56 \pm 0.01$ kcal mol^{-1}.

There is little quantitative information concerning the effect of deuterium substitution on hydrogen-bond formation, but the equilibrium isotope effect is thought to be small. Bell and Crooks [11] have, however, investigated the isotope effect on ion-pair formation. They observed no effect for 2,4-DNP with the strong bases triethyl-amine and piperidine, but a value of $K_H/K_D = 1.40$ was obtained for pyridine. This could be interpreted as indicating that there is no hydrogen-bonding in the ion-pair product with the stronger bases, but that a hydrogen-bonded ion-pair is formed with pyridine. Recently, however, the isotope effect for ion-pair formation between the phenolic acid indicator Bromophenol Blue and pyridine in toluene has been measured [15] with $K_H/K_D = 1.04(\pm 0.01)$ at 298 K. Clearly, it would be useful to have more data on these systems.

Table 1

Heats of mixing (ΔH^0) and equilibrium constants K for the reaction of substituted phenols with tri-n-butylamine in toluene at 298 K [12]

Acid	$-\Delta H^0$/kcal mol^{-1}	K/dm^3 mol^{-1}	Product
Phenol	7.0	13	H-bonded complex
p-Cl phenol	7.7	18.9	H-bonded complex
p-NO$_2$ phenol	8.5	120	H-bonded complex $\rightleftharpoons$ ion-pair
2,4-DNP	13.6	380	ion-pair
2,4,6-TNP	19.7	1075	ion-pair

DNP = dinitrophenol.
TNP = trinitrophenol.

We can conclude that the order of hydrogen-bond strengths, as measured by the equilibrium constant K (at 298 K) is:

Type of H-bond	O—H···N	O—H···O N—H···N	C—H···N
K/dm^3 mol^{-1}	100 $>$	20–50 $>$	1–10

5.2.3 Self-association

There is evidence for extensive self-association of phenols, carboxylic acids and certain amines in aprotic solvents at concentrations above 5×10^{-4} mol dm^{-3}. For example, imidazole at 10^{-3} mol dm^{-3} forms dimers or higher aggregates in inert solvents, and alcohols and

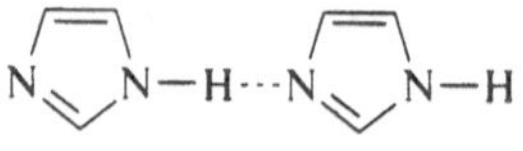

(Imidazole dimer)

carboxylic acids form both linear and cyclic oligomers. The complexity of the aggregates makes an unambiguous quantitative analysis difficult.

5.2.4 Solvent effects on hydrogen-bonding

There is a marked solvent effect on the self-association of benzoic acid in a variety of aprotic media [16], but recent data [17] for hydrogen-bond formation between the substituted phenol Magenta E and pyridine in a series of related aromatic aprotic solvents have not shown such pronounced effects (Table 2). The self-association results

Table 2

Thermodynamic data for hydrogen-bond formation in different solvents

System	Solvent	ΔG^0/kcal mol^{-1}	ΔH^0/kcal mol^{-1}
Benzoic acid[a]	Vapour	—	−8.1
dimerisation at	Cyclohexane	−4.71	−6.4
343 K	Carbon tetrachloride	−4.36	−5.5
	Benzene	−3.33	−3.8
Magenta E +	Toluene	−1.80	−6.1
pyridine at 298 K	Benzene	−1.90	−6.5
	Chlorobenzene	−2.00	−5.4
	o-Dichlorobenzene	−2.10	−7.6

[a] ΔH^0 per single H-bond.

the interaction of the phenolic hydroxyl group with the solvent acting with the π-electron cloud of benzene so that desolvation is required prior to dimerization. In the pyridine-Magenta E system, the interaction of the phenolic hydroxyl group with the solvent is similar in all the media studied. It seems that the strength of the

126

O—H$\cdots\pi$-electron bond (in aromatic solvents) is of the order of 2 kcal mol^{-1}. When very fast proton-transfer reactions are studied, weak interactions such as these may have to be taken into account in the kinetic analysis.

5.3 Interconversion of hydrogen-bonded complex and ion-pair

Depending on the solvent and the acid-base strength of the reactants, either a hydrogen-bonded complex (A—H$\cdots$B) or an ion-pair (A$^-\cdots$HB$^+$) will be the predominant species formed in media of low dielectric constant. There has, however, been much controversy as to whether (AH$\cdots$B) and (A$^-\cdots$HB$^+$) should be regarded (i) as limiting examples of a single product or (ii) as two distinct and separate species. In the first case, a potential-energy curve with a single minimum would be obtained. Although there is evidence for strong symmetrical hydrogen-bonds in the solid state, and hence for a single minimum potential, such structures have never been established in solution. This author believes that the evidence is strongly in favour of a double minimum potential (Figure 1), the most compelling evidence being provided by an analysis of infra-red [18] and visible spectra [19]. It is found that two peaks appear abruptly at

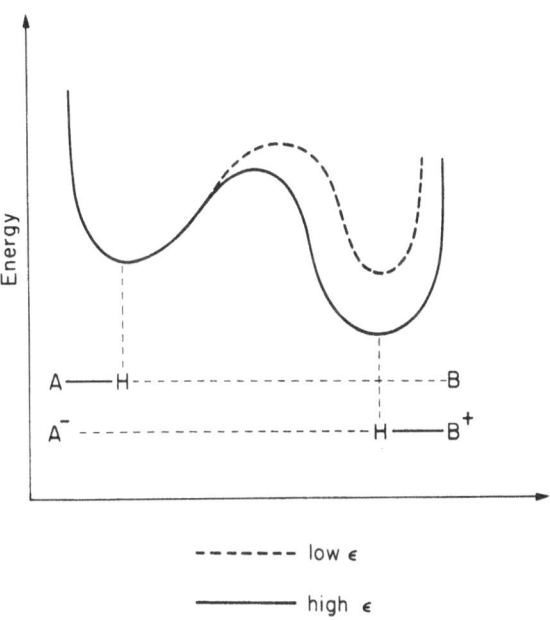

Figure 1. Double-minimum potential diagram for proton-transfer processes.

127

separate frequencies when a parameter affecting the tautomeric equilibrium (e.g. base species, solvent) is changed. Furthermore, a detailed thermodynamic analysis is consistent with the presence of two species in equilibrium (under the appropriate experimental conditions):

$$AH\cdots B \underset{k_{-3}}{\overset{k_3}{\rightleftharpoons}} A^-\cdots HB^+ \qquad K_3 = k_3/k_{-3} \qquad (3)$$

For a given acid-base pair, the degree of asymmetry of the double-minimum curve will depend primarily on the solvation of the ion-pair, the solvent polarity, and the effects of delocalization of charge. For solvents with $\varepsilon > 2$, increase in temperature generally favours the hydrogen-bonded complex, i.e. ΔH_3^0 is negative. However, in the excited state the double minimum potential is considerably more asymmetric, the ion-pair state having the lower energy [19].

The equilibrium constant K_3 correlates well with the value of ΔpK_a in water. Ion-pairs are more strongly favoured as the polarity of the solvent is increased; for example, in the case of the p-nitro-phenol-triethylamine system, ion-pairs are only formed in significant amounts when $\varepsilon > 6$. Further evidence for ion-pairing is shown by dipole moments [20]; moments of 10–18 D are observed when proton transfer is indicated, these values being of the same order as for tetra-alkylammonium salts, $R_4N^+X^-$ [21].

It is of interest to consider the factors determining the rate at which the proton is transferred between the tautomers, which is related to the barrier height in Figure 1. Proton transfer is associated with (i) solvent redistribution around the developing charge and (ii) possible delocalization of charge on A and B. The possibility of tunnelling in proton-transfer processes, as suggested by Bell in 1933, must also be considered since the barrier-width is narrow, owing to the preformed hydrogen-bond, and so the likelihood of tunnelling is increased. A full discussion of the factors influencing proton tunnelling and its experimental verification has been provided by Caldin [22] and calculations have been made on tunnelling through an Eckart-type barrier by Weiss [23]. For a realistic barrier width (0.08 nm), rate constants of the order of 10^{12} s^{-1} are calculated for barriers as high as 10 kcal mol^{-1}, with a corresponding primary isotope effect of around 6. Decreasing the barrier height to 2 kcal mol^{-1} increases the rate constant to 5×10^{13} s^{-1} and decreases the isotope effect to about 1.1. However, the difficulties with these model calculations are that the height of the barrier is difficult to estimate for real systems, and the calculations cannot allow adequately for the presence and role of the solvent and the effects of solvation. Such calculations are only meaningful when the solvent motion can be considered to be uncoupled from the actual step involving transfer of the proton, and in this situation it may be that the solvent reorganization is rate-limiting in the overall proton-transfer process. This point will be amplified in Section 7.

Some interesting results [24] have been obtained for the acetic acid-triethylamine interaction in chloroform, which is a system with a tendency for both acid dimerization and proton transfer. Infra-red line broadening was observed, suggesting the formation of a compromise 2:1 complex, with bonding of the protonated amine to two acid moieties, as in 5, with very rapid proton exchange in a time of the order of 10^{-14} s.

$$H_3C-C \overset{O\cdots H-O}{\underset{O^-\cdots H\cdots O}{<>}} C-CH_3$$

$$^+N(Et)_3$$

5

5.4 Role of the solvent in ion-pair formation

The ion-pair concept was introduced by Bjerrum [25] in 1926, and has more recently been reconsidered by Fuoss [26] and by Prue [27]. There is still some difficulty regarding a precise definition of an ion-pair since the role of the solvent in ion-pairing is not often considered [28].

Bjerrum's theory for ion-pair formation relates the equilibrium constant K_{assn} for

$$R^+ + X^- \rightleftharpoons R^+X^-$$

in a hypothetical continuous and structureless medium to the dielectric constant (ε) of that medium, the interionic separation, a, in the ion-pair, the ionic charges ze, and the temperature T; N is the Avogadro number. The equation is:

$$K_{assn} = \tfrac{4}{3}\pi Na^3 \exp(-z_+ z_- e^2/\varepsilon akT) \qquad (4)$$

Only one type of ion-pair can be envisaged on the basis of this approach. For a typical value of a (0.7 nm), it is found from equation 4 that free ions predominate over a wide concentration range when $\varepsilon > 40$; for $\varepsilon < 5$, the concentration of free ions is negligible, since K_{assn} is $> 10^8$ dm^3 mol^{-1}.

Since an ion-pair is necessarily the first-formed product of a proton-transfer reaction, and is the only product in a solvent of low dielectric constant, it is worth considering in some detail the ion-pair concept, with special emphasis on solvent involvement. In a real solvent with discrete molecular and dipolar structure, two types of ion-pair species can be visualized, which are usually termed 'contact' or 'solvent-separated' ion-pairs. (The latter is sometimes known as an 'outer-sphere complex').

In solvents of intermediate dielectric constant ($\varepsilon = 10$ to 30), the ions will interact strongly with the solvent, S and solvent-separated ion-pairs may then be the preferred species. This is because the gain in free energy required for desolvation may not be fully compensated by the loss in free energy obtained by minimizing the coulombic potential, which is achieved by bringing the ions into direct contact. Figure 2 shows the likely species involved.

$$
\begin{array}{cccc}
\text{S} \quad \text{S} & \text{S} \quad \text{S} & \text{S} \quad \text{S} & \text{S} \ \text{S} \\
\text{SR}^+\text{S} + \text{SX}^-\text{S} \overset{K_1}{\rightleftharpoons} & \text{SR}^+\text{SSX}^-\text{S} \overset{K_2}{\rightleftharpoons} & \text{SR}^+\text{SX}^-\text{S} \overset{K_3}{\rightleftharpoons} & \text{SR}^+\text{X}^-\text{S} \\
\text{S} \quad \text{S} & \text{S} \quad \text{S} & \text{S} \quad \text{S} & \text{S} \ \text{S}
\end{array}
$$

| Free-ions | Solvent-separated ion-pair (1) | Solvent-separated ion-pair (2) | Contact ion-pair |

Figure 2. Types of ion-pair and solvation patterns.

Values of K_1 can be calculated with reasonable confidence by the Bjerrum equation only when both cation and anion are strongly solvated with slow solvent-exchange about the ions. The evaluation of K_2 and K_3 is more difficult since several types of interaction are involved, such as the following: (i) There are short-range nearest-neighbour interactions. For ion-pairs involving acid-base systems, the most important of these is likely to be hydrogen-bonding in the contact ion-pair (for example, $R'C-O^- \cdots H-^+NR''$). (ii) There are specific solvation effects. Ions in contact are not likely to be as strongly solvating as free ions; because of charge compensation, fewer solvent molecules may be involved. These factors suggest that in poorly solvating aprotic solvents of low dielectric constant, such as benzene and toluene, contact ion-pairs are likely to be the only species formed. Precise calculations are difficult, but a thermodynamic approach has been discussed by Szwarc [29].

We may conclude that (i) if both ions are weakly solvated, contact ion-pairs predominate; (ii) if one ion solvates strongly and the other to a much lesser extent (because of charge delocalization, for instance), the ions in the ion-pair may be separated by only one solvent molecule linked with the more strongly solvating ion; (iii) if both ions strongly solvate, there will be two solvent molecules between the ions in the solvent-separated ion-pair (see Figure 2); (iv) the magnitude of K_1 is the result of the balance of thermal and coulombic energy. The sizes of K_2 and K_3 are consequences of the balance of coulombic plus short-range interactions versus desolvation.

The thermodynamics of ion-pair formation have been extensively investigated, the most useful techniques being uv-visible spectrophotometry, conductometry and electron spin resonance. Conversion of a contact ion-pair into a solvent-separated ion-pair is accompanied by a red (bathochromic) shift of about 20 nm in the visible or ultraviolet spectrum. Many indicator acids exhibit this effect in the conjugate base (e.g. the 2,4-dinitrophenolate ion), and equilibrium constants for ion-pair formation have been determined by means of this technique [30]. The importance of hydrogen-bonding in the contact ion-pair is shown by the data [31] in Table 3 for association constants of picrates in benzene and 1,1-dichloroethane. Hydrogen-bonding in the ion-pair is possible with trialkylammonium ions but not with tetra-alkylammonium ions, and may be responsible for the increase of $K(=K_1K_2K_3)$ by three orders of magnitude, though different solvation energies may also be involved.

Table 3

Association constants of picrates at 298 K

Salt	Solvent (ε)		$K/\mathrm{dm^3\,mol^{-1}}$
Am_4NPic			5.0×10^{16}
	C_6H_6	(2.4)	
Am_3NHPic			2.5×10^{20}
Bu_4NPic			2.2×10^4
	CH_3CHCl_2	(10.0)	
Bu_3NHPic			5.0×10^7

Conductivity measurements indicate that, in solvents of low ε, additional species in the form of triple ions ($R^+X^-R^+$ and $X^-R^+X^-$) can be produced [32] at concentrations above $10^{-5}\,\mathrm{mol\,dm^{-3}}$. Another mode of association is dimerization [33] (or higher aggregation) of ion-pairs, i.e. $(2R^+X^- \rightleftharpoons (R^+X^-)_2)$. The extent of association depends on the magnitude of the dipole moment and the ion-size, aggregations involving one small and one large ion being especially favoured. The available results on ion-pairing have recently been comprehensively reviewed by Davis [1, 2], and further insight into the dynamic aspects of the ion-pair concept is provided by the remarkable studies of Ritchie [34] on the reactivity of nucleophiles towards stable carbonium ions.

5.5 Kinetics of proton-transfer processes in aprotic solvents

5.5.1 Detailed reaction scheme

A bimolecular reaction in solution requires that the reacting molecules come together in the initial step to form an encounter complex (EC), in which no new bonds are formed and desolvation of the reactants has not yet occurred. This species may then undergo reaction in the direction of product formation, or the molecules forming the 'encounter' complex may separate again to re-form the reactants. Likely products are a hydrogen-bonded complex, or an ion-pair, depending on the acid-base strength. Which type of ion-pair is formed depends on the dielectric constant (ε).

A detailed reaction scheme for a proton-transfer process represented in general by equation 5 is:

$$\text{AH} + \text{B} \underset{k_{-1}}{\overset{k_1}{\rightleftharpoons}} \text{AH}\|\text{B} \underset{k_{-2}}{\overset{k_2}{\rightleftharpoons}} \text{AH}\cdots\text{B} \underset{k_{-3}}{\overset{k_3}{\rightleftharpoons}} \text{A}^-\cdots\text{HB}^+ \underset{k'_{-2}}{\overset{k'_2}{\rightleftharpoons}}$$

Acid Base	ECl	Hydrogen-bonded	Ion-pair
(a)	(b)	complex (c)	(d)

$$\text{A}^-\|\text{HB}^+ \underset{k'_{-1}}{\overset{k'_1}{\rightleftharpoons}} \text{A}^- + \text{HB}^+$$

Solvent separated   Free ions
ion-pair (EC2)   (f)   (5)
(e)

Concerning the actual proton-transfer process, (c) → (d), it is clear that this is only one step in a complex reaction sequence. Kinetic information can, in general, only be obtained for the overall reaction, with forward rate constant k_f and backward rate constant k_b. If proton transfer is rate-limiting, then a steady-state treatment shows that each of these observed rate constants is composed of contributions from the various equilibria preceding proton transfer, multiplied by the rate constant specifically for proton transfer, i.e.:

$$k_f = (k_1 k_2)(k_{-1}k_{-2})^{-1}k_3$$
$$k_b = (k'_1 k'_2)(k'_{-1}k'_{-2})^{-1}k_{-3}$$

The analysis of the overall process into elementary steps was performed by Eigen [35] as long ago as 1954, but the full implications of this treatment have only recently been realized. They are reflected in the Marcus [36] treatment for proton-transfer reactions and in the further refinements of Albery *et al.* [37] and Kreevoy *et al.* [38]. To date, nearly all the data refer to reactions in an aqueous medium. The field has recently been reviewed [39], and there is a detailed discussion by Crooks in Chapter 6 of this volume.

5.5.2 *Diffusion-controlled reaction processes*

The thermodynamics of steps (a) $\rightarrow$ (c) and steps (d) $\rightarrow$ (f) have already been considered in Sections 2 and 4 respectively. Concerning the kinetics of the overall process, it is possible to calculate theoretically with reasonable confidence only the rate constants k_1, k_{-1}, k'_1 and k'_{-1}. These reactions (by definition of an encounter complex) are controlled entirely by diffusion processes, and so k_1 and k'_1 represent limiting second-order rate constants. The simplest theory for these rate processes is due to Smoluchowski [40] and is based on Fick's Law of diffusion. Several important simplifications were made in the original treatment. These are: (i) the reactants were assumed to be rigid spheres moving in a homogeneous continuum subject to the macroscopic laws of diffusion; (ii) no short-range (reactant-solvent) interactions were considered; (iii) the Stokes–Einstein equation for diffusion coefficients was applied. The application of this equation to reactants comparable in size to solvent molecules is open to question. However, for uncharged molecules in non-polar solvents, we might expect the Smoluchowski treatment to be a good approximation. More refined treatments have recently been proposed by Noyes [41] and by Schmitz and Schurr [42], but the simple treatment is still useful for general discussion.

For uncharged reactants, the rate constant for diffusion-controlled bimolecular combination (k_D) is given by:

$$k_D = k_1 = 4\pi Na(D_{AH} + D_B) \times 10^{-3} \, \text{dm}^3 \, \text{mol}^{-1} \, \text{s}^{-1} \qquad (6)$$

where D is a diffusion coefficient (in $\text{cm}^2 \, \text{s}^{-1}$), a is the separation (in cm) in the encounter complex, and other symbols are as defined previously (4). For reactants of similar size, equation 6 can be further simplified, leading to (with the viscosity η in poise):

$$k_D = k_1 = 4RT\eta^{-1} \times 10^{-3} \, \text{dm}^3 \, \text{mol}^{-1} \, \text{s}^{-1} \qquad (7)$$

For charged reactants, as considered by Debye [43]:

$$k'_D = k'_1 = 4\pi Na(D_{A^-} + D_{BH^+})[\phi(e^\phi - 1)^{-1}] \times 10^{-3} \, \text{dm}^3 \, \text{mol}^{-1} \, \text{s}^{-1} \qquad (8)$$

where

$$\phi = z_A - z_{BH} + e^2/\varepsilon a k T \qquad (9)$$

Similar expressions can be derived [35] for the first-order rate constant for diffusional separation of molecules and ions. These are, for neutral species:

$$k_{-D} = k_{-1} = 3(D_{AH} + D_B)/a^2 \text{ s}^{-1} \qquad (10)$$

for charged species:

$$k'_{-D} = k'_{-1} = [3(D_{A^-} + D_{HB^+})/a^2][\phi e^\phi/(e^\phi - 1)] \text{ s}^{-1} \qquad (11)$$

Combining equation 6 with 10, and 8 with 11, the equilibrium constants characterizing encounter-complex formation are obtained. These are, for neutral species:

$$K = k_1/k_{-1} = \tfrac{4}{3}\pi N a^3 \times 10^{-3} \text{ dm}^3 \text{ mol}^{-1} \qquad (12)$$

for charged species:

$$K' = k'_1/k'_{-1} = \tfrac{4}{3}\pi N a^3 e^{-\phi} \times 10^{-3} \text{ dm}^3 \text{ mol}^{-1} \qquad (13)$$

It should be noted that (i) equation 13 corresponds to the Bjerrum equation 4, (ii) from equation 12, K is independent of solvent (if the value of a is independent of solvent). Some of the implications of these equations have been recently discussed by Hemmes [44]. Table 4 shows values of ϕ, and the corresponding rate and equilibrium constants, calculated from equations 10, 11, 12, 13 for $a = 0.5$ nm and $D_{AH} = D_{A^-} = D_{BH^+} = D_B = 2 \times 10^{-6} \text{ cm}^2 \text{ s}^{-1}$.

It can be seen from the table that the predictions of the Debye–Smoluchowski theory are as follows. (i) The formation and dissociation of an encounter-complex involving uncharged species is always a very rapid process. (ii) In solvents of high dielectric constant, such as water, charges have little effect on the rates, but in media of low dielectric constant both rate and equilibrium constants are drastically affected. (iii) In aprotic solvents of intermediate dielectric constant, the diffusion apart of ions is a slow process, and may be rate-limiting in an overall proton-transfer reaction. (iv) In media of low dielectric constant, free ions are not formed at the usual concentrations. In what follows, several of these predictions are used and tested in the interpretation of experimental data.

134

Table 4

Calculated rate and equilibrium constant values for encounter-complex formation

ε	$-\phi$	K	$10^{-9}k_1$	$10^{-9}k_{-1}$	K'	$10^{-9}k'_1$	k'_{-1}
		$\mathrm{dm^3\,mol^{-1}}$	$\mathrm{dm^3\,mol^{-1}\,s^{-1}}$	$\mathrm{s^{-1}}$	$\mathrm{dm^3\,mol^{-1}}$	$\mathrm{dm^3\,mol^{-1}\,s^{-1}}$	$\mathrm{s^{-1}}$
80(eg H_2O)	1.39	0.31	1.5	4.8	1.24	2.8	2.25×10^9
5(eg PhCl)	22.3	0.31	1.5	4.8	1.2×10^9	33	27.5
2(eg C_6H_6)	55.6	0.31	1.5	4.8	1.5×10^{24}	83	5.5×10^{-14}

5.6 Kinetics of proton-transfer processes from carbon acids

There have been surprisingly few kinetic studies in aprotic solvents as opposed to non-aqueous solvents such as the alcoholic and mixed protic–aprotic solvent systems. It is well known that the situation can be complicated in the latter cases by direct solvent participation in the proton-transfer reaction through a cooperative mechanism.

Interesting results have, however, been obtained recently by Caldin and Mateo [45], who used the stopped-flow method with a wide variety of aprotic solvents, to follow the reaction between 4-nitro-phenylnitromethane (NPNM) and the bases tetramethylguanidine (TMG) of formula $HN=C(NMe_2)_2$, tri-n-butylamine (TBA), and triethylamine (TEA). Some of the results are shown in Table 5, and may be compared with the predictions of theoretical treatments of isotope effects with and without tunnelling.

For a classical proton transfer, the highest possible value of k_H/k_D (when the zero-point energy difference is completely lost in the transition state) is about 17, with a corresponding value of $\log(A_D/A_H)$ of about 0.3. It would seem, therefore, that tunnelling is definitely shown by the reaction with tetramethylguanidine in toluene and chlorobenzene, i.e. the less polar solvents. In fact, the values of k_H/k_D are among the highest so far recorded, although large values have also been found for hydrogen-atom [47] and hydride-ion transfer processes [48] in similar solvents.

For the more polar solvents, methylene dichloride and acetonitrile, tunnelling is not unequivocally indicated. This may be because the solvent rearranges (at least in part) as the proton is being transferred; this results in the effective mass of the proton in the transition state (m_H) having a value greater than unity. (A value of $m_H = 1$ indicates proton motion uncoupled from any solvent motion.) The timing of solvation changes in proton-transfer reactions is the subject of much debate at the present time, and has been discussed in detail by Kurz and Kurz [49]. Although the observed enthalpy of activation ($\Delta H^{\ddagger}$) should strictly be split into contributions from (i) initial hydrogen-bond formation, (ii) solvent reorganization prior to proton transfer and (iii) the actual proton transfer itself, it is reasonable to suppose that in this case the hydrogen bond formed by the carbon acid will be weak (with ΔH^0 around -1 to $-3\,\text{kcal mol}^{-1}$) and will effectively compensate for any small enthalpy change due to prior solvent reorganization. The large negative value of $\Delta S_H^{\ddagger}$ is perhaps indicative of some preceding solvent rearrangement.

136

Table 5

Ion-pair formation between 4-nitrophenylnitromethane (NPNM) and bases in various solvents [46]

Kinetic primary isotope effect and derived parameters

	Base					
	TMG				TEA	
	Solvent				Solvent	
	PhCH$_3$	PhCl	CH$_2$Cl$_2$	CH$_3$CN	PhCH$_3$	CH$_3$CN
ε	2.40	5.60	9.10	37.50	2.40	37.50
$k_H/\mathrm{dm^3\,mol^{-1}\,s^{-1}}$, 298 K	2290	7080	5240	5890	132	12.2
$K_H/\mathrm{dm^3\,mol^{-1}}$, 298 K	160	—	6480	3500	4.20	137
k_H/k_D, 298 K	45	50.0	11.40	11.80	11.00	3.10
$\Delta H_H^{\ddagger}/\mathrm{kcal\,mol^{-1}}$	3.60	3.60	3.50	4.40	3.50	6.70
$\Delta H_D^{\ddagger} - \Delta H_H^{\ddagger}$	4.30	3.70	1.90	1.50	2.20	1.00
$\Delta S_H^{\ddagger}/\mathrm{cal\,K^{-1}\,mol^{-1}}$	−31.00	−29.00	−30.10	−26.60	−37.20	−22.10
$\log(A_D/A_H)$	1.50	1.03	0.35	0.00	0.38	0.26
m_H	1.00	1.00	1.24	1.27	1.00	1.39
$E_H/\mathrm{kcal\,mol^{-1}}$	8.60	8.55	4.85	5.85	5.30	7.70

TMG = tetramethylguanidine, HN=C(NMe$_2$)$_2$.

TEA = triethylamine, N(C$_2$H$_5$)$_3$.

These results provide a tentative hypothesis as to why tunnelling has rarely been detected for proton transfer in aqueous media. By extrapolation from the behaviour of polar solvents of low dielectric constant, it is to be expected that dipolar water molecules interact strongly with the reactants and are therefore responsive to the charge development accompanying proton migration; the result is synchronous proton transfer and heavy-atom rearrangement, so that the effective mass is much greater than that of the proton and there is little possibility of tunnelling in water.

It is interesting that tetramethylguanidine is the only base so far studied which clearly shows tunnelling. The ratios of k_H/k_D for triethylamine and tri-n-butylamine, while large, can be explained in terms of classical transfer. The different behaviour may be associated with the delocalization and change in geometry required in the protonated base.

From the limited data at present available, there appear to be interesting differences between proton transfer involving carbon acids and oxygen acids in aprotic solvents, but more systems must be studied before definite conclusions can be drawn. In particular, an attempt should be made to correlate the extent of proton transfer in the transition state (by means of a Brønsted-type plot) with the primary isotope effect, as has been done for aqueous media.

5.7 Kinetics of proton-transfer reactions from oxygen acids

To date, many kinetic determinations have been made using phenolic indicator acids, especially 2,4-dinitrophenol (2,4-DNP), Magenta E and Bromophenol Blue (BPB). While 2,4-dinitrophenol and Magenta E form ion-pairs with only the stronger aliphatic amine bases (such as tri-n-butylamine and triethylamine), Bromophenol Blue is a strong enough acid to form ion-pair complexes readily with the much weaker substituted pyridine bases [6, 50]. This property, together with the fact that the ion-pair product is formed after a ring-opening step, makes Bromophenol Blue a particularly interesting acid for study. From the available data it seems that (i) when the base is strong, proton transfer along the hydrogen bond can be very rapid ($k > 10^{10} \, \text{s}^{-1}$), (ii) when the base is weak, proton transfer can be slow (down to $k \sim 10^2 \, \text{s}^{-1}$), because of the need for extensive solvent reorganization and/or charge separation.

5.7.1 *Proton transfer to strong bases*

For these systems, the overall forward reaction is very fast ($k_f \sim 10^8$ to 10^9 dm^3 mol^{-1} s^{-1}) and the actual proton-transfer step is not rate-limiting; the slowest step appears to be the formation of some weak intermediate complex preceding the proton transfer. This is found for the reactions of 2,4-dinitrophenol and Magenta E with tertiary aliphatic amines [51], which have been studied by means of the microwave temperature-jump technique. Kinetic measurements where the base structure is systematically varied, using alkyl groups of different chain length, enable steric and inductive effects on the rate of formation of hydrogen-bonded complexes to be investigated. In addition, the solvent can be systematically varied, and theories for diffusion-controlled processes may be tested under conditions where the solvent is most likely to provide a pure medium effect [52].

Three facts suggest that at least one further intermediate formed after the encounter-complex and before the hydrogen-bonded complex should be considered: (i) the anomalous viscosity-dependence of the observed rate constant k_{obs}; (ii) the fact that k_{obs} is ten times smaller than that calculated for a simple diffusion-controlled process; and (iii) that $\Delta H^{\ddagger}_{obs}$ is not the same as that derived from the temperature-dependence of the solvent viscosity. This intermediate contains the reactants in contact, but in a situation where they can mutually rotate until they achieve the necessary orientation for further reaction; weak stabilization is thought to be provided by dispersion and dipole-dipole forces [52]. The requirement that any intra-molecular hydrogen bond in the acid must be broken before hydrogen-bonding to the nitrogen base can occur is also likely to be important, especially for 2,4-dinitrophenol.

These results are interesting, in that they provide information concerning the time scale and the species formed in the early stages of a proton-transfer reaction. A rigorous analysis is difficult since there are several discrete processes occurring in the same time range (such as desolvation, rotation, breaking of intra-molecular hydrogen-bond or of hydrogen-bond to solvent), so that isolating one single dominant factor may not be possible.

Other methods of studying the rate of formation of hydrogen-bonded complexes are ultrasonics [53] and dielectric relaxation [54]. It seems that the rate constants for formation are usually slightly less than for diffusion-control ($\sim 10^9$ dm^3 mol^{-1} s^{-1}), while the lifetime of a hydrogen-bonded complex is around 10^{-7} s. Special attention has been given to hydrogen-bond formation between nucleotide derivatives as models for base-pairing in DNA (Section 8).

5.7.2 Proton transfer to weak bases

Not all proton transfers between oxygen and nitrogen centres are rapid. Proton transfer from Bromophenol Blue (1) to pyridine bases can be measured in the millisecond time range [55]; both stopped-flow and laser temperature-jump techniques have been employed, giving identical results [56]. Thermodynamic and kinetic measurements suggest that the reaction can be most simply expressed as the formation of an ion-pair in chlorobenzene:

$$AH + B \underset{k_b}{\overset{k_f}{\rightleftharpoons}} A^- \cdots HB^+$$

However, the analysis in Table 6 shows that negative values of $\Delta H_f^{\ddagger}$ are obtained with the stronger bases. This indicates that the reaction is at least a two-step process, with an intermediate, which is probably a hydrogen-bonded complex (AH$\cdots$B):

$$AH + B \cdot \underset{k_{21}}{\overset{K_{12}}{\rightleftharpoons}} AH \cdots B \underset{k_{32}}{\overset{k_{23}}{\rightleftharpoons}} A^- \cdots HB^+$$

Thus the mechanism is:

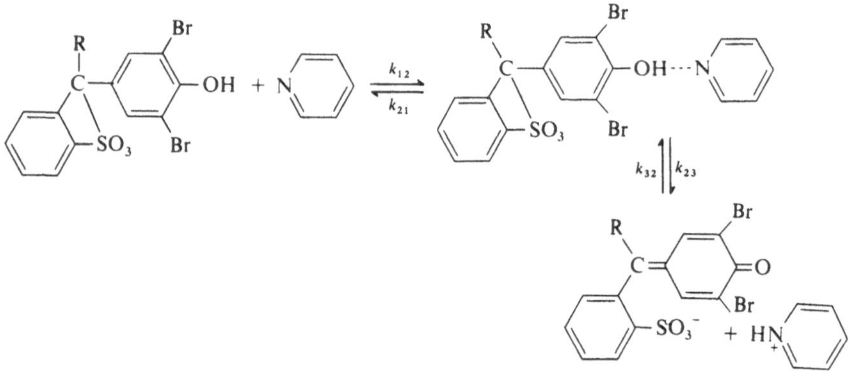

Fortunately, the pre-equilibrium (i.e. K_{12}) can be separately monitored, since Magenta E (2) forms only a hydrogen-bonded complex

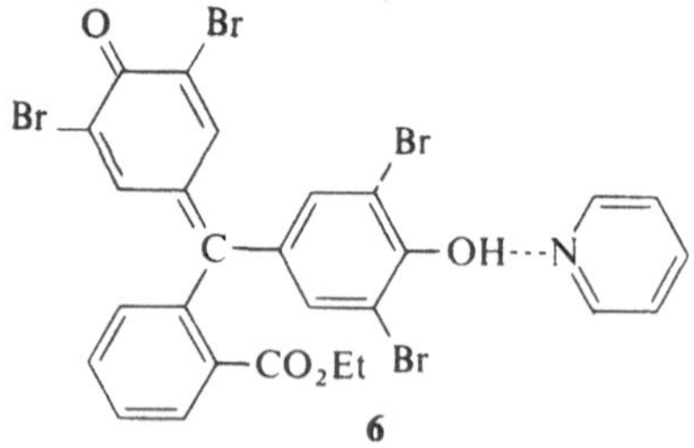

6

6 with pyridine, and this complex is evidently a very good model*
for the initial interaction of Bromophenol Blue with pyridine [17],
so that values of K_{12} could be taken from Table 2.

If the overall reaction can be described as a fast pre-equilibrium
followed by a slow unimolecular conversion, then in a temperature-
jump experiment two relaxation times are expected, the faster of
which is not observed, while the slower, which is observed and is of
the order of milliseconds, is given by:

$$k_{obs} = \tau^{-1} = K_{12}k_{23}([AH] + [B]) + k_{32}$$

By varying the concentrations and determining the first-order rate
constant, it is therefore possible to find values of $K_{12}k_{23}(=k_f)$
experimentally, and so to estimate k_{23}, with the aid of the values of
K_{12}. From the temperature-variation of $K_{12}k_{23}$ we can find the
value of $\Delta H_{12}^0 + \Delta H_{23}^{\ddagger}$, whence by using the value of ΔH_{12}^0 for the
model system we can estimate $\Delta H_{23}^{\ddagger}$. Thus k_{23} and $\Delta H_{23}^{\ddagger}$, which refer
specifically to the proton-transfer step, can both be derived, along with
related quantities. The results are given in Table 6.

It can be seen that values of $\Delta S_{23}^{\ddagger}$ are negative for all the bases;
this indicates (since the reaction is unimolecular) that some solvent
reorganization is required to form the transition state. This is supported by the value of $\Delta V^{\ddagger}$ of about $-16\,cc\,mol^{-1}$ recently determined [17] for the reaction with pyridine in chlorobenzene. This
value is very similar to that of ΔV^0 and indicates that the solvation
of the transition state resembles that of the ion-pair.

These conclusions are also consistent with the solvent dependence
[33] of k_f in solvents of dielectric constant 2 to 10. It is found that the
values of $\log k_f$ correlate well with the solvent polarity parameter E_T,
indicating that the proton-transfer process is dominated by solvent
reorganization rather than by the actual movement of the proton.

In a sophisticated discussion of the mechanism, it is necessary to
ask: (i) Does solvent motion precede, accompany or follow the actual
transfer of the proton? (ii) Is the ring-opening step synchronous with,
or subsequent to, the actual proton transfer step (i.e. is the charge
delocalized as it forms on the anion or not, or is the transition state
best represented by 7 or 8)? (iii) Is the proton transfer slow for these
systems solely because ring opening is involved?

In an attempt to answer these questions, the primary isotope effects
on both kinetics and equilibria have recently been measured [15].
For the reaction of Bromophenol Blue with pyridine in toluene,

---

* This model is better than the system 2,6-dibromophenol-pyridine in carbon tetrachloride
which was previously employed [55]; the results are similar.

Table 6

Thermodynamic and activation parameters for the reaction of Bromophenol Blue with aromatic amines in chlorobenzene (298 K)

Amine	$-\Delta H^{o}$	$\Delta H_f^{\ddagger}$	$-\Delta S_f^{\ddagger}$	$10^{-4}k_{23}$	k_{32}	$\Delta H_{23}^{\ddagger}$	$-\Delta S_{23}^{\ddagger}$
	kcal mol^{-1}	kcal mol^{-1}	cal mol^{-1}K^{-1}	s^{-1}	s^{-1}	kcal mol^{-1}	cal mol^{-1}K^{-1}
Pyridine	10.3	2.2	28.6	1.5	50	6.9	20.7
2-Methylpyridine	13.9	−1.2	35.7	19	44	3.8	21.9
2,6-Dimethylpyridine	13.2	−1.7	31.0	130	70	4.8	22.9
2,4,6-Trimethylpyridine	14.4	−3.6	33.3	890	98	1.9	20.2

7   8

K_H/K_D is 1.04($\pm$0.01), and all the kinetic isotope effect is found to reside in k_f. In view of the carbon-acid isotope effect (Section 6), this result is rather surprising, and can only indicate either (i) tunnelling coupled with much associated motion of heavy atoms, in solvent reorganization and opening of the sultone ring, or (ii) very fast tunnelling of both proton and deuteron in an essentially uncoupled motion, following rate-limiting solvent reorganization. A detailed consideration of all the available data seems to support the second hypothesis, the ring opening being a relatively fast step subsequent to proton transfer, with $k \sim 10^6$ s^{-1}. However, because of the limited number of systems so far studied, it would be unwise at this stage to generalize from these results.

It is clear, however, that such detailed analyses regarding mechanism can now be attempted, and it is to be hoped that as more results become available a coherent picture will emerge.

5.8 Hydrogen-bonding and proton-transfer reactions in DNA and related species

The genetic message is stored in double-helical DNA, which is stabilized both by hydrogen-bonding between nucleotide base-pairs and by stacking forces (hydrophobic and dispersion forces) between the adjacent nucleotides. The very high specificity found experimentally in the pairing of the purine-pyrimidine bases in DNA establishes the molecular basis for information storage. It is found, in accordance with Chargaff's rules, that adenine (A) pairs exclusively with thymine (T) linked through two hydrogen-bonds, while cytosine (C) is associated exclusively with guanine (G) through three hydrogen-bonds. (In RNA, thymine is replaced by the closely related pyrimidine uracil (U)). It follows that if one strand of DNA has a base sequence such as —CATAGAT—, then the complementary strand will be —GTATCTA—. The structures of the associated nucleotide bases are shown in Figure 3.

143

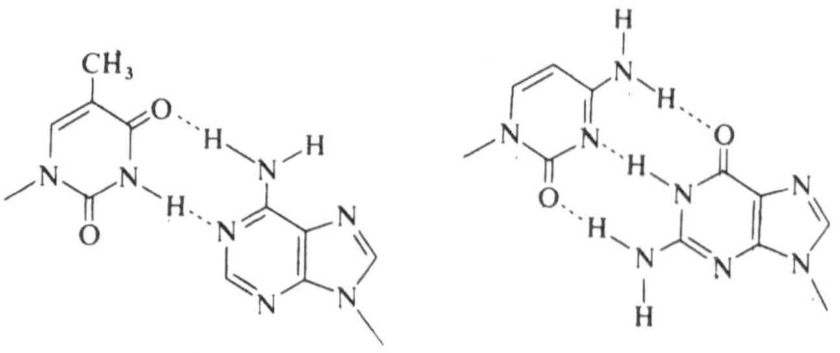

Figure 3. Purine-pyrimidine base-pairing structures as found in DNA (Watson–Crick configurations).

To understand the mechanisms of replication and code-reading and the evolution of the genetic apparatus [59], it is clearly desirable to investigate both thermodynamically and kinetically the nature of the elementary steps involved.

5.8.1 *Single base-pairing*

The hydrogen-bonded association of single bases is best studied in aprotic solvents of low dielectric constant, since the interior of the double helix is out of contact with water and may be considered to represent a low-polarity environment. In addition, hydrophobic effects, which in aqueous media lead to planar stacking (oligomerization) of the bases, are minimized in aprotic solvents, and competitive hydrogen-bonding to the solvent is much reduced. Infra-red measurements enable the equilibrium constants for hydrogen-bond formation between various permutations of the bases to be determined [58] (Table 7). It is clear that the very specific associations found in DNA do not result solely from a consideration of the thermodynamics of simple hydrogen-bond formation. However, there must have been

Table 7

Equilibrium constants for base-pairing in $CDCl_3$ at 298 K

(Values from Ref. 58. Where values are not quoted, K is very low)

Combination		$K/dm^3\,mol^{-1}$	Combination		$K/dm^3\,mol^{-1}$
Adenine (A) with	U (T)	590	Cytosine (C) with	G	10^4
	A	80		C	28
	C	—		U	—
	G	—	Uracil (U) with	U	115
Guanine (G) with	G	10^2–10^3		G	—

some stage in molecular evolution when enzymes did not exist that could direct the specificity of these processes. Consideration of Table 7 shows that the most discriminating base-pair is that between C and G, and indeed it would appear that more primitive life-forms contain a higher proportion of CG pairs, and that DNA containing relatively more of the AT pairs has emerged later in the course of molecular evolution.

The kinetics of base-pairing have been studied, in benzene and chloroform, by means of the techniques of dielectric relaxation [54] and ultrasonics [53]. The analysis can be complex because of dimerization coupled to hetero-association. Furthermore, it is not always clear which groups on a given base are acting as donors and acceptors, although the situation is clearer in DNA itself. There is the added possibility that the solvent (*e.g.* chloroform) can act as an hydrogen-bond donor, and so it is likely that association rate constants in such solvents may be slightly less than the diffusion-controlled limiting value, as is observed. Results are shown in Table 8 for the self-association of uracil and hetero-association with adenine.

Table 8

Rate constants for purine-pyrimidine hydrogen-bond association

	Method	Solvent	$10^{-9}k_f/dm^3 mol^{-1} s^{-1}$	$10^{-7}k_b/s^{-1}$	K	Ref.
$U^a :: U^a$	DR	Benzene	4.4	3.8	115	[54b]
$A :: U^a$	DR	Benzene	10.0	3.6	280	[54b]
$A :: U^b$	US	Chloroform	4.0	3.2	125	[53]

[a] 1-isobutyl-6-methyluracil.
[b] 1-cyclohexyl-uracil.

The ease with which the base-pair can be split (lifetime $\simeq 0.1\ \mu s$) allows a rapid separation of complementary strands in the replication process. However, this low thermodynamic stability, which is advantageous for rapid code-reading, necessitates that other factors, for example stacking, cooperativity and especially enzyme recognition, must undoubtedly be involved to account for the low error-rate in replication.

5.8.2 *Oligo-nucleotide formation*

Additional fundamental information can be obtained concerning the cooperative nature of the double-helix-to-coil transition in DNA (which involves coupled hydrogen-bond rupture) from model studies of helix formation by oligo-nucleotides of chain length 4–10. Examples are the interaction of oligo-A with oligo-A, and that of oligo-A with oligo-U(T)) [59, 60]. The reaction is shown in Figure 4.

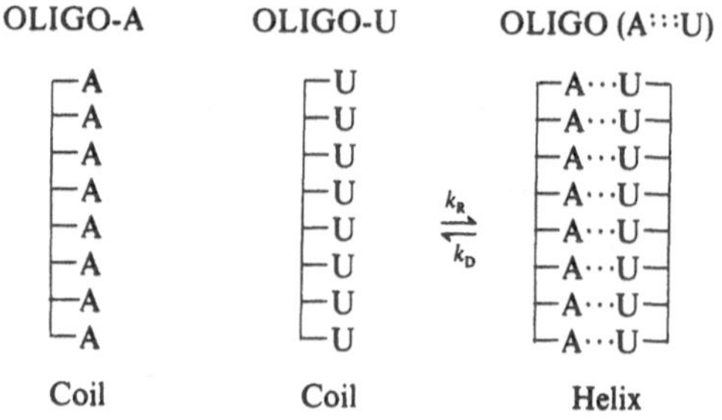

OLIGO-A	OLIGO-U		OLIGO (A∷∷U)
A	U		A···U
A	U		A···U
A	U		A···U
A	U	k_R	A···U
A	U	$\rightleftharpoons$	A···U
A	U	k_D	A···U
A	U		A···U
A	U		A···U
Coil	Coil		Helix

Figure 4. Oligo-nucleotide base-pair formation in water

'Melting curve' measurements show clearly that the conformational change is cooperative, and this may be ascribed to the gain in free energy from the stacking interaction between adjacent base-pairs. For such a system, the kinetic analysis described by Schwarz [61] is appropriate. The conformational transition is readily monitored by the hypochromic effect in the ultra-violet absorption spectrum at 260 nm, and so the dynamics of the change can be followed by the temperature-jump method within the transition (melting) range, which is of the order of 10–20 K. For short-chain nucleotides, a single relaxation process is observed, corresponding to an 'all or none' transition from double helix to coil, any intermediate hydrogen-bonded states being present in negligible concentration. The forward rate constant (k_R) is found to be independent of chain length and is of the order of $10^6 \, dm^3 \, mol^{-1} \, s^{-1}$, while the enthalpy of activation ($\Delta H_R^{\ddagger}$) is negative and in the region of 0 to $-20 \, kcal \, mol^{-1}$. The reverse rate constant (k_D) depends on chain length and has a high enthalpy of activation.

The results suggest that: (1) two to three base-pairs have to be formed in a nucleation stage for the association of an oligo-A molecule with another oligo-A or an oligo-U molecule before the transition state for double helix formation is reached; this may be relevant to the mechanism of codon-anticodon recognition during protein synthesis, which involves recognition of three bases on the transfer RNA; (2) following nucleation, base-pairing takes place at a rate of about 10^7 pairs per second; (3) the dynamics of the conformational transition for DNA are much more difficult to analyse since a spectrum of relaxation times is observed.

5.8.3 *Proton-transfer in DNA*

Löwdin has suggested [62] that proton-transfer may occur along the hydrogen bonds of the base pairs of DNA, leading to a coupled tautomerism in the purine-pyrimidine pair. It is expected that two protons would move simultaneously, in opposite directions, within a given base-pair to prevent charge development. A consequence of this hypothesis is that if the helix splits with the protons in the transferred state (as indicated by the arrows in Figure 5), forming the less likely

Normal forms:
Before proton transfer:

Tautomeric forms:
After proton transfer:

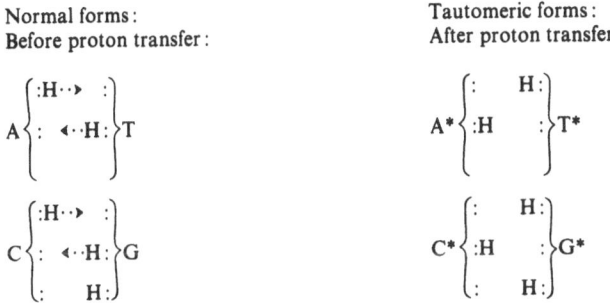

Figure 5. Diagrammatic view of base pairing.

enol and imine forms, rather than the normally stable keto- and amine forms, then errors can occur in the code-reading and replication processes. This is because the tautomers formed, A*, T*, C*, and G*, will tend to pair A*C, T*G, C*A and G*T, and so lead to an error or deletion in the coded message and a loss of genetic information. This is represented in Figure 6. It is difficult to provide an estimate of

Figure 6. Possible pairings of the tautomeric form produced after proton transfer.

the rate constant for tautomerism, since the barrier height and asymmetry cannot be estimated and tunnelling would have to be considered. Clearly we require more fundamental information before definite conclusions can be drawn, but the hypothesis is an interesting one.

An experimental approach based on induced proton-transfer may be more profitable. This could be effected either by irradiation techniques, since proton-transfer is facilitated in the excited state; or perhaps through the study of the effects of intercalated mutagens and carcinogens (such as acridine dyes or anthracene derivatives) into the double helical structure. It may be that the enhanced environmental mobility conferred on the otherwise rigid helical structure is one factor facilitating proton-transfer within the intact DNA, and this may help to explain the carcinogenic action of certain types of molecules.

5.9 Micellar catalysis of proton-transfer processes in aprotic solvents

There has been considerable interest recently in the field of micellar catalysis [63], both in aqueous and non-aqueous media. Micellar species form readily in aprotic solvents, and such species are characterized by having the charged or polar groups localized on the inside of the aggregated structure, forming a cavity out of contact with the solvent, while the hydrocarbon part of the surfactant forms an outer layer immediately adjacent to the solvent. For this reason, they are known as 'reversed' micelles, to distinguish them from micelles in water. The number of molecules in a micelle for a charged surfactant is small. In the case of the well-characterized long-chain alkylammonium carboxylates, values of 5 to 10 seem typical. Thus, these reversed micelles can be thought of as oligomeric ion-pair aggregates.

Micellar catalysis in water is not striking; rate enhancements greater than 10^2 have seldom been obtained, and micelles exhibit only limited substrate specificity (in contrast to enzymes). Of particular interest, therefore, is the remarkable rate enhancement which has been observed for the mutarotation of 2,3,4,6-tetramethyl-α-D-glucose (**8** in Figure 7) catalysed by alkylammonium carboxylate micelles

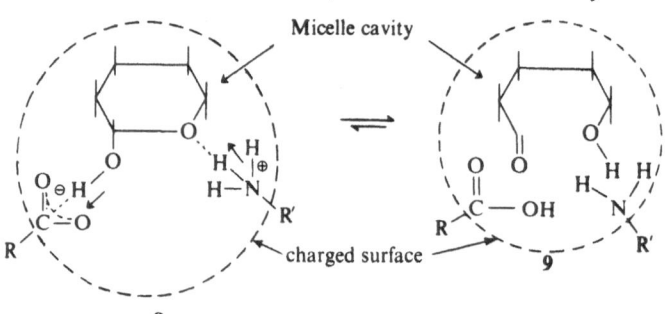

Figure 7. Proposed mechanism for micellar catalysis of the mutarotation of glucose in aprotic solvents (MicS-**8** → MicP-**9**).

(such as dodecylammonium propionate, DAP) in benzene and cyclohexane [64]. The plot of rate constant against surfactant concentration is sigmoidal, showing an upper limiting rate (k_{Mic}) at high concentrations of surfactant. The rate enhancement ratio (k_{Mic}/k_0) varies from 4×10^2 to 8×10^2 and is dependent on the anionic surfactant group (such as propionate, benzoate, butyrate) and on the solvent. The presence of moisture has only a slight effect on catalysis.

The kinetic scheme is analogous to the Michaelis-Menten scheme for enzyme catalysis. The solvent-dependence of the rate-enhancement ratio can be explained in terms of different values for the initial interaction between substrate and micelle (Mic) measured by the partition coefficient K_1. The full scheme for a transformation of substrate (S) to product (P) is:

$$ \text{Mic} + \text{S} \underset{\text{fast}}{\overset{K_1}{\rightleftharpoons}} \text{MicS} \underset{\text{slow}}{\overset{k_2}{\rightleftharpoons}} \text{MicP} \rightleftharpoons \text{Mic} + \text{P} $$

The site of solubilization (corresponding to an enzyme active site) is thought to be at the cavity of the micelle, in a hydrophilic microenvironment of high charge density. The complex MicS will in general be stabilized by hydrogen-bonding between the substrate and the charged surfactant groups. These specific interactions can facilitate ring opening by a bifunctional catalysis mechanism (Figure 7).

Similar large rate enhancements are observed for the decomposition of Meisenheimer complexes [66], such as sodium 1,1-dimethoxy-2,4,6-trinitro-cyclohexadienylide (**10** in Figure 8) in benzene containing 0.05 per cent added DMSO to dissolve the substrate. (Even at

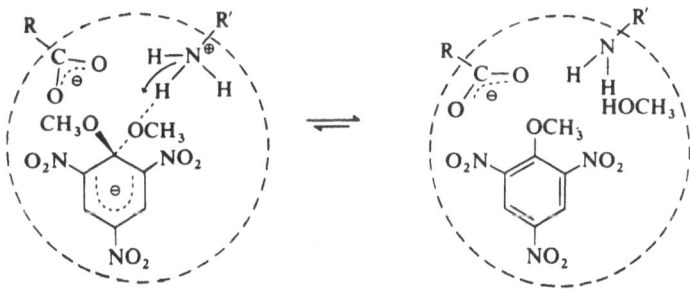

Figure 8. Proposed mechanism for catalytic decomposition of the 1,1-dimethoxy-2,4,6-trinitro-cyclohexadienylide ion (MicS-**10** → MicP).

these low concentrations, doubling the DMSO concentration halves the rate enhancement (through a decrease in K_1)). Both dodecylammonium benzoate ($k_{Mic}/k_0 = 6.3 \times 10^4$) and lecithin ($k_{Mic}/k_0 = 2 \times 10^3$) have been used as catalysts. The mechanism is thought to involve proton transfer from the ammonium group on the surfactant to the leaving methoxy-group. In this case, the anionic group in the surfactant is not directly involved.

It might be possible to elucidate the detailed mechanism from measurements of the primary isotope effect. Micellar catalysis offers the possibilities both of new methods of synthesis (in hydrophilic environments in aprotic solvents), and of the study of catalytic action at charged interfaces, and so opens up an exciting new area in the study of proton-transfer processes in aprotic solvents.

REFERENCES

[1] M. M. Davis, *Acid–Base Behaviour in Aprotic Organic Solvents*, National Bureau of Standards, Monograph 105 (1968).

[2] M. M. Davis, Brønsted Acid-Base Behaviour in 'Inert' Organic Solvents, in *The Chemistry of Non-Aqueous Solvents*, ed. J. J. Lagowski, Acad. Press, Vol. 13 (1970).

[3] J. N. Brønsted and R. P. Bell, *J. Amer. Chem. Soc.*, **53**, 2478 (1931).

[4] R. P. Bell, *Proc. Roy. Soc.*, A **143**, 377 (1934).

[5] R. P. Bell and E. F. Caldin, *J. Chem. Soc.*, 382 (1938).

[6] R. P. Bell and J. W. Bayles, *J. Chem. Soc.*, 1518 (1952).

[7] S. N. Vinogradov and R. H. Linnell, *Hydrogen-Bonding*, Van Nostrand, Reinhold (1971).

[8] G. C. Pimentel and A. L. McClellan, *The Hydrogen-Bond*, Freeman and Co. (1960).

[9] A. Allerhand and P. von R. Schleyer, *J. Amer. Chem. Soc.*, **85**, 866, 1233, 1715 (1963); L. Joris, P. von R. Schleyer and R. Gleiter, *J. Amer. Chem. Soc.*, **90**, 327 (1968).

[10] H. Baba and S. Suzuki, *J. Chem. Phys.*, **35**, 1118 (1961).

[11] R. P. Bell and J. E. Crooks, *J. Chem. Soc.*, 3513 (1962).

[12] D. Neerinck, A. Van Audenhaege, L. Lamberts, and P. Huyskens, *Nature*, **218**, 461 (1968).

[13] E. M. Arnett, L. Joris, E. Mitchell, T. S. S. R. Murty, T. M. Gorrie, and P. von R. Schleyer, *J. Amer. Chem. Soc.*, **92**, 2365 (1970).

[14] R. W. Taft, personal communication.

[15] B. H. Robinson and K. J. A. Hargreaves, to be published.

[16] G. Allen, J. G. Watkinson, and K. H. Webb, *Spectrochim. Acta*, **22**, 807 (1966).

[17] B. H. Robinson, C. J. Wilson, and T. Altinata, to be published.

[18] C. L. Bell and G. M. Barrow, *J. Chem. Phys.*, **31**, 300 (1959); **31**, 1158 (1959).

[19] R. Scott and S. Vinogradov, *J. Phys. Chem.*, **73**, 1890 (1969); H. Baba, A. Matsuyama and H. Kokubun, *Spectrochim. Acta*, **25A**, 1709 (1969).

[20] L. Sobczyk and Z. Pawetka, *J. Chem. Soc., Faraday Trans. I*, **76**, 832 (1974).

[21] K. Bauge and J. W. Smith, *J. Chem. Soc.* (*A*) 616 (1966).

[22] E. F. Caldin, *Chem. Rev.*, **69**, 135 (1969).

[23] J. J. Weiss, *J. Chem. Phys.*, **41**, 1120 (1964).

[24] D. L. DeTar and R. W. Novak, *J. Amer. Chem. Soc.*, **92**, 1361 (1970).

[25] N. Bjerrum, *Kgl. Danske Videnskab. Selskab*, 7, No. 9 (1926).

[26] R. M. Fuoss, *J. Amer. Chem. Soc.*, **80**, 5059 (1958).

[27] J. E. Prue, *J. Chem. Ed.*, **46**, 12 (1969).

[28] J. R. Jones in *Prog. Reaction Kinetics*, 7, 1 (1973).

[29] M. Szwarc, *Accounts Chem. Res.*, **2**, 87 (1969).

[30] T. E. Hogen-Esch and J. Smid, *J. Amer. Chem. Soc.*, **88**, 307 (1966); L. L. Chan and J. Smid, *J. Amer. Chem. Soc.*, **90**, 4954 (1968).

[31] C. A. Kraus, *J. Phys. Chem.*, **60**, 129 (1956).

[32] C. A. Kraus and R. M. Fuoss, *J. Amer. Chem. Soc.*, **55**, 21 (1933).

[33] G. Gammons, B. H. Robinson, and M. J. Stern, *J. Chem. Soc. Chem. Comm.*, 1157 (1972).

[34] C. D. Ritchie, *Accounts Chem. Res.*, **5**, 348 (1972).

[35] M. Eigen, *Z. Phys. Chem.* (*Frankfurt*), **1**, 176, (1954); *Angew. Chem., Int. Ed.*, **3**, 1 (1964).

[36] R. A. Marcus, *J. Amer. Chem. Soc.*, **91**, 7224 (1969); *J. Phys. Chem.*, **72**, 891 (1968).

[37] W. J. Albery, A. N. Campbell-Crawford, and J. S. Curran, *J. Chem. Soc. Perkin II*, 2206 (1972).

[38] M. M. Kreevoy and D. E. Konasewitch, *Adv. Chem. Phys.*, **21**, 243 (1971); M. M. Kreevoy and Sea-Wha Oh, *J. Amer. Chem. Soc.*, **95**, 4805 (1973).

[39] M. H. Davies, J. R. Keeffe, and B. H. Robinson, *Ann. Rep. Chem. Soc.* (A), pp. 123–171, (1973).

[40] M. V. Smoluchowski, *Z. Phys. Chem.*, **92**, 129 (1917).

[41] R. M. Noyes, *Prog. Reaction Kinetics*, 1, 129 (1961).

[42] J. M. Schurr, *Biophys. J.*, **10**, 700 (1971).

[43] P. Debye, *Trans. Electrochem. Soc.*, **82**, 235 (1942).

[44] P. Hemmes, *J. Phys. Chem.*, **94**, 75 (1972).

[45] E. F. Caldin and S. Mateo, *J. Chem. Soc. Chem. Comm.*, 854 (1973).

[46] E. F. Caldin and S. Mateo, *ibid.* (1973) and personal communication.

[47] E. F. Caldin and S. Mateo, *ibid.* (1973). Ref. 6 cited therein.

[48] E. S. Lewis and J. K. Robinson, *J. Amer. Chem. Soc.*, **90**, 4337 (1968).

[49] J. L. Kurz and L. C. Kurz, *J. Amer. Chem. Soc.*, **94**, 4451 (1972).

[50] O. Popovych, *J. Phys. Chem.*, **66**, 915 (1962).

[51] E. F. Caldin, J. E. Crooks, and D. O'Donnell, *J. Chem. Soc. Faraday Trans. I*, **69**, 993 (1973); K. J. Ivin, J. J. McGarvey, and R. Small, *Trans. Faraday Soc.*, **67**, 104 (1971); *J. Chem. Soc., Faraday Trans. I*, **69**, 1016 (1973).

[52] G. D. Burfoot, E. F. Caldin, and H. Goodman, *J. Chem. Soc., Faraday Trans. I*, **70**, 105 (1974).

[53] G. G. Hammes and A. C. Park, *J. Amer. Chem. Soc.*, **91**, 956 (1969).

[54] K. Bergmann, M. Eigen, and L. De Maeyer, *Ber. Bunsenges. Phys. Chem.*, **67**, 879 (1963); R. F. W. Hoppmann, *Ber. Bunsenges. Phys. Chem.*, **77**, 52 (1973).

[55] J. E. Crooks and B. H. Robinson, *Trans. Faraday Soc.*, **66**, 1436 (1970); **67**, 1707 (1971).

[56] B. H. Robinson and I. A. Watling, unpublished results.

[57] M. Eigen, *Naturwissenschaften*, **58**, 465 (1971); H. Kuhn, *Angew. Chem. Int. Ed.*, **9**, 798 (1972).

[58] E. Küchler and J. Derkosch, *Z. Naturforsch.*, **21B**, 209 (1966); Y. Kyogoku, R. C. Lord, and A. Rich, *J. Amer. Chem. Soc.*, **89**, 496 (1967).

[59] M. Eigen and D. Pörschke, *J. Mol. Biol.*, **53**, 123 (1970); **62**, 361 (1971).

[60] J. G. Hoggett and G. Maass, *Ber. Bunsenges. Phys. Chem.*, **75**, 45 (1971).

[61] G. Schwarz, *Rev. Mod. Phys.*, **40**, 206 (1968).

[62] P. O. Löwdin, *Adv. Quantum Chem.*, **2**, 213 (1966).

[63] E. H. Cordes and R. B. Dunlap, *Accounts Chem. Res.*, **2**, 329 (1969); E. J. Fendler and J. H. Fendler, *Adv. Phys. Org. Chem.*, **8**, 271 (1970).

[64] J. H. Fendler, E. J. Fendler, R. T. Medary, and V. A. Woods, *J. Amer. Chem. Soc.*, **94**, 7288 (1972).

[65] J. H. Fendler, *J. Chem. Soc. Chem. Comm.*, 269 (1972).

[66] J. H. Fendler, E. J. Fendler, and S. A. Chang, *J. Amer. Chem. Soc.*, **95**, 3273 (1973).

John E. Crooks

FAST AND SLOW PROTON-TRANSFER REACTIONS IN SOLUTION

6.1 Acids and pseudoacids

6.2 Fast proton-transfer processes

6.3 Slow proton-transfer processes

6.1 Acids and pseudoacids

It is immediately obvious from experiment that there is a profound difference in kinetic behaviour between those acids in which the acidic proton is bound to carbon and those acids in which the acidic proton is bound to oxygen or nitrogen. The neutralization of a micromolar aqueous solution of acetic acid by OH^-, for example, occurs as fast as the solutions can be mixed, whereas the corresponding neutralization of nitromethane, as followed by pH or u.v. absorption, requires several minutes at room temperature for completion. This was recognized by Hantzsch [1], who termed nitromethane and other carbon acids 'pseudo-acids'. He considered that their low rate of neutralization was due to the need for molecular rearrangement to occur before deprotonation:

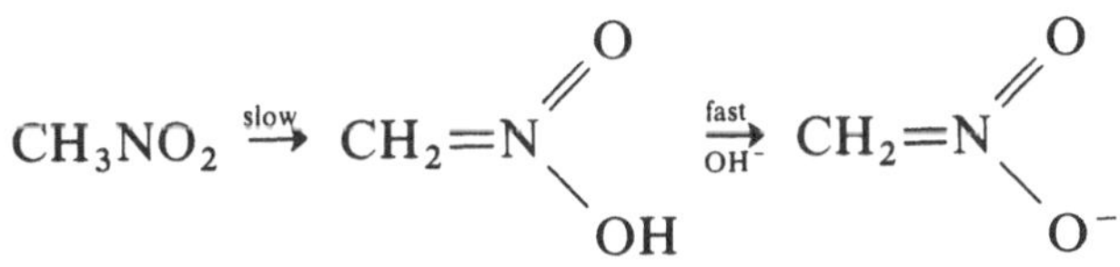

$$CH_3NO_2 \xrightarrow{slow} CH_2{=}N\diagup^{\textstyle O}_{\textstyle OH} \xrightarrow[OH^-]{fast} CH_2{=}N\diagup^{\textstyle O}_{\textstyle O^-}$$

Although we now believe that the departing proton comes from the carbon atom rather than the oxygen atom, Hantzsch's original concept, that there is a complicating factor in the proton-transfer reactions of carbon acids which is not present for other acids, is still valid. It would appear, however, that it is more fruitful to consider this complicating factor as an effect on the protonation of the anion rather than an effect on the deprotonation of the acid.

The ionization of an acid may be simply represented by:

$$HA \underset{k_b}{\overset{k_f}{\rightleftharpoons}} H^+ + A^-; \qquad K = \frac{[H^+][A^-]}{[HA]} = k_f/k_b \qquad (1)$$

The contrast between normal acids and pseudo-acids is neatly shown by an example from a review by Albery [2]. For acetic acid as HA, in aqueous solution,

$$K = 1.7 \times 10^{-5} \, \text{mol dm}^{-3}; \qquad k_f = 9 \times 10^5 \, \text{s}^{-1};$$
$$k_b = 5 \times 10^{10} \, \text{dm}^3 \, \text{mol}^{-1} \, \text{s}^{-1}$$

whereas for the pseudo-acid trifluoroacetylacetone,

$$K = 2.0 \times 10^{-5} \, \text{mol dm}^{-3}; \qquad k_f = 1.5 \times 10^{-2} \, \text{s}^{-1};$$
$$k_b = 7.5 \times 10^2 \, \text{dm}^3 \, \text{mol}^{-1} \, \text{s}^{-1}$$

There are two ways of looking at these values. One can say, after Hantzsch, that it is strange that trifluoroacetylacetone ionizes so much more slowly than acetic acid, even though it has the same acidic strength. On the other hand, one can say that it is strange that trifluoroacetylacetone anion reprotonates so much more slowly than the acetate ion, thereby, incidentally, making trifluoroacetylacetone as strong an acid as acetic acid by virtue of the relationship $K = k_f/k_b$. The second approach is more profitable and will be followed here.

6.2 Fast proton-transfer processes

6.2.1 The limiting rate for a proton-transfer reaction

The simplest assumption to make in the calculation of the rate of reaction between anions of weak acids and hydrogen ions is that, as the equilibrium strongly favours the reprotonation process, reaction occurs whenever a pair of ions are adjacent, *i.e.* are in the same encounter complex in the solution. The rate of the reaction is then the rate at which the ions diffuse towards each other to form encounter complexes. Such a calculation was originally performed by Smoluchowski [3] in order to find the rate of coagulation of colloidal particles. The calculation refers to featureless spherical particles diffusing through a homogeneous isotropic medium, an approximation more valid in the original calculation than for reagent molecules of similar size to solvent molecules. The rate of the bimolecular reaction for uncharged particles is found to be

$$k = 4\pi N (D_A + D_B) r_{AB}$$

154

where D_A and D_B are the diffusion coefficients (stoke) of reagent molecules A and B, r_{AB} is the intermolecular distance at which reaction occurs, i.e. the sum of the radii of A and B considered as spheres, and N is the Avogadro Number. Substitution of the Stokes–Einstein equation

$$D_A = kT/6\pi\eta r_A$$

where η is the viscosity of the solvent (poise), gives

$$k = 8RT/3\eta \tag{2}$$

If the diffusing particles are charged, a further factor must be included. Particles of unlike charge will tend to diffuse together more rapidly than neutral particles, and vice versa for particles of like charge. Debye [4] has calculated this factor to be q,

$$q = \frac{z_A z_B\, e^2}{kT\varepsilon r_{AB}\{1 - \exp(-z_A z_B/kT\varepsilon r_{AB})\}^{-1}} \tag{3}$$

where z_A is the algebraic value of the charge on A, e is the charge on the electron, and ε is the bulk dielectric constant of the solvent. The factor q is not very large, being around 2 to 3 for average size singly-charged ions, although it has a significant effect in a few extreme cases; for instance, the rate of reaction of OH^- with $HP_3O_{10}^{4-}$ is two orders of magnitude slower than predicted by equation 2. Substitution of numerical values into equation 2 gives a value of about $10^{10}\ dm^3\ mol^{-1}\ s^{-1}$ for the diffusion-controlled reaction between H^+ and the anion of a weak acid. As we have already seen, a rate constant of this order has been found for acetate anion, and this is typical for the anions of acids in which the acidic proton is located on oxygen or nitrogen. A large number of these rates have been measured by relaxation and electrochemical techniques and by n.m.r. [5, 6]. Rates of proton-transfer reactions of electronically-excited acids have been measured by the fluorescence-quenching technique. Some illustrative examples are given in Table 1. The reaction

$$H^+ + OH^- \rightarrow H_2O$$

is particularly noteworthy, as it is the fastest known bimolecular reaction in aqueous solution. The rate of this reaction has been measured by various techniques, and values obtained range from 7.3×10^{10} $dm^3\ mol^{-1}\ s^{-1}$ (at 15°C) [7] to $4.1 \times 10^{11}\ dm^3\ mol^{-1}\ s^{-1}$ (at 25°C) [8]. The most likely value at 25°C is between 1.0 and 1.4×10^{11} $dm^3\ mol^{-1}\ s^{-1}$, as determined by E-jump [9] and T-jump [10] techniques. As can be seen from Table 1, the rates of reprotonation of

Table 1

Rates of protonation of bases in water at 25°C

Data from M. Eigen, *Angew. Chemie, Internat. Ed.*, **3**, 1 (1964)

Base	pK of conjugate acid	$\log k_b/\mathrm{dm^3\,mol^{-1}\,s^{-1}}$
F^-	3.15	11.0
HCO_3^-	3.77	10.7
Benzoate	4.20	10.5
Acetate	4.74	10.7
Imidazole	6.95	10.2
p-nitrophenolate	7.14	10.6
$(CH_3)_3P$	8.80	9.7
NH_3	9.25	10.6
$(CH_3)_3N$	9.79	10.4
OH^-	15.70	11.2

the anions of oxygen and nitrogen acids are, in general, diffusion-controlled. Similar experiments performed on a wide range of oxygen and nitrogen acids show that the rates of proton transfer to OH^- from these acids are also diffusion-controlled.

6.2.2 Steric effects

The rate of protonation of an anion may be reduced by steric requirements. A simple ion such as F^- has spherical symmetry and has no preferred orientation for reaction with an approaching proton. The trimethylacetate ion, on the other hand, can only be protonated on one side, and k_b has a correspondingly lower value [11] of $1.5 \times 10^{10}\,\mathrm{dm^3\,mol^{-1}\,s^{-1}}$ (cf. Table 1). Weller [12] has given a detailed calculation of steric effects, considering the spatial angles on acid and base from and to which a proton is donated and accepted. Some values of fractional spatial angles assigned are unity for H_3O^+, OH^- and NH_4^+, $\frac{1}{2}$ for acetate and β-naphthol, $\frac{1}{3}$ for acridine and $\frac{1}{4}$ for trimethylamine and 3-acetylaminopyrene 5,8,10-trisulphonate. Values for the proton-transfer rate constants calculated by applying these steric factors to the rates calculated by the Debye-Smoluchowski equation, equations 2 and 3, agree with those measured by the fluorescence-quenching technique. For example, the rate of proton transfer from 3-acetylaminopyrene 5,8,10-trisulphonate to trimethylamine is calculated to be $6 \times 10^8\,\mathrm{dm^3\,mol^{-1}\,s^{-1}}$, which compares quite well with the observed rate of $2 \times 10^8\,\mathrm{dm\,mol^{-1}\,s^{-1}}$. Šolc and Stockmayer [13] give a formal mathematical proof of Weller's hypothesis that the rate calculated by the Debye-Smoluchowski equation must be multiplied by a factor involving the fractional surface area of the molecules available for reaction. Weller's calculation has been questioned by Caldin *et al.* [14] on the grounds that the Debye-Smoluchowski equation gives the rate of formation of encounter complexes, rather than the rate of collision of reactant molecules. There are many collisions in each encounter, so that the probability of a collision with the correct mutual orientation does not affect the observed rate, as long as one such collision occurs during the lifetime of the encounter. In a later theoretical study, Šolc and Stockmayer [15] predict that, for the same diffusion-controlled reaction carried out in a range of solvents, a plot of the reciprocal of the observed forward rate constant versus the solvent viscosity will have an intercept related to the rate at which the correctly-aligned encounter complex reacts to give products, and a slope related to the relative surface area of the reactant molecules available for reaction. Schmitz and Schurr [16] also show that a plot of the reciprocal rate versus viscosity is linear, whatever the degree of steric hindrance. Caldin *et al.* [14] find such a linear relationship for the proton-transfer reaction between the indicator acid Magenta E and tri-n-butylamine in a range of six aprotic solvents. They postulate a three-step mechanism;

$$AH + B \underset{k_{-1}}{\overset{k_1}{\rightleftharpoons}} AH, \ B \underset{k_{-2}}{\overset{k_2}{\rightleftharpoons}} AH \cdots B \underset{k_{-3}}{\overset{k_3}{\rightleftharpoons}} A^- \cdots HB^+$$

Here AH, B represents AH and B trapped together within the same solvation sphere, forming a weak complex, $AH \cdots B$ represents a hydrogen-bonded complex in which the $O-H \cdots N$ hydrogen bond is formed, and $A^- \cdots HB^+$ represents the final product, an ion-pair formed by motion of the proton across the hydrogen bond. The transformation of AH, B to $AH \cdots B$ is by rotational diffusion. For the particular reaction studied by Caldin *et al.*, the observed variation of the rate with temperature does not require that the rate of the re-orientation process within the encounter complex is limited by solvent viscosity. This is in agreement with the theoretical treatment by Šolc and Stockmayer [13], who find that rotational diffusion constants appear only in higher terms in their general formula for the rate of a diffusion-controlled reaction.

6.2.3 *Intramolecular hydrogen bonding*
Rates of ionization of monoanions of substituted malonic acids are much less than the diffusion-controlled limit, even though the equilib-

rium constant favours formation of the dianion [17]. Rate constants for the reaction

$$AH^- + OH^- \underset{k_b}{\overset{k_f}{\rightleftharpoons}} A^{2-} + H_2O; \qquad K = k_f/k_b$$

Table 2

The effect of intramolecular hydrogen bonding on $AH + OH^- \rightleftharpoons A^- + H_2O$

AH	log K	log k_f/dm^3 mol^{-1} s^{-1}	$\Delta H^\dagger$/ kcal mol^{-1}	$\Delta S^\dagger$/cal K^{-1} mol^{-1}	Bond	Ref.
Diethylmalonate anion	6.94	8.4	4.5	−5	O$^-\cdots$H—O	[17]
Ethyl-n-butylmalonate anion	6.75	8.2	5.5	−3	O$^-\cdots$H—O	[17]
Ethyl-iso-amylmalonate anion	6.69	8.3	5.5	−3	O$^-\cdots$H—O	[17]
Di-n-propylmalonate anion	6.66	7.5	6	−1	O$^-\cdots$H—O	[17]
Ethyl-iso-propylmalonate anion	5.97	7.7	7	0	O$^-\cdots$H—O	[17]
Di-iso-propylmalonate anion	5.42	7.7	7.5	+2	O$^-\cdots$H—O	[17]
Alizarin Yellow R	3.10	7.6	—	—	O$^-\cdots$H—O	[18]
Tropaeolin O	1.85	5.9	—	—	N$\cdots$H—O	[18]
Clayton Yellow	1.0	4.3	—	—	N$\cdots$H—O	[18]
Dimethylanthranilate anion	5.6	7.1	—	—	O$^-\cdots$H—O	[2]
2,4-Dihydroxyazonitrobenzene anion	2.0	5.7	—	—	N$\cdots$H—O	[2]

are given in Table 2. This effect is attributed to the proton in the monoanion being held in a strong intramolecular hydrogen bond [6],

which must be broken before a hydrogen bond to the attacking base, along which proton transfer occurs, can form. The finite value of $\Delta H^\ddagger$ is a measure of the energy required to break the intramolecular hydrogen bond, and the values found are typical of hydrogen bonds. Even smaller values of k_f have been found for some indicator acids [18], although, unfortunately, values of $\Delta H^\ddagger$ were not obtained. It is noteworthy that for all the O—H$\cdots$O hydrogen bonds in Table 2, the accepted oxygen bears a partial negative charge. If the acceptor

158

oxygen is neutral, as in o-nitrophenol or salicyclic acid [6], k_f is of the order of 10^{10} dm^3 mol^{-1} s^{-1}. This type of hydrogen bond does not appear to affect the kinetics in aqueous solution, even though it is well characterized in aprotic solvents [19]. By contrast, the O—H$\cdots$N bond has a pronounced kinetic effect.

6.2.4 *Eigen's description of the generalized proton-transfer process*

Eigen has shown [5] that in general, a proton-transfer process may be written as follows (the charges are omitted for generality).

$$AH + B \underset{k_{21}}{\overset{k_{12}}{\rightleftharpoons}} AH\cdots B \underset{k_{32}}{\overset{k_{23}}{\rightleftharpoons}} A\cdots HB \underset{k_{43}}{\overset{k_{34}}{\rightleftharpoons}} A + HB$$

Here A$\cdots$HB and A$\cdots$HB are encounter complexes, probably involving hydrogen-bonding, with, for some reactions, a solvent molecule interposed between the reactants as a hydrogen-bonded bridge for proton-transfer; k_{12} is the rate at which AH and B diffuse together, and k_{23} is the rate at which proton transfer occurs within the encounter complex. The overall forward and reverse rate constants, k_f and k_b, may be calculated by the steady-state approximation, in which it is assumed that the encounter complexes are present in small but constant concentrations. If at least one of the conditions:

$$k_{21} \ll k_{23}; \qquad k_{34} \ll k_{32}$$

is fulfilled, *i.e.* if the rates of proton transfer within the encounter complex are much faster than diffusion, a further approximation can be made to give:

$$k_f = \frac{k_{12}}{1 + (k_{21}k_{32}/k_{23}k_{34})}; \tag{4a}$$

$$k_b = \frac{k_{43}}{1 + (k_{34}k_{23}/k_{32}k_{21})} \tag{4b}$$

Values of k_{23} and k_{32} can sometimes be evaluated [20]. The width of a line in the Raman spectrum of aqueous trichloracetic buffer has been attributed to the proton exchange within the encounter complex H$^+\cdots$A$^-$. Values of the order 10^{12} to 10^{13} s^{-1} are thus found for k_{23} for the reaction

$$H^+\cdots A^- \rightarrow HA$$

whereas the value for the diffusion-controlled dissociation of H$^+\cdots$A$^-$ to free H$^+$ and A$^-$ is of the order of 10^{10} s^{-1}.

The overall equilibrium constant K is given by

$$K = K_{12}K_{23}K_{34} = \frac{k_{12}k_{23}k_{34}}{k_{21}k_{32}k_{43}}$$

If it is assumed that the rate at which AH and B diffuse together, k_{12}, is equal to the rate at which A and HB diffuse together, k_{43}, then equations 4a and 4b give:

$$k_f = \frac{k_{12}}{1 + K^{-1}} \tag{5a}$$

$$k_b = \frac{k_{43}}{1 + K} \tag{5b}$$

If the equilibrium favours the formation of A and HB, i.e. if $K \gg 1$, then the forward reaction is diffusion-controlled, since equation 5a reduces to $k_f = k_{12} = k_D$. If on the other hand $K \ll 1$, the backward reaction is diffusion-controlled, since equation 5b reduces to $k_b = k_{43}$ and therefore $k_b = k_D$.

If K has the value of unit, k_f and k_b are both equal to half the diffusion-controlled rate. K is related to the difference in acidity between AH and BH:

$$\log K = pK_{HB} - pK_{AH} = \Delta pK$$

Thus a plot of $\log k_f$ against ΔpK for a series of pairs of normal acids AH and HB will consist of a straight line of gradient zero if $pK_{HB} > pK_{AH}$, and a straight line of gradient unity if $pK_{HB} < pK_{AH}$. The converse holds for $\log k_b$. There are curved transition regions where ΔpK is around zero. These theoretical predictions are shown in Figure 1, together with experimental results for a range of oxygen, nitrogen, sulphur [21] and carbon [22] acids. The agreement between theory and experiment is fairly good for proton transfers between oxygen and nitrogen acids and bases, not so good for sulphur acids and bases, and very bad for the typical carbon acid acetylacetone. The values listed in Table 1 are in accordance with this scheme, as H_3O^+ is a much stronger acid than any of those listed. The slow rate of protonation of carbanions is now seen to be one manifestation of a general effect. Even for oxygen acids and bases there is a discrepancy between theory and experiment if ΔpK is around zero. For example, ΔpK for AH as acetic acid and B as propionate is -0.12, but $\log k_f$ is only 8.6 rather than 10, $\log k_b$ being 8.5. Equation 5a shows that k_f is half the diffusion-controlled rate if ΔpK is zero, but the reduction is too great to be explained by this term. There must be some small activation energy barrier to proton transfer if ΔpK is small. Proton transfer to sulphur centres is intermediate in character between that to oxygen and that to carbon. At sufficiently high ΔpK, the protonation of sulphur anions is diffusion-controlled,

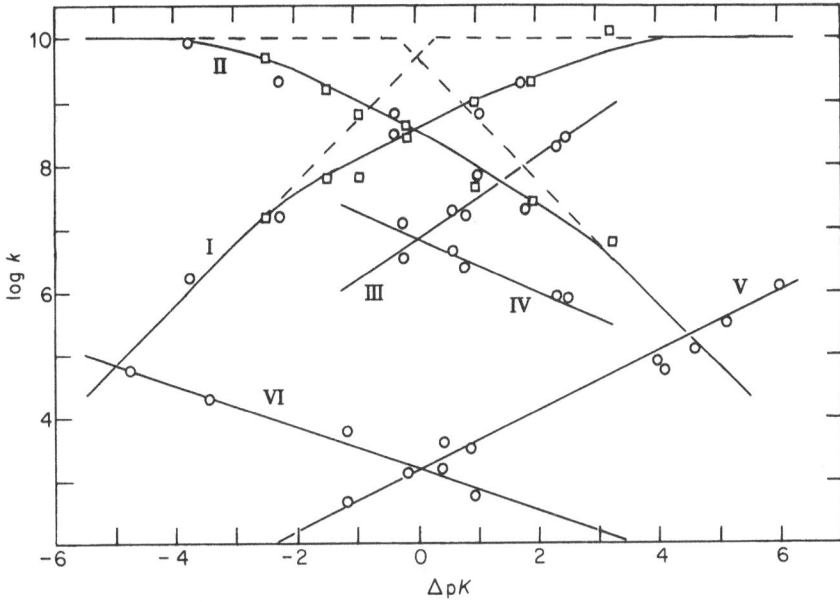

Figure 1.

Line I: squares; proton transfer to acetate from a series of oxygen acids. Circles; proton transfer to imidazole from a series of protonated amines.
Line II: the reverse reactions to those for Line I.
Line III: proton transfer to 2-hydroxyethylthiolate from a series of sulphur acids.
Line IV: the reverse reactions to those for Line III.
Line V: proton transfer from acetylacetone to a series of oxyanions.
Line VI: the reverse reactions to those for Line V.
Broken lines: theoretical predictions from Reference 5.
Data for Lines I, II, III, IV, Reference 21: data for Lines V and VI, Reference 22.

as found by the pulse radiolysis technique [23] for the radical anions of glutathione and lipoic acid. There is a correlation between ease of hydrogen-bonding between solvent and proton-acceptor and the deviation from the theoretical curve. It is sometimes stated that it is this hydrogen bonding which determines whether or not an acid is kinetically 'normal'. This point is discussed further in Section 6.3.1.

6.3 Slow proton-transfer processes

6.3.1 Hydrogen bonding and carbanions

A representative list of rates of protonation of anions of pseudo-acids is given in Table 3. These rates were obtained by dividing the observed rate of ionization, as measured by halogen uptake, by the observed acid dissociation constant. It is noteworthy that the site of

Table 3

Rates of protonation of carbanions in water at 25°C

Conjugate acid	$\mathrm{p}K_a$ of conjugate acid	$\log k_b/\mathrm{dm^3\,mol^{-1}\,s^{-1}}$
$CH_2(NO_2)_2$	3.6	3.5
$CH_3COCH_2NO_2$	5.1	3.6
$C_2H_5O_2CCH_2NO_2$	5.8	3.6
$(CH_3CO)_2CHBr$	7.0	5.4
$C_2H_5NO_2$	8.6	1.2
CH_3NO_2	10.2	2.8
$CH_3COCHC_2H_5CO_2C_2H_5$	12.7	7.6
$(C_2H_5CO_2)_2CHC_2H_5$	15.0	8.3

Data from Reference 25.

protonation in these ions is not the same as the site of the negative charge, which is typically delocalized on oxygen atoms adjacent to the protonation site. Protonation thus involves a reorganization of the ion, a transfer of negative charge from oxygen to carbon. The easy transmission of a proton by the proton-jump mechanism which occurs for an oxy-anion, such as acetate:

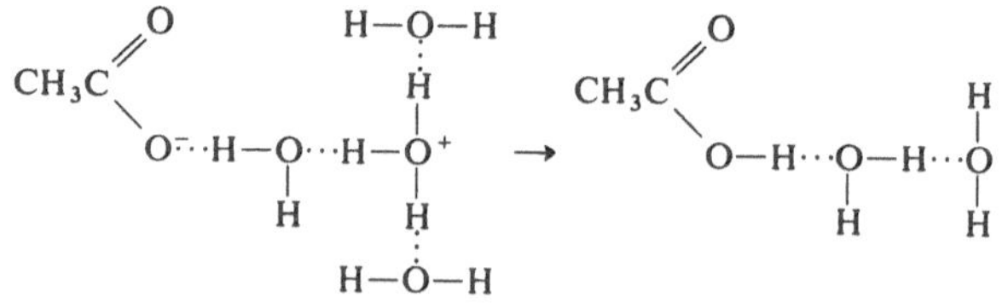

cannot occur for a carbanion, such as that of nitromethane, because of the need for this shift of charge:

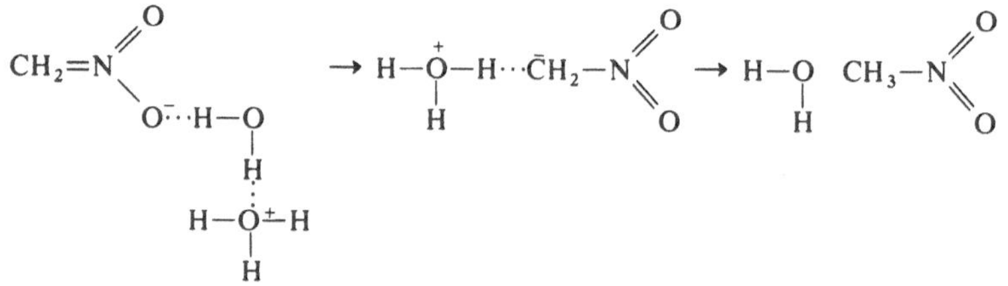

The transfer of charge and the corresponding disruption of the solvation pattern require energy which appears as an activation energy and reduces the rate below the diffusion-controlled limit.

On the scheme, as written above, proton transfer to acetate anion is thus from a $H_9O_4^+$ species, whereas that to nitromethane anion is from a H_3O^+ species. Albery [2] has reviewed the evidence for this distinction, and has given schematic energy surfaces illustrating the two types of process. Evidence for protonation of oxy-anions by $H_9O_4^+$ species comes from temperature-jump and n.m.r. studies. The rate of reaction between solvated proton and hydroxide ion in water agrees with that predicted by the Debye-Smoluchowski equation if the solvated proton has a diameter of 0.8 nm and the medium dielectric constant is that of bulk water. This suggests that H^+ has the kinetic behaviour of $H_9O_4^+$. N.m.r. studies on proton exchange between amines and ammonium ions, as described in the article by Grunwald and Eustace in this volume, show that the exchange occurs via a water molecule between the two nitrogen centres. Evidence for the protonation of carbanions by H_3O^+ comes from studies of the rates of acid-catalysed reactions in mixed H_2O-D_2O solvents.

If the activation energy found for the protonation of carbanions is due to the need for reorganization of the solvent structure in the inner solvation sphere, then a change in solvent should produce a dramatic change in the rate of reaction. These solvation effects are observed and are discussed in Section 3.3. Hydrogen bonding of solvent to proton-acceptor is not a necessary prerequisite for an exothermic proton-transfer reaction to be diffusion-controlled. Many proton-transfer reactions in aprotic solvents have been shown to be diffusion-controlled, as discussed in reference 14 and references quoted therein. The fundamental condition for diffusion-controlled proton-transfer is that no drastic change in the solvation pattern be required for proton-transfer to occur. If such a drastic change is required, as is found even for a few reactions in aprotic solvents, then the rate of the reaction is much reduced. The rate of reaction between Bromophenol Blue and aromatic amines in aprotic solvents is many orders of magnitude less than the diffusion-controlled limit, and very solvent-dependent [24]. The significance of hydrogen-bonding to the proton-acceptor in aqueous solution is that this sets up a solvent structure which for carbon acids is disrupted during proton transfer, but is not for oxygen and nitrogen acids.

Some carbanions are known in which the negative charge appears to be mostly localized on the carbon atom, and these are protonated at rates approaching the diffusion-controlled limit. Results for the disulphones [26, 27] and several cyano-compounds [28] are shown as an Eigen plot in Figure 2. Values for the disulphone and most of

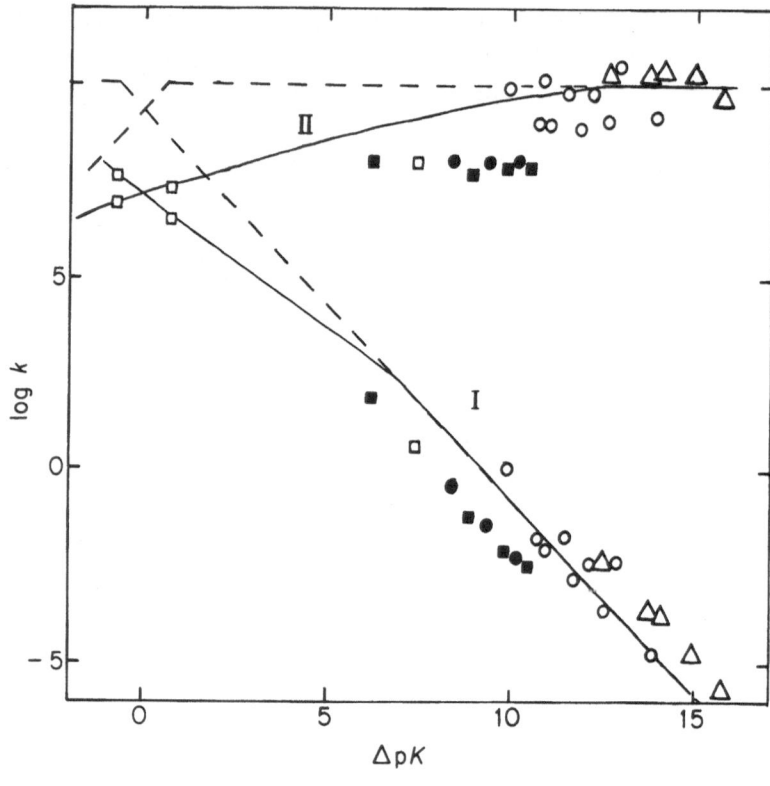

Figure 2.

Line I: filled squares; proton transfer from $(EtSO_2)_2$ CHMe to a series of oxyanions. Data from Reference 26. Filled circles; proton transfer from tert. butyl malononitrile to a series of oxyanions. Data from Reference 28. Open squares; proton transfer from malononitrile to morpholine, phosphate and bicarbonate. Data from Reference 28. Open circle; proton transfer from 1,4-dicyanobutene to a series of amines. Data from Reference 28. Open triangles; proton transfer from a series of disulphones to H_2O. Data from Reference 27.
Line II: the reverse reactions to those for Line I.
Broken Lines: theoretical predictions from Reference 5.

the cyano-compounds are calculated from the rate of proton-loss, as observed by halogenation or tritium exchange, and the measured acid dissociation constant. Values for malononitrile were obtained directly from temperature-jump experiments. Although the protonation rate reaches a limiting value independent of ΔpK, this is, for some compounds, rather lower than that predicted by theory, even making allowance for steric effects. However, the measurement of the acid dissociation constant for these extremely weakly acidic species is very difficult, and extrapolation of the results from the concentrated alkali solutions used to that appropriate to the conditions used in the kinetic experiments is dubious. It is not advisable to put too much reliance on the absolute values of the carbanion protonation rates, but their constancy for a range of proton donors is significant enough. The rates for the reaction $AH + OH^- \rightleftharpoons A^- + H_2O$ were found to be several orders of magnitude too low to fit the lines in Figure 2. This anomalously slow behaviour for the OH^--H_2O pair is often found for pseudo-acids. Kresge and Lin [29] suggest that this is due to the strong solvation of OH^-, causing the formation of an encounter complex between OH^- and a pseudo-acid to require more energy than for other bases. However if the proton donor is a normal acid, it can itself hydrogen-bond to the hydrated OH^-, and proton transfer then requires the minimum activation energy, which is the overall enthalpy increase. p-Nitrobenzoyl cyanide [28], a cyano-compound for which the negative charge on the anion can be delocalized, was found to be a typical pseudo-acid, the protonation of the anion being comparatively slow and dependent on the acidity of the proton donor.

Chloroform is a carbon acid whose structure suggests minimal electronic rearrangement on ionization. The rate of ionization has been measured by the tritium exchange method [30], but as no independent measurement of pK_A has been made, it is not possible to calculate the rate of protonation of the anion. Margolin and Long produce indirect evidence for chloroform being a normal acid. The ionization is subject to OH^- catalysis only, indicating that the rate of proton transfer to OH^- is much higher than to other bases. As discussed above, proton transfer from pseudo-acids to OH^- is typically anomalously slow. Increasing the basicity of OH^- by the admixture of DMSO with the solvent water increases the rate. A plot of $\log k$ versus the logarithmic activity of OH^- is a straight line of gradient unity, as is observed for normal acids on an Eigen plot.

The rate of proton-transfer from phenylacetylene to a range of amine bases [29], as measured by tritium exchange, gives a plot of $\log k$ versus ΔpK which is a straight line of gradient unity. If the generally accepted value for the pK_A of phenylacetylene, 21, is used, the rate of protonation of the anion by ammonium ion can be calculated to be around 10^{10} dm^3 mol^{-1} s^{-1}, suggesting that phenylacetylene is a normal acid. However, phenylacetylene resembles pseudo-acids, in that the rate of proton loss to OH^- is anomalously low, missing the line for the amine bases by a factor of 10^2. This is presumably because of the poor hydrogen-bond forming powers of phenylacetylene: chloroform, by contrast is a better hydrogen-bond donor, which accounts for the observations of Margolin and Long.

Rather surprisingly, pulse radiolysis studies [31] have shown that $C(NO_2)_3^-$ protonates at a rate approaching the diffusion-controlled limit ($3.6 \times 10^8 \ dm^3 \ mol^{-1} \ s^{-1}$), even though one would expect the negative charge to be delocalized over the three NO_2 groups.

6.3.2 The Brønsted equation

When there is a finite activation energy for both reactions setting up the equilibrium

$$AH + B \rightleftharpoons A^- + HB^+$$

it follows that k_{-1} is dependent on the nature of A^- and B, even though the equilibrium constant is much greater than unity, so that a plot of $\log k_1$ versus ΔpK is not necessarily a straight line of gradient unity. However if A or B is not varied too drastically, a linear relationship may still be observed, the gradient of the line being less than unity. In Figure 3 the course of the reaction is shown as the overlapping

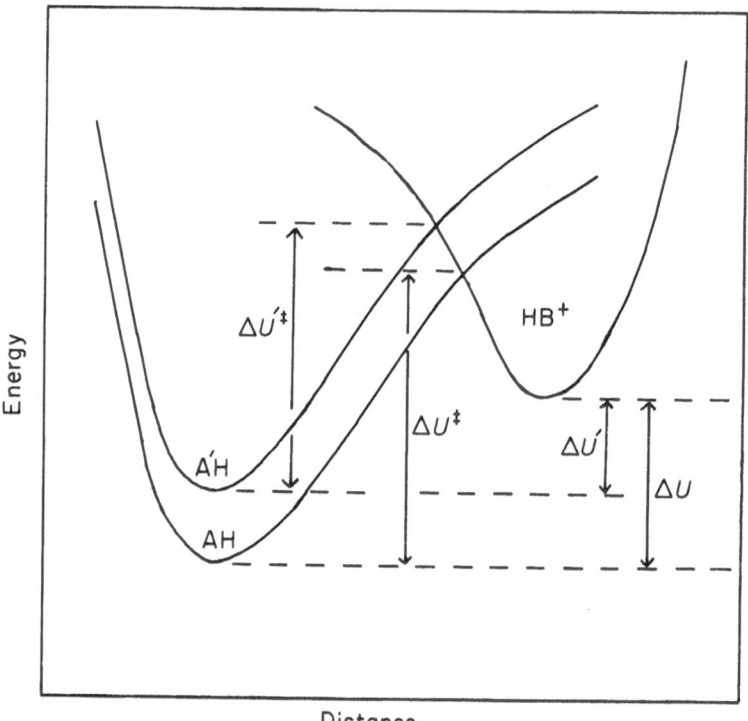

Distance

Figure 3.

Schematic diagram of energy versus distance along reaction co-ordinate for proton transfer.

of two potential energy curves of Morse type [32]. The minima correspond to AH + B and A^- + HB^+ respectively, and the point of intersection of the curves is the transition state $A\cdots H\cdots B$. Also shown in Figure 3 is the potential energy curve for another species A'H, where A' is similar enough to A to keep the curve the same shape, although with a different energy minimum. If the curves are assumed to be straight lines in the region of interest, simple geometry shows that

$$\Delta U'^{\ddagger} - \Delta U^{\ddagger} = \frac{s_1}{s_1 + s_2}(\Delta U' - \Delta U) \qquad (6)$$

where s_1 and s_2 are the gradients of the curves for AH (equal to that for A'H) and HB^+, both considered positive. If a quantity α is defined by

$$\alpha = \frac{s_1}{s_1 + s_2} \qquad (7)$$

and if the entropy terms are insignificant, or cancel out, equations 6 and 7 can be written in the form

$$\log k_1 = \alpha \log K_{AH} \qquad (8)$$

to give the rate of the proton transfer to B from a series of acids AH. Equation 8 was first derived from experimental data by Brønsted [33]. A large number of sets of proton-transfer reactions have been found to fit the Brønsted equation. Similarly, for a range of bases B accepting protons from a reference acid,

$$\log k_1 = -\beta \log K_{HB} \qquad (9)$$

when K_{HB} is the dissociation constant of the conjugate acid HB^+, and on the assumption that the entropy terms cancel, the potential-energy curves give

$$\beta = \frac{s_2}{s_1 + s_2} \qquad (10)$$

It follows from equations 7 and 9 that

$$\alpha + \beta = 1$$

The Brønsted equation can be seen to fit a region of the Eigen plot. If ΔpK is varied over too wide a region, α is no longer a constant, as Brønsted himself originally cautioned.

If the activation energy for the forward reaction is very low, the HB^+ curve intersects the AH curves in a region where s_1 is very small.

Thus, from equation 7, α approaches zero. Conversely, if the activation energy for the back reaction is very low, then s_2 is very small, and α approaches unity. These extremes correspond to the ideal kinetics shown in the Eigen plot (Figure 1). If the activation energy for the forward reaction is low, then the transition state may be taken to resemble AH + B. This is a consequence of the Hammond postulate [34], which proposes that if two states occur consecutively during a reaction process, and have nearly the same energy, their interconversion involves only a small reorganization of molecular structure. In other words, a low value of the Brønsted coefficient α may be, and often is, interpreted as meaning that the transition state resembles the reactants, whereas a value of α approaching unity means that the transition state resembles the products. The magnitude of α thus gives the position of the transition state along the reaction co-ordinate. There are further subtleties to this argument, which are discussed further in Section 3.4.

6.3.3 The effect of solvation

As Caldin [35] has pointed out, there are two ways in which solvation can affect the rate of a reaction. Solvated molecules cannot approach close enough to react unless some desolvation occurs, and this desolvation may require considerable energy if a reactant is ionic (cf. Chapters 2 and 3). Secondly, there may be changes in orientation of molecules within the solvation shell as the transition state is formed. These two effects may be distinguished as reactant, or initial-state, and transition-state solvation respectively. The objection is sometimes raised that, as the transition state only exists for a mathematically infinitesimal time, there is not enough time for solvent reorganisation around the transition state to occur, so that transition-state solvation cannot be significant [36]. This is not necessarily so; 'transition-state solvation' is a convenient term to describe the solvation pattern which must be set up before the transition state can occur. There is, for example, some evidence from values of $\Delta S^{\ddagger}$ for proton transfer to methoxybenzenes that transition-state solvation must exist [37].

A dramatic change in solvation can be made by the use of dimethylsulphoxide (DMSO) as solvent rather than water. DMSO is a dipolar aprotic solvent, of high dielectric constant ($\varepsilon = 46.4$), which solvates cations much more strongly than anions. The sulphone oxygen atoms have a partial negative charge, but the positive pole of the dipole is diffused in the methyl group in such a way that attraction to anions is small. DMSO is a good hydrogen-bond acceptor, but cannot act as a hydrogen-bond donor. The rates of a large number of S_N2 reactions with anionic nucleophiles have been shown [38] to be much faster in DMSO than in methanol, presumably because as the attacking anion is not solvated, preliminary desolvation of the reactant is not required. Ritchie and his co-workers [38] have carried out extensive studies of proton-transfer reactions of polycyclic aromatic hydrocarbons in DMSO. The results for proton transfer from these acids to anionic oxygen bases [39] are shown as an Eigen plot in Figure 4, and compared with results [40, 41] for proton-transfer

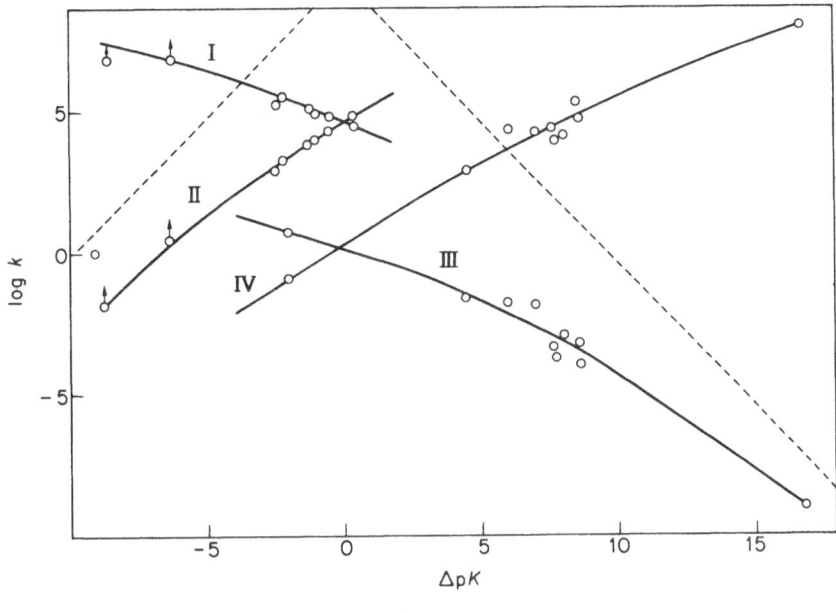

Figure 4.

Line I: proton transfer from aromatic polycyclic hydrocarbon acids to oxyanion bases in DMSO.
Line II: the reverse reactions to those for Line I.
Line III: proton transfer from aromatic polycyclic hydrocarbon acids to methoxide ion in methanol.
Line IV: the reverse reactions to those for Line III.
Broken Lines: theoretical predictions from Reference 5.
Data for Lines I, II, III and IV from Reference 40.

reactions from similar hydrocarbon acids to methoxide ion in methanol solution. The kinetic data for methanol solution were obtained by the tritium-exchange technique, whereas the reactions in DMSO were so much faster that the stopped-flow technique had to be employed. For three systems, indicated by arrows in Figure 4, the rates were so high that the rate constants could not be measured, and the points indicate minimum values. The rate constants for the back reactions were calculated using observed acidity constants in the appropriate solvents. The kinetics may be discussed in terms of the Eigen mechanism

$$AH + B^- \underset{k_{21}}{\overset{k_{12}}{\rightleftharpoons}} AH\cdots B^- \underset{k_{32}}{\overset{k_{23}}{\rightleftharpoons}} A^-\cdots HB \underset{k_{43}}{\overset{k_{34}}{\rightleftharpoons}} A^- + HB$$

where AH is the hydrocarbon and B^- the anionic oxygen base. The forward reaction is slower in methanol solution because B^- is strongly solvated, whereas AH does not form hydrogen bonds with the solvent. The formation of the encounter complex in which acid and base are adjacent, not separated by a solvent molecule, thus requires considerable activation energy, and k_{12} is low. In DMSO, on the other hand, B is not strongly solvated, and k_{12} can approach the diffusion-controlled limit. It is not likely that the slowness of the back reaction in methanol, as compared with DMSO, is due to reactant solvation. Values of hydrocarbon acidity constants in methanol and in DMSO show that hydrocarbon anions are more solvated by DMSO than by methanol [42]. The dispersion interactions of the delocalized anions with the polarisable DMSO are more important than the sum of the dispersion and hydrogen-bonding stabilization in methanol. In terms of the 'hard and soft acids and bases' principle [38], hydrocarbon anions are 'soft' bases, and so interact well with 'soft', i.e. polarisable, DMSO whereas methanol is a 'hard' acid and interacts well with 'hard' bases, such as OH^- or Cl^-. The contrast between the rates of hydrocarbon anion protonation must arise from the relative values of k_{32} for the two solvents. The amount of energy required to reorganize DMSO during the transition from $A^-\cdots HB$ to $AH\cdots B^-$ (and vice versa) is less than for methanol because of the importance of oriented hydrogen bonds in methanol. The methanol solvation structure must undergo a radical reorganization as protonation occurs, and this gives rise to a finite free energy of activation, which may have both enthalpy and entropy terms.

It is unfortunate that the data concerning the relative contributions of enthalpy and entropy terms to solvent reorganization around the transition state for proton-transfer is not only scanty but contradictory. The rate of proton-transfer from (+)-2-methyl-3-phenyl propionitrile to methoxide, as followed by the optical activity change [43], is accelerated by a factor of 10^9 in 90 per cent DMSO as compared with pure methanol solvent. This increase in rate is due entirely to a decrease in $\Delta H^{\ddagger}$ from 30.9 to 24.4 k cal mol^{-1}. The value of $\Delta S^{\ddagger}$ also decreases, from 10.6 to 6.0 cal K^{-1} mol^{-1}, so tending to reduce the rate in DMSO-rich solution. On the other hand, proton transfer from di(-4-nitrophenyl)methane to ethoxide is only twenty times faster in 5 per cent ethanol/95 per cent acetonitrile (a dipolar aprotic solvent similar to DMSO) than in 80 per cent ethanol/20 per cent acetonitrile [44]. This small change in $\Delta G^{\ddagger}$ is the net result of comparatively large changes in $\Delta H^{\ddagger}$ and $\Delta S^{\ddagger}$ acting against each other. The value of $\Delta H^{\ddagger}$ increases from 9.8 to 13.1 kcal mol^{-1} on going to the acetonitrile-rich solvent, and the observed increase in rate is due to the increase of $\Delta S^{\ddagger}$ from -22.6 cal K^{-1} mol^{-1} to -6.7 cal K^{-1} mol^{-1}. Data for mixed solvents are most difficult to interpret, as one of the solvent species may preferentially solvate various solute species to an unknown extent.

6.3.4 *The Marcus equation*

Marcus [45] has applied equation 11, originally developed by him for the interpretation of the rates of electron-transfer reactions, to proton-transfer reactions.

$$\Delta G^{\ddagger} = w_r + w_p + \lambda(1 + \Delta/\lambda)^2/4 \qquad (11)$$

where

$$1 > \Delta/\lambda > -1$$

Here $\Delta G^{\ddagger}$ is the free energy of activation, w_r is the free energy required to bring the reactants into an encounter complex, with reactants and solvation spheres correctly oriented for reaction, w_p is the corresponding term for the separation of the products, $\lambda/4$ is the free-energy barrier for proton transfer within the encounter complex if ΔG^0 is zero, and $\Delta = \Delta G^0 - w_r - w_p$.

If $\Delta > -1$, $\Delta G^{\ddagger} = w_r$. Qualitatively, w_r is related to the maximum rate for a reaction strongly favoured by ΔG^0, and λ is the parameter which governs the curvature of the Brønsted plot. The Marcus equation thus relates $\Delta G^{\ddagger}$ to ΔG^0, as does the Brønsted equation, but is more explicit about the various factors affecting the rate. The need for reactant and solvent reorganization is recognized by the inclusion of the term w_r. The equation, as originally applied to electron-transfer reactions, was derived for weak-overlap interactions, and its application to atom- and proton-transfer reactions, which involve strong-overlap interactions, requires justification by successful application.

Differentiation of equation 11 gives a value [46] for the Brønsted α:

$$\alpha = \left(\frac{\partial \Delta G^{\ddagger}}{\partial \Delta G^0}\right)_{\lambda, w_r, w_p} = \frac{1}{2}\left(1 + \frac{D}{\lambda}\right) \qquad (12)$$

For a series of similar acids, such as carboxylic acids, λ, w_r and w_p may be expected to be constant [47]. For a symmetrical reaction, Δ is zero, so that $\alpha = \frac{1}{2}$. For a reaction in which the products are favoured by the equilibrium constant, $\Delta < 0$ and so $\alpha < \frac{1}{2}$. If the reaction is diffusion-controlled, the treatment is no longer valid. For a reaction in which the reagents are favoured by the equilibrium constant, $\Delta > 0$ and so $\alpha > \frac{1}{2}$. This variation of α is in accordance with the Hammond postulate [34]. Differentiation of equation 12 gives

$$\frac{d\alpha}{d\Delta G^0} = \frac{1}{2\lambda} \qquad (13)$$

so that a value of λ can be calculated from the curvature of the Brønsted plot. Koeppel and Kresge construct an intersecting-parabola model for a simple hydrogen atom-transfer reaction, in which the point of intersection gives the activation energy [54]. If the two parabolas are of the same curvature, i.e., if AH and BH have the same force constant, and a constant horizontal displacement, then the expression for the activation energy reduces to

$$E_a = E_{a0}(1 + \Delta E/4\Delta E_{a0})^2$$

where E_{a0} is the activation energy if ΔE, the energy difference between the two minima, is zero. This is of the same form as equation 11, but the restrictions for this approximation to be valid are unlikely to be fulfilled. If force constants and interatomic distances are allowed to vary with ΔE in a realistic manner, a sigmoid dependence of α upon $\Delta E/\Delta E_{a0}$ is obtained, rather than the linear dependence predicted by equation 12. This can lead to a discrepancy of a factor of two in the value of λ calculated from observed values of α.

Much work on the experimental testing of the Marcus equation has been carried out on the hydrolysis of diazoacetate ion [46]. The rate-determining step

$$N_2CHCO_2^- + HA \underset{k_{-HA}}{\overset{k_{HA}}{\rightleftharpoons}} {}^+N_2 CH_2CO_2^- + A^-$$

is followed by a series of fast irreversible steps. For a wide range of acids HA, including phenols, carboxylic acids and alkylammonium ions, a plot of $\log k_{HA}$ versus $\log K_{HA}$, where K_{HA} is the acid dissociation constant of HA, is a curve [48], of slope approaching unity

172

for the weakest acids and 0.6 for the strongest acids. The standard free energy change for the reaction, ΔG_R^0, is the sum of w_r, Δ and w_p. Since ΔG_R^0 must vary with HA in the same way as ΔG_{HA}^0, the standard free energy of dissociation of HA, it follows that, if w_r and w_p are constant, then:

$$\Delta = \Delta G_{HA}^0 + C \qquad (14)$$

where C is a constant. The combination of equations 11 and 14 gives a relationship between the observable quantities $\Delta G^{\ddagger}$ and ΔG_{HA}^0, and the parameters $\lambda/4$, w_r and C. Kreevoy and Oh obtain values of the parameters to fit their results by a least-squares curve-fitting procedure. They also use this procedure to obtain the corresponding parameters for other slow proton-transfer reactions. The lines calculated for acetylacetone reactions, for example, exactly correspond to lines VII and VIII on Figure 1. A trio of adjustable parameters make for a great deal of latitude in curve-fitting, so it is probably not advisable to base arguments on the exact numerical values, but it is clear that the Marcus formulation can be used to account for the general features of a wide range of proton-transfer reactions.

The kinetic hydrogen isotope effect on the hydrolysis of diazo-acetate can also be successfully treated by equation 11. It is reasonable to suppose that the kinetic isotope effect is concentrated in which λ, w_r and Δ are little affected by isotopic substitution. From equation 11 we obtain:

$$\Delta G_{DA}^{\ddagger} - \Delta G_{HA}^{\ddagger} = \tfrac{1}{4}(\lambda_D - \lambda_H) + \tfrac{1}{4}\Delta^2(\lambda_D^{-1} - \lambda_H^{-1})$$

This equation describes very well the variation of $\log (k_{HA}/k_{DA})$ with $\log K_{HA}$. The only new adjustable parameter is λ_D, the others all being those calculated from the proton-transfer reactions.

The ionization of nitroalkanes presents some interesting data for interpretation. The rate of proton loss from a series of arylnitroalkanes to a base (OH$^-$, morpholine, or 2,4-lutidine) has been found to be increased by electron-withdrawing ring substitution much more than expected from the increase in acidity constant K_{AH} [49]. This has the effect of producing values of α greater than unity (1.54, 1.29 and 1.30 for OH$^-$, morpholine and 2,4-lutidine respectively). For a series of bases deprotonating any one arylnitroalkane, 'sensible' values of β are obtained, around 0.5. On the other hand, increasing methyl substitution increases both K_{AH} and the rate of proton loss to OH$^-$ in the nitroalkane series CH_3NO_2, $MeCH_2NO_2$, Me_2CHNO_2. This corresponds to a negative value for α (-0.8). Similar behaviour is found for the series $(CH_3CO)_2CH_2$, $CH_3COCH_2COCF_3$, $(CF_3CO)_2CH_2$, for which α is -0.15. Values of α outside the range from zero to unity are impossible according to the theories discussed in Section 3.2, and present a challenge to the Marcus theory. Kreevoy and Oh [48] find that the data for these reactions cannot be fitted to equations 11 and 14 with constant values of λ and w_r; this suggests that either λ or w_r (or both) vary within the series. Marcus [51] and Kresge [52] discuss variations in λ. Kresge considers the transition-state interaction between base and nitro group, and shows that a value of α greater than unity will be obtained if the substituent group on the nitro-alkane interacts more strongly with the base than with the nitro group, as occurs for arylnitroalkanes. For the nitroalkane series interaction between the base and substituent methyl groups destabilizes the transition state, thereby increasing the rate, whereas the methyl groups stabilize the carbanion, thus increasing K_{AH}, and α is thus negative. Albery [46] splits the w_r parameter into two, namely w_{ro}, the energy required for initial formation of the encounter complex, and w_s, the energy required to produce the correct solvation pattern for proton transfer to occur. Variations in w_s with ring substitution on arylnitroalkanes will cause α to be greater than unity. Albery classifies carbon acids as 'weak' or 'strong'. The strong acids, such as nitro- and diazo-compounds, are strong because they have ex-tensive π systems in the anion, which lead to large values of w_s, sensitive to substitution effects. The weak acids, such as ketones and cyano-compounds, have small values of w_s, little affected by substitu-tion. Fluoroacetylacetones seem to count as strong rather than weak.

Murdoch [53] derives an equation similar to the Marcus equation, equation 11, by the application of Hammond's postulate to the Eigen mechanism

$$AH + B^- \underset{k_{21}}{\overset{k_{12}}{\rightleftharpoons}} (AH\cdots B)^- \underset{k_{32}}{\overset{k_{23}}{\rightleftharpoons}} (A\cdots HB)^- \underset{k_{43}}{\overset{k_{34}}{\rightleftharpoons}} A^- + HB$$

Murdoch's equations enable the effect of finite activation energy for the second step on the plot of $\log k_f$ versus ΔpK to be predicted. The greater the activation energy the larger the transition region in which the slope changes from zero to unity, just as shown in Figure 1. Murdoch distinguishes between the observed value of the Brønsted exponent, α_{exp}, and the 'true' value, which gives the position of the transition state along the reaction co-ordinate. These are defined as follows:

$$\alpha_{exp} = d(\log k_f)/d(\log k_f/k_b)$$
$$\alpha = d(\log k_{23})/d(\log k_{23}/k_{32})$$

The relationship between α_{exp} and α thus depends on the relative rate constants of the various diffusive processes, i.e., $k_{12}, k_{21}, k_{34}, k_{43}$. For example, substitution of the numerical values of 14 kcal mol^{-1} for $\Delta G^{\ddagger}$, 10^3 dm^3 mol^{-1} s^{-1} for k_{12}, 10^{11} s^{-1} for k_{21}, 10^6 s^{-1} for k_{34} and 10^9 dm^3 mol^{-1} s^{-1} for k_{43} gives a value of α_{exp} of 0.38 at zero ΔpK, whereas α is 0.5. The calculated lines for these values are an exact fit for Ritchie's [40] data on proton-transfer reactions of hydrocarbons. No significance should be placed on the exact values of the parameters used to produce this fit, since they can be altered by an order of magnitude without drastically changing the features of the curve, but the effect of the diffusive steps on the value of α_{exp} is clearly shown. To quote Murdoch: 'Rate-equilibria relationships have a certain mesmerizing quality', and it is clear that they will continue to fascinate chemists for a long time to come.

REFERENCES

[1] A. Hantzsch, *Ber.*, **58**, 953 (1925).

[2] W. J. Albery, *Prog. Reaction Kinetics*, **4**, 355 (1967).

[3] A. Smoluchowski, *Z. Phys. Chem.*, **4**, 129 (1917).

[4] P. Debye, *Trans. Electrochem. Soc.*, **82**, 265 (1942).

[5] M. Eigen, *Angew. Chemie, Internat. Ed.*, **3**, 1 (1964).

[6] M. Eigen, W. Kruse, G. Maass, and L. De Maeyer, *Prog. Reaction Kinetics*, **2**, 285 (1964).

[7] G. C. Barker and D. C. Sammon, *Nature*, **213**, 65 (1967).

[8] C. L. Greenstock, P. C. Shragge, and J. W. Hunt, *J. Phys. Chem.*, **77**, 1624 (1973).

[9] M. Eigen and L. De Maeyer, *Z. Elektrochem.*, **59**, 986 (1955).

[10] G. Ertl and H. Gerischer, *Z. Elektrochem.*, **65**, 629 (1961).

[11] W. J. Albery and R. P. Bell, *Proc. Chem. Soc.*, 169 (1963).

[12] A. Weller, *Prog. Reaction Kinetics*, **1**, 189 (1961).

[13] K. Šolc and W. H. Stockmayer, *J. Chem. Phys.*, **54**, 2981 (1971).

[14] G. D. Burfoot, E. F. Caldin, and H. Goodman, *J. Chem. Soc., Faraday Transactions I*, **70**, 105 (1974).

[15] K. Šolc and W. H. Stockmayer, *Internat. J. Chem. Kinetics*, **5**, 733 (1973).

[16] K. S. Schmitz and J. M. Schurr, *J. Phys. Chem.*, **76**, 534 (1972).

[17] E. M. Eyring, *Adv. Chem. Phys.*, **21**, 237 (1971).

[18] M. C. Rose and J. Stuehr, *J. Amer. Chem. Soc.*, **90**, 7205 (1968).

[19] S. Singh, A. S. N. Murthy, and C. N. R. Rao, *Trans. Faraday Soc.*, **62**, 1056 (1966).

[20] M. M. Kreevoy and C. A. Mead, *Disc. Faraday Soc.* **39**, 166 (1965).

[21] M. L. Ahrens and G. Maass, *Angew. Chem. Internat. Edn.* **7**, 818 (1968).

[22] M. L. Ahrens, M. Eigen, W. Kruse, and G. Maass, *Ber. Bunsengesell. Phys. Chem.*, **74**, 380 (1970).

[23] M. Z. Hoffman and E. Hayon, *J. Amer. Chem. Soc.*, **94**, 7950 (1972).

[24] G. Gammons, B. H. Robinson, and M. J. Stern, *J. Chem. Soc. Chem. Comm.*, 1157 (1972).

[25] R. G. Pearson and R. L. Dillon, *J. Amer. Chem. Soc.*, **75**, 2439 (1953).

[26] R. P. Bell and B. G. Cox, *J. Chem. Soc.*, (*B*), 652 (1971).

[27] F. Hibbert, *J. Chem. Soc., Perkin II*, 1289 (1973).

[28] E. A. Walters and F. A. Long, *J. Amer. Chem. Soc.*, **91**, 3733 (1969); F. Hibbert, F. A. Long, and E. A. Walters, *J. Amer. Chem. Soc.*, **93**, 2829 (1971); F. Hibbert and F. A. Long, *J. Amer. Chem. Soc.*, **94**, 2647 (1972).

[29] A. J. Kresge and A. C. Lin, *J. Chem. Soc. Chem. Comm.*, 761 (1973).

[30] Z. Margolin and F. A. Long, *J. Amer. Chem. Soc.*, **95**, 2757 (1973).

[31] S. A. Chaudhri and K.-D. Asmus, *J. Phys. Chem.*, **76**, 26 (1972).

[32] R. P. Bell and O. M. Lidwell, *Proc. Roy. Soc.*, *A*, **176**, 114 (1940).

[33] J. N. Brønsted and K. J. Pedersen, *Z. Physikal. Chem.*, **108**, 185 (1924).

[34] G. S. Hammond, *J. Amer. Chem. Soc.*, **77**, 334 (1955).

[35] E. F. Caldin, *J. Chem. Soc.*, 3345 (1959).

[36] M. M. Kreevoy and R. A. Kretchmer, *J. Amer. Chem. Soc.*, **86**, 2435 (1964); E. Grunwald and E. Price, *J. Amer. Chem. Soc.*, **86**, 2965 (1964); R. P. Bell, *Discussions Faraday Soc.*, **39**, 16 (1965).

[37] A. J. Kresge, Y. Chiang, and Y. Sato, *J. Amer. Chem. Soc.*, **89**, 4418 (1967).

[38] A. J. Parker, *Chem. Rev.*, **69**, 1 (1969).

[39] C. D. Ritchie, *J. Amer. Chem. Soc.*, **91**, 6749 (1969).

[40] C. D. Ritchie and R. E. Uschold, *J. Amer. Chem. Soc.*, **90**, 3415 (1968).

[41] A. Streitwieser, A. P. Marchand, and A. H. Pudjaatmaka, *J. Amer. Chem. Soc.*, **89**, 693 (1967).

[42] C. D. Ritchie and R. E. Uschold, *J. Amer. Chem. Soc.*, **90**, 2821 (1968).

[43] D. J. Cram, B. Rickburn, C. A. Kingsbury, and B. Haberfield, *J. Amer. Chem. Soc.*, **83**, 3678 (1961).

[44] J. H. Kim and K. T. Leffek, *Can. J. Chem.*, **51**, 2805 (1973).

[45] R. A. Marcus, *J. Phys. Chem.*, **72**, 891 (1968); A. O. Cohen and R. A. Marcus, *J. Phys. Chem.*, **72**, 4249 (1968).

[46] W. J. Albery, A. N. Campbell-Crawford, and J. S. Curran, *J. Chem. Soc., Perkin II*, 2206 (1972).

[47] M. M. Kreevoy and D. E. Konasewich, *Adv. Chem. Phys.*, **21**, 243 (1971).

[48] M. M. Kreevoy and S. W. Oh, *J. Amer. Chem. Soc.*, **95**, 4805 (1973).

[49] F. G. Bordwell and W. J. Boyle, Jr., *J. Amer. Chem. Soc.*, **94**, 3907 (1972).

[50] J. R. Jones and S. P. Patel, *J. Amer. Chem. Soc.*, **96**, 574 (1974).

[51] R. A. Marcus, *J. Amer. Chem. Soc.*, **91**, 7224 (1969).

[52] A. J. Kresge, *J. Amer. Chem. Soc.*, **92**, 3210 (1970).
[53] J. R. Murdoch, *J. Amer. Chem. Soc.*, **94**, 4410 (1972).
[54] G. W. Koeppl and A. J. Kresge, *J. Chem. Soc.*, *Chem. Comm.*, 371 (1973).

7

A. Jerry Kresge

THE BRØNSTED RELATION: SIGNIFICANCE OF THE EXPONENT

7.1 The hypothesis

The Brønsted relation correlates the effectiveness of acids and bases as catalysts in proton transfer reactions with their acid or base strength. It is shown in equation 1 for the case of acid catalysis, where k_A

$$k_A = G_A(K_A)^\alpha \tag{1}$$

is the specific rate constant for catalysis by the acid whose acid dissociation constant is K_A, G_A is a proportionality constant, and α is the Brønsted exponent.

This relation was first proposed by Brønsted and Pedersen in 1924 [1]. In the years since, much evidence has accumulated in support of its essential validity, and it is now generally recognized that all reactions which show general acid or general base catalysis must conform to the Brønsted relation. The relation has seen widespread use, not only as a means of summarizing data and making predictions, but also as an indicator of reaction mechanism. Perhaps the most interesting aspect of the latter application is the idea that Brønsted exponents provide information about transition state structure, and in particular that the numerical value of an exponent is equal to the fractional extent of proton transfer at the reaction's transition state.

The hypothesis that Brønsted exponents measure transition state structure in this way may be traced back to the Leffler principle [2]. This principle assumes that some perturbation on a reacting system, say a substituent change in one of the reactants, will change the free energy of its transition state by an amount intermediate between the changes produced in the free energy of the reactants and the free energy

179

of the products. This proposition may be stated mathematically as in equation 2, where δ_R is an operator denoting the effect of the

$$\delta_R G^{\ddagger} = (1 - \alpha)\delta_R G_r^0 + \alpha\delta_R G_p^0 \tag{2}$$

substituent change on the quantity which it precedes, and $G^{\ddagger}$, G_r^0, and G_p^0 are the standard free energies of the transition state, the reactants, and the products, respectively. The parameter α, which is constrained to lie between the limits zero and one, measures the degree to which the transition state resembles the reactants or the products: when α is near zero, the substituent effect on $G^{\ddagger}$ is similar to that on G_r^0, which implies a structural similarity between the transition state and the reactants; and when α is near unity, the substituent effect on $G^{\ddagger}$ is similar to that on G_p^0, which implies a structural similarity between the transition state and the products. Rearrangement of equation 2 and substitution of $\Delta G^{\ddagger}$ for $(G^{\ddagger} - G_r^0)$ and ΔG^0 for $(G_p^0 - G_r^0)$ gives equation 3. This provides a useful

$$\alpha = \delta_R \Delta G^{\ddagger}/\delta_R \Delta G^0 \tag{3}$$

definition of α as a ratio of substituent effects, namely, the substituent effect on the free energy of activation divided by the substituent effect upon the standard free energy change of the reaction.

The Leffler principle relates the free energy of activation of a reaction to the standard free energy change of the same process, and it therefore deals with the rate and equilibrium constants of a single reaction. The Brønsted relation as normally formulated, on the other hand, correlates rate constants of a catalysed reaction with equilibrium constants of a different process, the acid ionization reaction of the catalyst. The two relationships are nevertheless completely equivalent, for variation of the acid has exactly the same effect on the standard free energy change of the catalysed reaction as on the standard free energy change of the acid ionization. This may be seen by considering the consequences of the change $HA_1 \rightarrow HA_2$ on the pairs of reactions in equations 4 and 5. In each case, the variation

$$\left.\begin{array}{l} HA_1 + S \rightarrow HS^+ + A_1^- \\ HA_2 + S \rightarrow HS^+ + A_2^- \end{array}\right\} \tag{4}$$

$$\left.\begin{array}{l} HA_1 \rightarrow H^+ + A_1^- \\ HA_2 \rightarrow H^+ + A_2^- \end{array}\right\} \tag{5}$$

is limited to the same two species, HA and A^-, and the standard free energy change of each process must therefore be altered by the same amount. Taking the variation $HA_1 \rightarrow HA_2$ to be a substituent change then converts equation 1 into equation 3, and the exponent of the Brønsted relation may be seen to be entirely equivalent to the parameter α in the Leffler principle.

Brønsted exponents may therefore be regarded as ratios of substituent effects: $\alpha = \delta_R \Delta G^{\ddagger}/\delta_R \Delta G^0$. The magnitude of these substituent effects is determined by the structural changes which occur as the reaction proceeds. If these structural changes parallel the changes in bonding to the proton being transferred, then Brønsted exponents will measure the degree of proton transfer at the reaction's transition state. We shall now examine some evidence to determine whether or not this is actually so.

7.2 Qualitative evidence

Some evidence of a qualitative nature which bears on the hypothesis that Brønsted exponents measure the extent of proton transfer at the transition state of proton transfer reactions comes from application of the Hammond postulate [3]. This postulate states that transition states for exothermic reactions will tend to be reactant-like and those for endothermic processes, product-like. Proton transfer has, of course, not yet taken place in the reactants of a proton transfer reaction, and it is complete in the products of such a process. It follows, then, that the degree of proton transfer should be low, with α generally less than 0.5, in the reactant-like transition states of exothermic reactions; and that it should be high, with α generally greater than 0.5, in the product-like transition states of endothermic reactions. A graphic illustration of such behavior is provided by proton transfer between oxygen and nitrogen donors and acceptors, i.e. between species which Eigen has called 'normal' acids and bases [4]. Here Brønsted exponents are zero for proton transfers which are exothermic by more than a few kcal mol^{-1} and are unity for transfers endothermic by more than a similar small amount; in the narrow space between these two regions, the exponent changes smoothly from one to the other of its limiting values and is approximately 0.5 at $\Delta G^0 = 0$.

Similar systematic changes in Brønsted exponent have also been found in proton transfer reactions to and from carbon, but here the variation is usually more gradual than the sharp changes which occur in systems consisting only of normal acids and bases [5]. Variation of this kind is in itself evidence that Brønsted exponents measure the extent of proton transfer, for the structure of transition states, and therefore the degree of proton transfer at the transition state, cannot be expected to remain constant if ΔG^0 is varied over a sufficiently large range. This follows from the Hammond postulate [3, 6], and it is also required by Marcus rate theory [7] as well as by certain other theoretical models of the proton transfer process, such as those based upon two intersecting parabolic potential energy functions [8], or the BEBO method of estimating reaction barriers [9], or a modified Sato potential energy surface [10].

A particularly well documented example of systematic variation in Brønsted exponent for proton transfer involving carbon, and the curved Brønsted relation which results, is provided by the reaction of carbonyl compounds with bases (equation 6). The data, gathered over many years in a number of laboratories, have recently been

$$\underset{\substack{|}}{\overset{O}{\underset{|}{\overset{\|}{C}}}}-\underset{|}{\overset{|}{C}}-H + B \rightarrow \underset{}{\overset{O}{\overset{\|}{C}}}=\bar{C}\diagup \quad + BH^+ \tag{6}$$

summarized by R. P. Bell in the form of a graph [11], which is reproduced here as Figure 1. The Brønsted exponent for this system varies from 0.4 at $\log K = 5$ $(\Delta G^0 = -7 \, \text{kcal mol}^{-1})$ to 0.9 at $\log K = -20$ $(\Delta G^0 = 27 \, \text{kcal mol}^{-1})$ and is 0.5 at $\Delta G^0 = 0$.

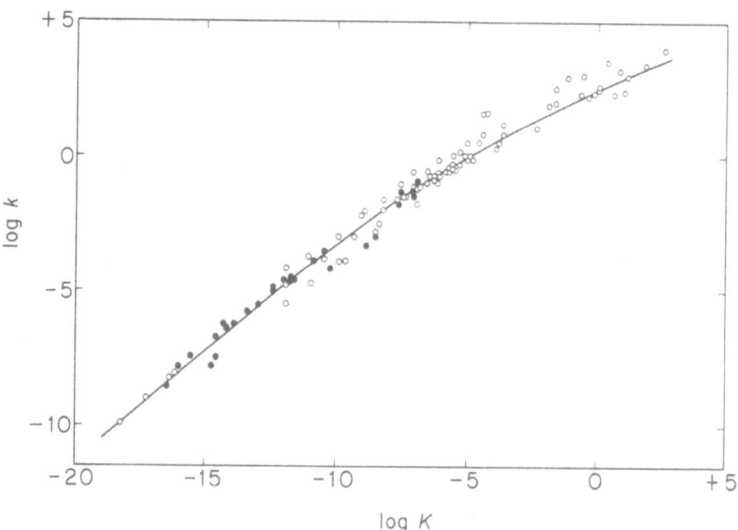

Figure 1. Curved Brønsted plot for the reaction of carbonyl compounds with bases. [Reproduced by permission from *The Proton in Chemistry* (second edition), Cornell University Press, Ithaca, New York (1973), p. 203.]

Additional qualitative support for the hypothesis that Brønsted exponents measure transition state structure comes from kinetic isotope effects. It is well known that the magnitude of primary hydrogen isotope effects changes systematically with transition state structure. This behaviour was originally thought to be a force-constant effect, determined by the relative strengths of the partial bonds to the proton being transferred [12], but there is now evidence that it may be the result of proton tunnelling [13]. Be that as it may, both explanations require the isotope effect to pass through a maximum value when the proton is half-transferred and to fall off as the degree of transfer becomes significantly less or significantly greater than 0.5.

A good example of behaviour of this kind is provided by the ionization of carbonyl compounds (equation 6), to which additional data for proton transfer from nitroalkanes to various bases may be added [13, 14]. The isotope effect on these reactions rises to a maximum of $k_H/k_D \simeq 10$ at $\Delta G^0 = 0$, just where the Brønsted exponent is one-half, and it falls off to considerably lower values on either side of this point. The endothermic side of $\Delta G^0 = 0$ is particularly well documented: here k_H/k_D drops to about 3 when $\Delta G^0 \simeq 25$ kcal mole^{-1} and the Brønsted exponent becomes $ca.$ 0.9. Aromatic hydrogen exchange shows a similar correspondence between isotope effect [15] and Brønsted exponent [16], and additional examples may be found in the hydrolysis of vinyl ethers [17] and diazocompounds [18], as well as in the diazo-coupling reaction [19].

Several different lines of evidence thus suggest that Brønsted exponents do measure transition state structure in at least a qualitative way. However, the hypothesis being examined goes beyond this: it proposes that α is numerically equal to the degree of proton transfer at the transition state. To test this aspect of the idea, quantitative evidence must be examined.

7.3 *Quantitative evidence*

It is necessary, before proceeding to an investigation of the idea that Brønsted exponents measure transition state structure in a quantitative way, to formulate a fairly precise definition of 'degree of proton transfer.' It would seem logical to use bond order for this purpose, and to set the degree of proton transfer equal to the order of the bond being formed between the proton in motion and the proton acceptor.

It is difficult, unfortunately, to determine the order of the forming bond, or in fact to obtain any other detailed information about the structure of transition states: these species cannot be observed directly, and their properties must therefore be inferred by indirect means. Secondary isotope effects provide one source of such indirect information. It is known that secondary isotope effects are produced by changes in the force constants of the isotopically substituted bonds [20], and the logarithms of these isotope effects are, to a good degree of approximation, linear functions of the force constant changes [20, 21]. Since force constants are themselves approximately linear functions of bond order [22], logarithms of secondary isotope effects then become linear functions of bond order changes. It remains only for the bond order changes in the isotopically substituted 'non-reacting' bonds, which cause the secondary isotope effects, to parallel the bond order changes in the reacting bonds in order that the logarithm of a secondary isotope effect on a proton transfer reaction might measure the order of the forming bond, and therefore the extent of proton transfer, at the reaction's transition state. In situations where the limiting value of the secondary isotope effect is known, perhaps from the equilibrium measurements, an isotopic exponent, α_i, may then be defined as the ratio $[\log(k_H/k_D)]/[\log(k_H/k_D)_{lim}]$. This, by hypothesis, will be numerically equal to the degree of proton transfer.

The secondary isotope effect on proton transfer from the hydronium ion, $(k_{H_3O^+}/k_{D_3O^+})_{II}$, is the most frequent source of isotopic exponents of this kind. As shown in equation 7, removal of a proton from H_3O^+ converts the two non-reacting bonds of this species into $O-H$ bonds of water; the limiting value of this isotope effect may therefore be expressed in terms of l, the known D-H fractionation factor of the hydronium ion $[l = (D/H)_{hydronium\ ion}/(D/H)_{water}]$, as l^{+2}. This leads to the relationship $\alpha_i = [\log(k_{H_3O^+}/k_{D_3O^+})_{II}]/[\log(l^2)]$. This expression

$$H_3O^+ + S \rightarrow (H_2\overset{\delta+}{O}\cdots H\cdots \overset{\delta+}{S})^{\ddagger} \rightarrow H_2O + HS^+ \qquad (7)$$

may be used in conjunction with directly measured values of this isotope effect, obtained either by comparing rate and product isotope effects [23], or, in the case of isotope exchange reactions, by comparing rates of tritium uptake from H_2O and D_2O solutions [24]. Values of α_i are more commonly determined, however, by the analysis of rate measurements made in H_2O-D_2O mixtures [25].

Some results for systems in which Brønsted exponents are also available are listed in Table 1. In a few cases, duplicate values of α_i have been determined by different methods; these differ by 0.03 to 0.14, which is probably no greater than the combined uncertainties of the individual measurements.

The correspondence between α_i and the Brønsted exponents α_B, may be seen to be considerably better for the first seven entries of this table than for the last four. The latter, however, all refer to examples of diazo-compound hydrolysis, which is a reaction known to give

Table 1

Comparison of Brønsted, α_B, and isotopic, α_i, exponents

Reaction	α_i	α_B	Reference
Aromatic hydrogen exchange in 1,3,5-trimethoxybenzene	0.70[a] 0.67[b]	0.56[d] 0.59[e]	[16], [26], [27], [28]
Aromatic hydrogen exchange in 1,3-dimethoxybenzene	0.70[a]	0.62[e]	[16], [26]
Hydrolysis of ethyl vinyl ether	0.56[c], 0.65[b]	0.70[e]	[17b], [29] [30]
Hydrolysis of 2-dichloromethylene-1,3-dioxolane	0.42[b]	0.49[e]	[31]
Hydrolysis of cyanoketene dimethyl acetal	0.5–0.65[b]	0.62[e]	[32]
Acid cleavage of allylmercuric iodide	0.65[c], 0.79[b]	0.67[e]	[33, 34]
Acid cleavage of isobutenylmercuric bromide	0.69[b] 0.83[c]	0.69[e]	[35]
Hydrolysis of diazoacetate ion	0.30[c]	0.51[e]	[18a], [36]
Hydrolysis of 3-diazo-2-butanone	0.27[b]	0.61[e]	[18b], [37]
Hydrolysis of ethyl 2-diazopropionate	0.29[b]	0·59[e]	[18b], [37]
Hydrolysis of p-nitrophenyldiazomethane	0.49[b]	0.69[e]	[38]

[a] By comparison of rates of tritium uptake in H_2O and D_2O solution.
[b] From rate measurements in H_2O–D_2O mixtures.
[c] By comparison of rate and product isotope effects.
[d] By differentiation of $\Delta G^{\ddagger}$–ΔG^0 plot.
[e] From carboxylic acid catalytic coefficients.

rather sharply curved Brønsted plots [18d]. In these cases, therefore, the extent of proton transfer from a strong acid, such as the hydronium ion used to determine α_i, should be less than the degree of proton transfer from the weak carboxylic acids upon which α_B is based. It is significant in this respect that in all four cases the disagreement between the two kinds of exponent is indeed in the direction $\alpha_i < \alpha_B$.

For the first seven entries in Table 1, the difference between α_i and α_B is never greater than 0.14 and its average value is only 0.08. This agreement, however, is somewhat deceptive, for the range covered by each set of exponents is rather small. Statistical analysis of the data gives a least-squares relationship with rather large uncertainties in the slope and intercept parameters, $\alpha_i = (0.20 \pm 0.26) + (0.73 \pm 0.42)\alpha_B$, and a correlation coefficient, $r = 0.466$, which indicates that only 36 per cent ($r^2 \times 100$) of the variation in α_i may be attributed to a linear dependence upon α_B. There is, in addition, some uncertainty that α_i based upon this isotope effect really measures the degree of proton transfer accurately: model calculations of isotope effects on proton transfer from the hydronium ion [39] show that, although α_i does increase regularly with the order of the forming bond, it tends to lag behind the bond order, sometimes by as much as 0.2.

Better agreement between α_B and α_i has been obtained by using exponents based upon other secondary isotope effects. The reaction investigated was the hydrolysis of 3-diazo-2-butanone catalysed by formic and acetic acids [40]. Kinetic isotope effects were measured with the catalyst pairs $HCO_2H\text{-}DCO_2H$ and $CH_3CO_2H\text{-}CD_3CO_2H$ and the isotope effects on the equilibrium ionization of these acids were used as limiting values of the kinetic effects. The results obtained, $\alpha_i = 0.64 \pm 0.07$ for formic acid and $\alpha_i = 0.64 \pm 0.06$ for acetic acid, are in excellent agreement with the Brønsted exponent $\alpha_B = 0.61 \pm 0.03$, also based upon formic and acetic acids.

Another system examined in much the same way involves proton transfer from 2-nitropropane to hydroxide and acetate ions [41]. In this case, however, isotopic substitution was made in the substrate rather than in the catalysts: rate and equilibrium isotope effects for the pair $(CH_3)_2CHNO_2$ and $(CD_3)_2CHNO_2$ were compared. The results here, $\alpha_i = 0.43 \pm 0.02$ for hydroxide ion and $\alpha_i = 0.38 \pm 0.03$ for acetate ion, are each significantly different from $\alpha_B = 0.55 \pm 0.02$, based upon acetate and chloroacetate ion catalysis [42]; moreover, the values of α_i for a strong catalyst (hydroxide ion) and a weak one (acetate ion) stand in the wrong order. These discrepancies may be understood on the reasonable assumption that two factors contribute to these isotope effects and that one of the factors does not change monotonically along the reaction coordinate. As will be seen below, a similar explanation can be applied to some unusual Brønsted exponents given by this same reaction under certain conditions.

Quantitative estimates of the degree of proton transfer at the transition state of proton transfer reactions have also been made using medium effects on rates and equilibria [16], [43]. The reaction to which this method was applied is carbon protonation of weak aromatic bases, such as trimethoxybenzene (equation 8). The study was conducted

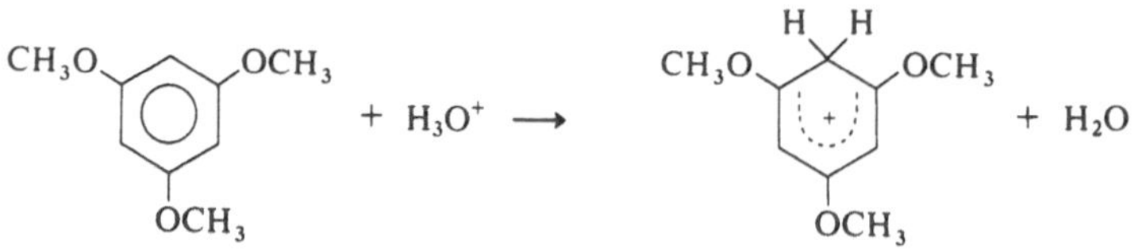

$$CH_3O\text{-}C_6H_3(OCH_3)_2 + H_3O^+ \longrightarrow [CH_3O\text{-}C_6H_4(OCH_3)_2]^+ + H_2O \qquad (8)$$

in concentrated aqueous perchloric acid, in which both rates and equilibria could be measured, and in which changes in the concentration of the acid could also supply the medium effect. Although medium effects in concentrated acids such as this are not yet completely understood, it is likely that they are governed largely by the charges on the species involved. Insofar as this is correct, the ratio $\delta_M \Delta G^{\ddagger} / \delta_M \Delta G^0$, where $\delta_M \Delta G^{\ddagger}$ is the medium effect on the free energy of activation for the protonation of a given aromatic substrate and $\delta_M \Delta G^0$ is the medium effect on the standard free energy of reaction, will measure the amount of positive charge transferred to the substrate at the transition state for proton transfer. Since the transfer of positive charge is here accomplished by proton transfer, this ratio will also measure the extent of proton transfer at the transition state.

Such a ratio has been called an acidity exponent, α_A. Values for a number of aromatic substrates, together with the corresponding Brønsted exponents, are listed in Table 2. The Brønsted exponents

Table 2

Comparison of Brønsted, α_B, and acidity, α_A, exponents [16]

Kinetic substrate	α_A	α_B[a]
Azulene	0.55	0.47
1,3,5-Trimethoxybenzene	0.55	0.56
1,3,5-Triethoxybenzene	0.54	0.55
1,3-Dihydroxy-2-methylbenzene	0.54	0.59
1,3-Dimethoxybenzene	0.64	0.62
1,3-Dimethoxy-2-methylbenzene	0.65	0.62
Anisole	0.71	0.72
Benzene	0.93	0.87

[a] By differentiation of $\Delta G^{\ddagger}$-ΔG^0 plot.

were obtained by differentiating a Brønsted plot constructed with free energies of activation and standard free energies of reaction for the protonation reactions themselves, and they therefore, just like the acidity exponents, refer to the hydronium ion as the donor acid.

The agreement between these two kinds of exponent may be seen to be good. Statistical analysis gives a least-squares relationship, $\alpha_A = (-0.01 \pm 0.09) + (1.04 \pm 0.14)\alpha_B$, whose intercept and slope parameters are close to the zero and unit values expected for a one-to-one correspondence between the two variables, and whose correlation coefficient, $r = 0.949$, indicates that 90 per cent of the variation in α_A may be attributed to a linear dependence upon α_B.

There is thus a strong indication that α_A and α_B measure the same thing. Whether or not this is the degree of proton transfer at the transition state depends, of course, upon the validity of the argument made above concerning medium effects. Since there seems to be no reason to doubt its essential correctness, it would appear that Brønsted exponents do measure transition state structure in more than a qualitative way.

7.4 Anomalous exponents

Quite recently an apparently severe blow was dealt to the idea that Brønsted exponents measure transition state structure in even a qualitative way, when systems were discovered for which the exponents are negative and others in which they are greater than unity [44]. These results are clearly at odds with the hypothesis that α is equal to the degree of proton transfer at the transition state, inasmuch as negative transfer and more than complete transfer can have no real meaning. However, the factors responsible for this anomalous behaviour are now well understood, and the circumstances under which they are likely to play an important role can be fairly well predicted.

The systems which produce these anomalous Brønsted exponents all involve proton transfer from nitro compound pseudo-acids; a particularly simple example is given by the reaction of compounds of the series CH_3NO_2, $CH_3CH_2NO_2$, $(CH_3)_2CHNO_2$ with the hydroxide ion (equation 9). The effect of the progressive methyl-

$$R_2CHNO_2 + HO^- \rightarrow R_2C{=}NO_2^- + H_2O \qquad R = H \text{ or } CH_3$$

$$(9)$$

substitution here is to lower the specific rate of reaction but raise the equilibrium constant of the process; this gives a negative Brønsted exponent, $\alpha = -0.5 \pm 0.1$ [45]. Moreover, since the Brønsted exponents for the forward and reverse directions of a single reaction must add up to unity: $\alpha + \beta = 1$, the exponent for reprotonation of the nitronate ions becomes greater than unity: $\beta = 1.5 \pm 0.1$.

The rate-retarding effect of successive methyl substitution on the forward reaction along this series may be recognized as the normal polar (inductive) interaction of methyl in a situation where negative charge is being generated. The methyl group effect on the equilibrium, on the other hand, has for some time been attributed to hyperconjugative stabilization of the $C{-}N$ double bond in the nitronate anions [46]. These two interactions are of opposite sign, but that in itself is not enough to produce the inversion of overall substituent effect observed in going from the kinetics to the equilibria; the two interactions must also develop at different rates as the system moves along the reaction coordinate. There is good reason to believe, however, that hyperconjugative stabilization will lag behind the polar effect. Hyperconjugation requires the presence of unsaturation, but the π-system which provides this in the final state is not yet fully developed at the transition state: the p-orbital of the α-carbon atom, which is an integral part of this π-system, is still being used to form a partial bond to the departing proton. This incomplete conjugation will also impair delocalization of negative charge onto the nitro group. Charge will therefore build up on the α-carbon atom, and thus the polar effect will be enhanced at the expense of hyperconjugation. It is in fact possible to reproduce the experimental Brønsted exponent [45], $\alpha = -0.5$, by making quantitative estimates of the above effects and of one other, an intermolecular interaction to be discussed below. The agreement may, moreover, be extended to other anomalous systems with $\alpha > 1$ [44], in which hyperconjugation cannot be a factor, simply by leaving out the hyperconjugative effect.

An important feature of this analysis is the presence of several interactions between the substituent and the rest of the system which develop at different rates. If the substituent interacts with the system in just one way, or if several interactions operate but they all develop at the same rate, then the total substituent effect on the rate, $\delta_R \Delta G^{\ddagger}$, cannot be greater than, or of different sign from, that on the equilibrium, $\delta_R \Delta G^{\circ}$, and α must lie between the limits zero and one. If, on the other hand, several interactions of different sign are operative, and if these develop at different rates, then it is a simple matter to have α less than zero or greater than unity.

The position in the system at which the substituent is introduced has an important bearing on whether or not some interactions will lag behind or lead others. It was seen above that negative charge builds up on the α-carbon atom in the transition state of a nitro-alkane ionization reaction because the π-system through which this charge will ultimately be delocalized is not yet adequately developed; a substituent on the α-carbon atom thus produces an inordinately large transition-state interaction. Substitution in the base, on the other hand, will have a different effect, for the π-system there will either already be present, as in carboxylate or phenolate ions, or else the charge will remain undelocalized, as is the case with amines. It is significant in this respect that Brønsted relations, for series of nitroalkane ionization reactions in which the nitro compound is kept the same but the base is varied, all have exponents which are non-anomalous [47].

These considerations could be generalized by saying that variation of the pseudo-acid (or pseudo-base) in a proton transfer reaction between such a species and a normal oxygen or nitrogen base (or acid) might lead to anomalous behaviour, but that change of the other reaction partner is unlikely to do so. The same conclusion has also been reached on the basis of entirely different reasoning [48], and there is some additional experimental evidence to support it. In the carbonyl-compound ionizations shown in equations 10,

$$A: X-C_6H_4 \cdot CH_2 \cdot CH(CO \cdot CH_3)_2 + RCO_2^-$$

$$\rightarrow X-C_6H_4 \cdot CH_2 \cdot C(CO \cdot CH_3)_2^- + RCO_2H$$

$$B: X-C_6H_4 \cdot CH_2 \cdot CH(CO \cdot CH_3)(CO_2Et) + RCO_2^-$$

$$\rightarrow X-C_6H_4 \cdot CH_2 \cdot C(CO \cdot CH_3)(CO_2Et)^- + RCO_2H$$

$$(10)$$

variation of the base by change of R gives a Brønsted exponent of 0.45 in both Series A and Series B; variation of the carbonyl compound by change of X, on the other hand, gives 0.58 for Series A and 0.70 for Series B [49]. The difference produced by shifting the site of substitution into the pseudo-acid is in each case in the direction expected if there is lagging of negative-charge delocalization; in neither case, however, is the abnormality great enough to produce an anomalous exponent. Similar effects could be operating in some of the reactions that show scatter from the correlation of Figure 1, for the variation there includes changes in the carbonyl compound as well as in the base accepting the proton. Negative Brønsted exponents have also been reported recently for the ionization of a series of fluorine-substituted β-diketones, but it is not clear just how much of the effect there is due to abnormal behaviour and how much is caused by complications associated with extensive hydration of the ketone groups [50].

Variation of the pseudo-base in the aromatic protonation reactions of Table 2, on the other hand, seems not to produce abnormal behaviour: none of the Brønsted exponents in this system is anomalous, and the set as a whole shows good agreement with degrees of proton transfer estimated in another way. Of course, the substituent changes here were made at some distance away from the reaction centre, and any effects due to unusual transition state charge distributions are therefore likely to be attenuated. It is possible, on the other hand, that even if the substituents had been introduced at the carbon atom undergoing protonation, little or no abnormal behaviour would have resulted, because this reaction differs in charge type from nitroalkane and carbonyl compound ionization; this difference has an important bearing on the presence or absence of unusual effects.

It is convenient, for the purpose of seeing how this difference in charge type operates, to view the aromatic protonation reaction in the reverse direction. The process then becomes, just like the ionization of a nitroalkane or of a carbonyl compound, a proton transfer from saturated carbon situated next to some group Z into which the electron pair left behind can delocalize (equation 11). In the reaction

$$B + H-\overset{|}{\underset{|}{C}}-Z \rightarrow BH^+ + \overset{\diagdown}{\underset{\diagup}{C}}=Z^- \tag{11}$$

of a nitroalkane or a carbonyl compound, Z is initially neutral, and localization of electron density on the α-carbon atom in the transition state occurs at the expense of giving Z some negative charge. This is illustrated in 12, where a transition state for this kind of system, arbitrarily chosen as one in which the proton is half-transferred and shown without the base for the sake of simplicity, is represented as a resonance hybrid of charge-localized, **1**, and charge-delocalized, **2**,

$$\left\{ \text{H}\cdots\overset{\tfrac{1}{2}-}{\underset{\underset{X}{|}}{C}}-Z \leftrightarrow \text{H}\cdots\overset{\tfrac{1}{2}-}{\underset{\underset{X}{|}}{C}}\!=\!Z \right\}^{\ddagger} \tag{12}$$

$$\qquad\quad \mathbf{1} \qquad\qquad \mathbf{2}$$

structures. Structure **1** makes a relatively greater contribution to the resonance hybrid here than does its counterpart in the case of a fully-formed nitronate ion, and the substituent thus sees an inordinately large negative charge on the carbon atom to which it is attached. This gives an abnormally large transition state interaction, and that leads to an abnormal Brønsted exponent.

In a system such as aromatic deprotonation, on the other hand, Z is initially positive. That makes the charge-localized structure of the transition state resonance hybrid, **3** in 13, a charge-separated form, which raises its energy and lowers the contribution it can make to

$$\left\{ \text{H}\cdots\overset{\tfrac{1}{2}-}{\underset{\underset{X}{|}}{C}}-\overset{+}{Z} \leftrightarrow \text{H}\cdots\overset{\tfrac{1}{2}-}{\underset{\underset{X}{|}}{C}}\!=\!Z \right\}^{\ddagger} \tag{13}$$

$$\qquad\quad \mathbf{3} \qquad\qquad \mathbf{4}$$

the hybrid. There is thus a stronger driving force for delocalization in a transition state of this type, and a negative charge is therefore less likely to build up on the α-carbon atom. Such build up of negative charge, moreover, requires additional positive charge to accumulate on Z, and that serves to strengthen another interaction which already functions to offset any interaction involving negative charge.

Some experimental support for these ideas comes from a study of the effect of chlorine substitution on the addition of ethanol to the double bond of ethyl vinyl ether [51]. This reaction occurs through rate-determining proton transfer to the β-carbon atom of the ether, equation 14, and successive chlorine substitution at this position

$$CX_2{=}CHOEt + H^+ \underset{slow}{\rightleftharpoons} CHX_2{\cdot}CHOEt^+ \overset{EtOH}{\underset{fast}{\rightarrow}} CHX_2{\cdot}CH(OEt)_2$$

$$(14)$$

X = H or Cl

retards the process considerably. The effect on $\Delta G^{\ddagger}$ is 5 kcal mole^{-1} per chlorine atom, which is similar in magnitude to the influence of β-chlorine substitution on the full positive charge generated in the limiting solvolysis of tertiary carbinyl halides [52]. The similarity of these two effects argues against the occurrence of any significant build-up of negative charge on the chlorine-substituted carbon atom which is being protonated in reaction 14. The contribution of charge-localized resonance structure (similar to 3 in 13) to the structure of this transition state must therefore be minimal. This conclusion is reinforced by the fact that protonation of the double bond in these chlorine-substituted vinyl ethers is strongly endothermic; deprotonation must then be strongly exothermic, which gives it an early transition state. This is just the situation in which the double bond is least likely to be well-developed at the transition state, and where a build-up of negative charge would therefore have the strongest tendency to occur.

There is still another effect which might contribute to anomalous behaviour, and this is the intermolecular interaction of proton donor and proton acceptor which can occur as these two come together to form a transition state [53]. Steric hindrance is a common manifestation of this effect [54]. It is well-known, for example, that, although alkyl substitution at the 2- and 6-positions of pyridines raises the basicity of these substances, it often lowers their ability to function as proton acceptors [55]. Hydrophobic interactions can also be significant, and the fact that pivalic acid is sometimes a better acid catalyst than its pK_a would suggest has been attributed to this effect [56]. Polar interactions are another source of the intermolecular effect. In the acid-catalysed hydrolysis of ethyl vinyl ether, for example, the rate constants for positively charged amino acids fall about a factor of two below a Brønsted relation based upon neutral carboxylic acids [57], and much larger deviations due to differences in charge type can be found in the base-catalyzed decomposition of nitramide [57, 58].

Intermolecular interactions of this kind will lead to anomalous behaviour only when systems in which they are present are combined with others in which they are not, i.e. when charged and uncharged acids are used together in a single reaction series or when sterically hindered bases are combined with unhindered species. Of course, this is not usually done, and special care is in fact generally exercised to assure a maximum amount of structural consistency among the substances used in a Brønsted correlation. In this circumstance, intermolecular interactions contribute the same amount to each value of $\Delta G^{\ddagger}$ along the series, and these unwanted effects then cancel out in the differences $\delta_R \Delta G^{\ddagger}$, and cannot enter into the ratio $\delta_R \Delta G^{\ddagger}/\delta_R \Delta G^0$ $(= \alpha)$. An example of this may be seen in the hydrolysis of ethyl vinyl ether, where, as mentioned above, charged amino acids deviate from a Brønsted relation based upon neutral carboxylic acids; the charged acids nevertheless provide a correlation of their own, parallel to that for the neutral catalysts, and the exponents of the two relations show no significant difference from one another: $\alpha = 0.65 \pm 0.04$ and 0.70 ± 0.03 [57].

There is some evidence that dipolar interactions along a reaction series which is homogeneous in charge type may also differ enough to produce deviations from a Brønsted plot [59], but the effects here are small and are not likely in general to have a substantial influence on the magnitude of Brønsted exponents. An interesting exception to this conclusion is provided by the reaction of nitroalkanes (CH_3NO_2, $CH_3CH_2NO_2$, $(CH_3)_2CHNO_2$) with hydroxide ion. In this system, the intermolecular interaction of the charge remaining on the hydroxide ion and the dipole of the α-methyl group has been estimated to be about 0.6 kcal mole^{-1}. This interaction is about half as large as that between the residual hydroxide charge and the negative charge which builds up on the α-carbon atom, and it is sufficiently great to make a significant contribution to the anomalous exponent, $\alpha = -0.5$. But this is an unusual situation, inasmuch as the substitution occurs directly at the reaction site. More remote substitutions will be much more common, and these will certainly produce relatively less significant effects.

In summary then, Brønsted exponents that do not reflect the extent of proton transfer in the transition state may result when the reaction partner which is varied is a neutral pseudo-acid, especially if the variation is accomplished by making changes directly at the reaction center. Variation of a neutral pseudo-base, on the other hand, is less likely to give an abnormal exponent, and judicious variation of a normal acid or base in a reaction which also involves a pseudo-species will almost certainly give a Brønsted exponent that is a reliable index of transition state structure.

7.5 A mechanistic application

Knowledge of the degree of proton transfer at the transition state of a proton transfer reaction has many useful applications. An especially interesting one, due to Gravitz and Jencks, uses only changes in the Brønsted exponent [60], and is therefore free of any uncertainty produced by lingering doubts that Brønsted exponents do in fact measure transition state structure in a quantitative way. This work deals with the mechanism of formation of the N,O-trimethylene-phthalimidium cation, 5, from the corresponding N-acyl ester aminals (equation 15).

$$\text{(15)}$$

This reaction is acid-catalysed, and it therefore involves proton transfer in addition to C—O bond-breaking. In this respect, it resembles the slow stage in the hydrolysis of ortho esters (equation 16), and, as in ortho-ester hydrolysis, two mechanistic possibilities exist: (a) proton transfer and C—O bond breaking could take place in separate, consecutive steps, 17, or (b) they could occur together in a single concerted process, 18 [61]. Resolution of this mechanistic problem in the case of ortho esters has received much recent attention, for it bears upon, among other things, the mechanism of lysozyme action, and Gravitz and Jencks' work on the formation of a phthalimidium ion therefore takes on added significance.

$$\text{(16)}$$

194

$$\text{HA} + \underset{\underset{R}{\overset{|}{O}}}{\overset{}{\underset{}{C}}}\overset{\overset{|}{N-}}{\underset{O-}{}} \rightarrow A^- + \underset{\underset{R}{\overset{|}{H-O}}}{\overset{}{\underset{}{C}}}\overset{\overset{|}{N-}}{\underset{O-}{}} \rightarrow A^- + \text{HOR} + -C^{\overset{+}{}}\overset{\overset{|}{N-}}{\underset{O-}{}} \quad (17)$$

$$\text{HA} + \underset{\underset{R}{\overset{|}{O}}}{\overset{}{\underset{}{C}}}\overset{\overset{|}{N-}}{\underset{O-}{}} \rightarrow \left\{\underset{\overset{\delta-}{A\cdots H\cdots O}\;\overset{|}{\underset{R}{}}\;\overset{\delta-}{O-}}{C^+}\overset{\overset{|}{N-}}{}\right\}^{\ddagger} \rightarrow A^- + \text{HOR} + -C^{\overset{+}{}}\overset{\overset{|}{N-}}{\underset{O-}{}} \quad (18)$$

Two different predictions may be made for these two mechanisms concerning the effect of changes in the structure of the leaving alcohol on the behaviour of the system. These predictions use a method of analysis which is based upon the 'tilting' potential energy surfaces first applied to proton transfer reactions by Albery [62] and later generalized by More O'Ferrall [63]. The argument, in brief, depends upon the fact that proton transfer makes up essentially all of the reaction coordinate of the first step in the step-wise mechanism, 17, and changes in the basicity of the oxygen atom to which the proton is being transferred will therefore affect the system according to the Hammond postulate, i.e. an increase in basicity will make this step less endothermic, shifting its transition state to an earlier position along the reaction coordinate and decreasing the degree of proton transfer. In the concerted mechanism, 18, on the other hand, the reaction coordinate is complex, consisting of several different kinds of atomic motion, and changes in basicity of the proton-accepting site affect the system in a direction perpendicular to the reaction coordinate. This leads to 'anti-Hammond postulate behaviour' [64], with the opposite relationship between oxygen basicity and extent of proton transfer, i.e. an increase in basicity will now produce more proton transfer at the transition state.

The experimental results, summarized in Table 3, show that introduction of electron-withdrawing substituents near the oxygen atom of the leaving alcohol reduces the Brønsted exponent for phthalimidinium ion formation by a considerable amount. This structural change lowers the basicity of the proton-accepting site, as shown, for example, by an increase in the acid strength of the leaving alcohol. According to the Hammond postulate, this change should raise the degree of proton transfer at the transition state, but the drop in α actually observed indicates that the opposite has in fact taken place.

195

Since anti-Hammond postulate behaviour is predicted for the concerted process, and not for the step-wise path, this reaction must therefore occur via the concerted mechanism.

The acid catalysts which Jencks used in this study form a rather mixed group: they include, in addition to several neutral carboxylic acids, the positively charged hydronium ion and the negatively charged dihydrogen phosphate ion. Brønsted exponents obtained from such a set of catalysts may not be accurate quantitative measures of transition state structure, and, as Table 3 shows, exponents calculated only from the carboxylic acid data do differ somewhat from

Table 3

Brønsted exponents for phthalimidium ion formation [60]

Leaving alcohol	pK_a of alcohol	α All acids	α Carboxylic acids only
Ethanol	16.0	0.74	—
Methanol	15.5	0.71	0.83
β-Chloroethanol	14.3	0.64	0.71
Propargyl alcohol	13.6	0.60	0.62
β,β-Dichloroethanol	12.9	0.53	0.60
β,β,β-Trifluoroethanol	12.4	0.49	0.51

those based on all of the acids. The order in which the exponents fall, however, is exactly the same in both sets, and use of a non-homogeneous group of catalysts has therefore not upset the mechanistic conclusion.

REFERENCES

[1] J. N. Brønsted and K. Pedersen, *Z. Phys. Chem.*, **108**, 185 (1924).
[2] J. E. Leffler, *Science*, **117**, 340 (1953); J. E. Leffler and E. Grunwald, *Rates and Equilibria of Organic Reactions*, John Wiley, New York (1963), p. 156.
[3] G. S. Hammond, *J. Amer. Chem. Soc.*, **77**, 334 (1955).
[4] M. Eigen, *Angew. Chem., Intl. Edn.*, **3**, 1 (1964).
[5] A. J. Kresge, *Chem. Soc. Rev.*, **2**, 475 (1973).
[6] J. R. Murdoch, *J. Amer. Chem. Soc.*, **94**, 4410 (1972).
[7] R. A. Marcus, *J. Phys. Chem.*, **72**, 891 (1968).
[8] G. W. Koeppl and A. J. Kresge, *J.C.S. Chem. Comm.*, 371 (1973).

[9] G. W. Koeppl and A. J. Kresge, to be published.

[10] G. W. Koeppl, unpublished work.

[11] R. P. Bell, *The Proton in Chemistry* (second edition), Cornell Univ. Press, Ithaca, New York (1973), p. 203.

[12] F. H. Westheimer, *Chem. Rev.*, **61**, 265 (1961); J. Bigeleisen, *Pure Appl. Chem.*, **8**, 217 (1964).

[13] R. P. Bell, W. H. Sachs, and R. L. Tranter, *Trans. Faraday Soc.*, **67**, 1995 (1971).

[14] R. P. Bell, *The Proton in Chemistry* (second edition), Cornell Univ. Press, Ithaca, New York, 1973, pp. 262–266; J. E. Dixon and T. C. Bruice, *J. Amer. Chem. Soc.*, **92**, 905 (1970); F. G. Bordwell and W. J. Boyle, Jr., *J. Amer. Chem. Soc.*, **93**, 512 (1971).

[15] J. L. Longridge and F. A. Long, *J. Amer. Chem. Soc.*, **89**, 1292 (1967).

[16] A. J. Kresge, S. G. Mylonakis, Y. Sato, and V. P. Vitullo, *J. Amer. Chem. Soc.*, **93**, 6181 (1971).

[17] (a) A. J. Kresge, D. S. Sagatys, and H. L. Chen, *J. Amer. Chem. Soc.*, **90**, 4174 (1968); (b) A. J. Kresge, H. L. Chen, Y. Chiang, E. Murrill, M. A. Payne, and D. S. Sagatys, *J. Amer. Chem. Soc.*, **93**, 413 (1971).

[18] (a) M. M. Kreevoy and D. E. Konasewich, *Adv. Chem. Phys.*, **21**, 243 (1972); (b) W. J. Albery, A. N. Campbell-Crawford, and K. S. Hobbs, *J.C.S. Perkin II*, 2180 (1972); (c) W. J. Albery, A. N. Campbell-Crawford, and R. W. Stevenson, *J.C.S. Perkin II*, 2198 (1972); (d) M. M. Kreevoy and S. Oh, *J. Amer. Chem. Soc.*, **95**, 4805 (1973).

[19] S. B. Hanna, C. Jermini, and H. Zollinger, *Tetrahedron Letters*, 4415 (1969).

[20] M. Wolfsberg and M. J. Stern, *Pure Appl. Chem.*, **8**, 325 (1964); *J. Chem. Phys.*, **45**, 2618 (1966).

[21] J. Bigeleisen and M. Wolfsberg, *Adv. Chem. Phys.*, **1**, 15 (1958).

[22] H. S. Johnston, *Gas Phase Reaction Rate Theory*, Ronald Press, New York, 1966, p. 82.

[23] (a) M. M. Kreevoy and R. A. Kretchmer, *J. Amer. Chem. Soc.*, **86**, 2435 (1964); (b) M. M. Kreevoy, P. J. Steinwand, and W. V. Kayser, *J. Amer. Chem. Soc.*, **86**, 5013 (1964); (c) J. M. Williams and M. M. Kreevoy, *Adv. Phys. Org. Chem.*, **6**, 63 (1968).

[24] A. J. Kresge and D. P. Onwood, *J. Amer. Chem. Soc.*, **86**, 5014 (1964).

[25] A. J. Kresge, *Pure Appl. Chem.*, **8**, 243 (1964).

[26] A. J. Kresge, D. P. Onwood, and S. Slae, *J. Amer. Chem. Soc.*, **90**, 6982 (1969).

[27] A. J. Kresge, Y. Chiang, and R. A. More O'Ferrall, to be published.

[28] A. J. Kresge, S. Slae, and D. W. Taylor, *J. Amer. Chem. Soc.*, **92**, 6309 (1970).

[29] M. M. Kreevoy and R. Eliason, *J. Phys. Chem.*, **72**, 1313 (1968).

[30] A. J. Kresge and Y. Chiang, *J. Chem. Soc. B*, 58 (1967).

[31] V. Gold and D. C. A. Waterman, *J. Chem. Soc. B*, 849 (1968).

[32] V. Gold and D. C. A. Waterman, *J. Chem. Soc. B*, 839 (1968).

[33] M. M. Kreevoy, P. J. Steinwand, and W. V. Kayser, *J. Amer. Chem. Soc.*, **88**, 124 (1966).

[34] M. M. Kreevoy, T. S. Straub, W. V. Kayser, and J. L. Melquist, *J. Amer. Chem. Soc.*, **89**, 1201 (1967).

[35] M. M. Kreevoy and R. A. Lanholm, *Int. J. Chem. Kinet.*, **1**, 157 (1969).

[36] M. M. Kreevoy and D. E. Konasewich, *J. Phys. Chem.*, **74**, 4464 (1970).

[37] W. J. Albery and A. N. Campbell-Crawford, *J.C.S. Perkin II*, 2190 (1972).

[38] G. Diderich and H. Dahn, *Helv. Chim. Acta*, **55**, 1 (1972); H. Dahn and G. Diderich, *Helv. Chim. Acta*, **54**, 1950 (1971).

[39] R. A. More O'Ferrall, G. W. Koeppl, and A. J. Kresge, *J. Amer. Chem. Soc.*, **93**, 9 (1971).

[40] W. J. Albery, J. R. Bridgeland, and J. S. Curran, *J.C.S. Perkin II*, 2203 (1972).

[41] M. H. Davies, *J.C.S. Perkin II*, 1018 (1974).

[42] R. P. Bell and D. M. Goodall, *Proc. Roy. Soc. A*, **294**, 273 (1966).

[43] A. J. Kresge, R. A. More O'Ferrall, L. E. Hakka, and V. P. Vitullo, *Chem. Comm.*, **46** (1965).

[44] F. G. Bordwell, W. J. Boyle, Jr., J. A. Hautala, and K. C. Yee, *J. Amer. Chem. Soc.*, **91**, 4002 (1969); M. Fukuyama, P. W. K. Flanagan, F. T. Williams, Jr., L. Frainier, S. A. Miller, and H. Schechter, *J. Amer. Chem. Soc.*, **92**, 4689 (1970); F. G. Bordwell, W. J. Boyle, Jr., and K. C. Yee, *J. Amer. Chem. Soc.*, **92**, 5926 (1970); F. G. Bordwell and W. J. Boyle, Jr., *J. Amer. Chem. Soc.*, **93**, 511 (1971); **94**, 3907 (1972).

[45] A. J. Kresge, *Can. J. Chem.*, **52**, 1897 (1974).

[46] C. K. Ingold, *Structure and Mechanism in Organic Chemistry*, Cornell Univ. Press, Ithaca, New York, 1953, p. 559; A. Streitwieser, Jr. and J. H. Hammons, *Prog. Phys. Org. Chem.*, **3**, 43 (1965); A. J. Kresge, D. A. Drake and Y. Chiang, *Can. J. Chem.*, **52**, 1889 (1974).

[47] R. G. Pearson and F. V. Williams, *J. Amer. Chem. Soc.*, **76**, 258 (1954); M. J. Gregory and T. C. Bruice, *J. Amer. Chem. Soc.*, **89**, 2327 (1967); J. E. Dixon and T. C. Bruice, *J. Amer. Chem. Soc.*, **92**, 905 (1970); D. J. Barnes and R. P. Bell, *Proc. Roy. Soc. A*, **318**, 421 (1970); F. G. Bordwell and W. J. Boyle, Jr., *J. Amer. Chem. Soc.*, **94**, 3907 (1972).

[48] R. A. Marcus, *J. Amer. Chem. Soc.*, **91**, 7224 (1969).

[49] R. P. Bell, personal communication.

[50] J. R. Jones and S. P. Patel, *J. Amer. Chem. Soc.*, **96**, 574 (1974).

[51] A. Kankaanperä, P. Salomaa, P. Juhala, R. Aaltonen, and M. Mattsen, *J. Amer. Chem. Soc.*, **95**, 3618 (1973).

[52] A. Streitwieser, Jr., *Solvolytic Displacement Reactions*, McGraw-Hill, New York (1962), p. 124.

[53] A. J. Kresge, *J. Amer. Chem. Soc.*, **92**, 3210 (1970).

[54] V. Gold in *Progress in Stereochemistry*, Vol. 3, P. B. D. de la Mare and W. Klyne, Eds., Butterworths, London (1962), p. 191.

[55] R. P. Bell, M. H. Rand, and K. M. A. Wynne-Jones, *Trans. Faraday Soc.*, **52**, 1093 (1956); F. Covitz and F. H. Westheimer, *J. Amer. Chem. Soc.*, **85**, 1773 (1963); J. A. Feather and V. Gold, *J. Chem. Soc.*, 1752 (1965).

[56] R. P. Bell, E. Gelles, and E. Moeller, *Proc. Roy. Soc. A*, **198**, 308 (1949).

[57] A. J. Kresge and Y. Chiang, *J. Amer. Chem. Soc.*, **95**, 803 (1973).

[58] R. P. Bell, *Acid-Base Catalysis*, Oxford Univ. Press, London, 1941, p. 85.

[59] A. J. Kresge, H. L. Chen, Y. Chiang, E. Murrill, M. A. Payne, and D. S. Sagatys, *J. Amer. Chem. Soc.*, **93**, 413 (1971).

[60] N. Gravitz and W. P. Jencks, *J. Amer. Chem. Soc.*, **96**, 507 (1974).

[61] Y. Chiang, A. J. Kresge, P. Salomaa, and C. I. Young, *J. Amer. Chem. Soc.*, **96**, 4494 (1974); E. H. Cordes and H. G. Bull, *Chem. Rev.*, **74**, 581 (1974); T. Fife, *Accts Chem. Research*, **5**, 264 (1972); R. H. DeWolfe, *Carboxylic Ortho Acid Derivatives*, Academic Press, New York (1970), p. 134 ff.

[62] W. J. Albery, *Progress in Reaction Kinetics*, Pergamon Press, Oxford (1967), Vol. 4, p. 353.

[63] R. A. More O'Ferrall, *J. Chem. Soc. B*, 274 (1970).

[64] E. R. Thornton, *J. Amer. Chem. Soc.*, **89**, 2915 (1967).

8

Rory A. More O'Ferrall

SUBSTRATE ISOTOPE EFFECTS

8.1 Introduction

This chapter considers hydrogen isotope effects upon rates of proton transfer for reactions in solution not involving solvent isotope effects. Practically speaking, examples are confined to proton transfer to or from carbon, and the point of view most commonly apparent in experimental investigations is that of an organic chemist interested in reactivity. As with reaction rates, the emphasis of interpretations has been less upon the absolute magnitudes of isotope effects than upon correlations describing their variation from one reaction to another. In this regard Westheimer's suggestion [1], that the size of the isotope effect depends upon the structure of the transition state and that a maximum value should be found for a transition state in which the hydrogen is symmetrically bonded to the atoms between which it is being transferred, has been particularly fruitful. Made in 1961, this proposal has since stimulated a good deal of experimental work and discussion, and has led to wide use of measurements of primary isotope effects as criteria of the extent of proton transfer in the transition state.

The Westheimer effect and its implications form the main subject of this chapter, and no attempt is made to consider for example the more traditional application of primary isotope effects to the study of reaction mechanisms. However, a further point that is emphasized is that interpretations of isotope effects may be appreciated without resort to calculations, and before discussing kinetic effects some time is spent in considering, from a qualitative standpoint, the origins of hydrogen isotope effects and the isotopic properties of stable molecules and equilibria. In this preliminary review previous accounts of hydrogen isotope effects are extensively used [2–9] and among these

201

the monographs by Bell [2] and Johnston [3] should especially be mentioned.

8.2 Isotope effects upon stable molecules and equilibria

The starting point for most discussions of isotope effects is the implication of the Born-Oppenheimer approximation that nuclear substitution leads to no appreciable change in electronic energy. Of the other factors leading to energy differences between molecules, it is a feature of primary hydrogen isotope effects that zero-point energy changes are usually much more important than differences in isotopic partition functions. It is useful to begin therefore with an outline of the characteristics of isotopic zero-point energies in stable molecules. This will be followed by a briefer account of isotope effects upon partition functions and of the relative importance of zero-point energies and partition functions in isotopic equilibria.

8.2.1 Zero point energy

Diatomic molecules. For a diatomic molecule, in the harmonic approximation, the zero point energy ε_0 is related in a familiar manner [10] to the frequency of molecular vibration v_0 and, through the velocity of light c, to its corresponding wave number ω_0:

$$\varepsilon_0 = \tfrac{1}{2}hv_0 = \tfrac{1}{2}hc\omega_0 \tag{1}$$

The zero point energy is controlled by two factors; the bond force constant f, and the effective mass of the vibration m':

$$\varepsilon_0 = (h/4\pi)(f/m')^{\frac{1}{2}} \tag{2}$$

The force constant is a characteristic of the potential energy curve for the molecule (Figure 1), and for the small displacements (Δr) from the equilibrium internuclear distance (r^0) that occur in most molecular vibrations the curve is well approximated by the parabola

$$V - V_0 = \tfrac{1}{2}f\Delta r^2 \tag{3}$$

and the force constant determines the change in potential energy accompanying displacement. The force constant measures the stiffness of a chemical bond and corresponds to the curvature of the potential energy curve at its minimum:

$$f = (\partial^2 V/\partial \Delta r^2)_{r^0} \tag{4}$$

Potential energy curves for molecules with small and large force constants are compared in scheme 1.

202

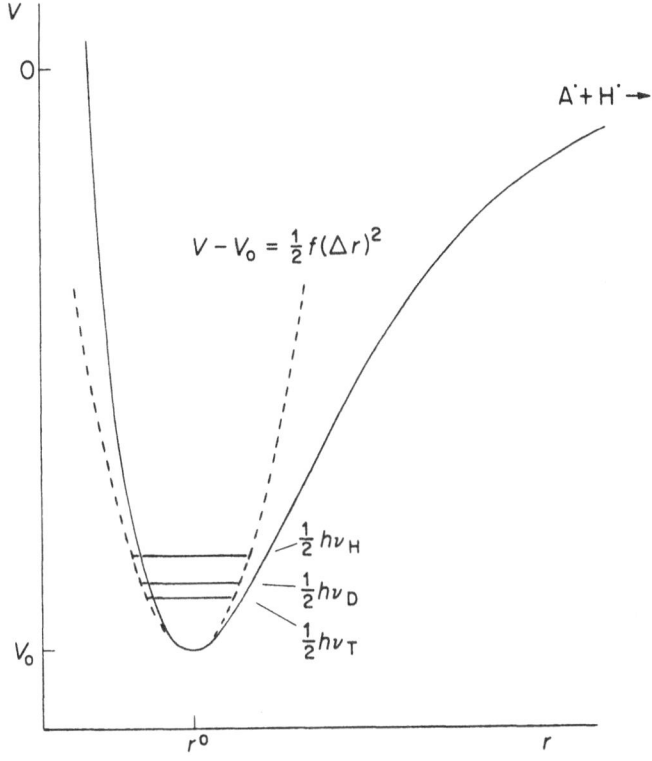

Figure 1. Potential energy curve for a diatomic hydride A–H. The zero point energies and, hence, anharmonicities of the vibrations are exaggerated.

Within the Born-Oppenheimer approximation the potential energy and hence force constants of a molecule are unaffected by isotopic substitution. However the zero-point energy is affected because there is a change in effective mass. The effective mass is determined by the masses of the atoms in motion in the vibration and for a diatomic molecule is equal to the reduced mass μ. For a diatomic hydride A—H,

$$\frac{1}{\mu} = \frac{1}{m_A} + \frac{1}{m_H} \tag{5}$$

and a simplifying feature of hydrogen isotope effects is that, except in the case that A itself is hydrogen, the mass of A is normally sufficiently greater than that of the isotopes of hydrogen (H, D, and T) that in the vibrations of the isotopic hydrides A—H, A—D, and A—T the atom A hardly moves and the effective masses are to a good approximation m_H, m_D and m_T respectively. Since the isotopic masses are 1, 2 and 3, the zero-point energy differences between

203

molecules take a very simple form:

$$\Delta\epsilon(H, D) = \tfrac{1}{2}h(\nu_H - \nu_D) \cong (hf^{\frac{1}{2}}/4\pi)(1 - 2^{-\frac{1}{2}}) \tag{6}$$

$$\Delta\epsilon(H, T) = \tfrac{1}{2}h(\nu_H - \nu_T) \cong (hf^{\frac{1}{2}}/4\pi)(1 - 3^{-\frac{1}{2}}) \tag{7}$$

As is illustrated on the (common) potential energy curve of Figure 1, individual zero-point energies fall in the order

$$A-H > A-D > A-T$$

Although force constants are unaffected by isotopic substitution they control the magnitude of the zero-point energy change taking place when substitution occurs, and it is the possibility of a difference in force constants between reactants and products, or between reactants and transition state, that leads to isotope effects upon rates and equilibria. This is shown in scheme 1 for the equilibrium

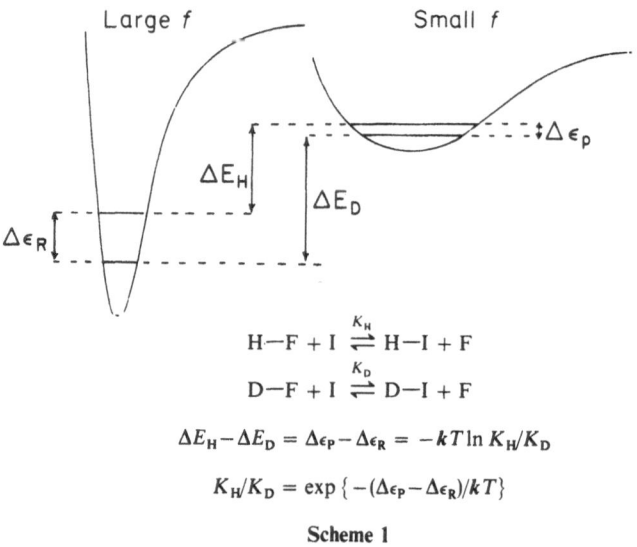

$$H-F + I \underset{K_D}{\overset{K_H}{\rightleftharpoons}} H-I + F$$

$$D-F + I \underset{}{\overset{}{\rightleftharpoons}} D-I + F$$

$$\Delta E_H - \Delta E_D = \Delta\epsilon_P - \Delta\epsilon_R = -kT\ln K_H/K_D$$

$$K_H/K_D = \exp\{-(\Delta\epsilon_P - \Delta\epsilon_R)/kT\}$$

Scheme 1

HF + I$^-$ $\rightleftharpoons$ HI + F$^-$ where formation of a product with a diminished force constant leads to a decrease in equilibrium constant on replacing hydrogen by deuterium ($K_H/K_D > 1$). The smaller force constant gives a smaller zero-point energy change for isotopic substitution and the zero-point energy difference in the product ($\Delta\epsilon_P$) fails to offset that in the reactant ($\Delta\epsilon_R$) favouring greater reaction of the lighter isotopic reactant. Thus a decrease in force constant gives an isotope effect (K_H/K_D) greater than unity and an increase in force constant gives one of less than unity.

204

The relation between force constants and zero-point energy changes can be written explicitly as follows

$$\Delta\epsilon_P - \Delta\epsilon_R \cong (h/4\pi)(f_P^{\frac{1}{2}} - f_R^{\frac{1}{2}})(1 - 2^{-\frac{1}{2}}) \qquad (8)$$

where f_R and f_P are the force constants in the reactant and product. The significance of this relation is emphasized by Stern and Wolfsberg's dictum that there are no quantum-mechanical isotope effects where there are no force constant changes [11]. For equilibrium effects this applies specifically to zero-point energies, and although subject to some qualifications [11] the importance of force constants is certainly not exaggerated.

An interesting implication of equation 1 pointed out by Swain and coworkers [12] is that the relative magnitudes of different hydrogen isotope effects *are* independent of force constant changes. Replacing deuterium by tritium alters the expression only in changing $2^{\frac{1}{2}}$ to $3^{\frac{1}{2}}$, and it is quickly shown that:

$$K_H/K_T = (K_H/K_D)^{(1 - 1/3^{\frac{1}{2}})/(1 - 1/2^{\frac{1}{2}})} = (K_H/K_D)^{1.442} \qquad (9)$$

which should be true independently of the nature of the chemical reaction occurring. Notwithstanding its derivation for equilibria between diatomic molecules, this relationship between deuterium and tritium isotope effects is remarkably general and applies widely to proton-transfer reactions [13]. This is an illustration of how simple considerations of hydrogen isotope effects often carry over to complex molecules and reactions.

Polyatomic molecules. In many respects hydrogen vibrations in polyatomic molecules are similar to those of diatomics. When, as is normally the case, hydrogen is monovalent and attached by a single bond to a heavy atom, there is usually a single hydrogen stretching frequency with an effective mass m_H and a force constant equal to that of the bond to hydrogen. This frequency shifts by about $1/2^{\frac{1}{2}}$ upon deuteration, while other vibrations, whose effective masses are insensitive to the mass of hydrogen, remain unaffected.

Polyatomic molecules possess bending as well as stretching vibrations, and the most common situation, especially in organic molecules, is that hydrogen is involved in two bending vibrations. For a linear molecule such as HCN the bends are degenerate and for the general case A—B—H the frequency is given by [14]

$$\nu_H = \frac{1}{2\pi}\left\{ f_b\left(\frac{1}{m_A r_1^2} + \frac{1}{m_H r_2^2} + \frac{1}{m_B}\left[\frac{1}{r_1} + \frac{1}{r_2} \right]^2 \right) \right\}^{\frac{1}{2}} \qquad (10)$$

where r_1 and r_2 are the bond lengths and f_b is the bending force constant.

$$A \overset{r_1}{\underset{\displaystyle f_b}{\frown}} B \overset{r_2}{\frown} H$$

Provided that m_H, $m_D \ll m_A$, m_B,

$$\Delta\epsilon = \tfrac{1}{2}h(\nu_H - \nu_D) \cong (f_b^{\frac{1}{2}}/4\pi)(1 - 1/2^{\frac{1}{2}}) \tag{11}$$

and again the isotopic shift in frequency is $1/2^{\frac{1}{2}}$. For molecules larger than triatomic, additional bending vibrations are isotopically insensitive.

In practice isotopic shifts in frequencies are usually a little less than $1/2^{\frac{1}{2}}$. As an example of a molecule with two bends and one stretch sensitive to substitution, the vibration frequencies [14a] of $CHCl_3$ and $CDCl_3$ are listed in Table 1, and it is clear that the ratios of isotopic

Table 1

Vibration frequencies (cm^{-1}) of liquid $CHCl_3$ and $CDCl_3$[a]

$CHCl_3$[b]	$CDCl_3$[b]	ω_H/ω_D
262 (2)	262 (2)	1.0
366	367	1.0
668	651	1.03
761 (2)	738 (2)	1.03
1216 (2)	908 (2)	1.34
3019	2256	1.34

[a] R..W. Wood and D. A. Rank, *Phys. Rev.*, **42**, 386 (1932).
[b] Degeneracy indicated in brackets.

frequencies ω_H/ω_D are only 1.34. Apart from small anharmonic effects, which should be less important for zero-point energy differences than for vibrational transitions, there are two reasons for this. The first and most obvious is that the effective masses of the vibrations are not exactly m_H and m_D. The second and often more important is that the normal vibrations of the molecule do not precisely correspond to stretching and bending of bonds to hydrogen, but involve some degree of mechanical coupling with other atomic motions, leading to a spreading of isotopic sensitivity between vibrations. On this point Stern and Wolfsberg [15] have shown in detail that irrespective of such interactions isotopic zero-point energy changes are to a very very good approximation determined solely by the stretching and bending force constants directly affecting the motion of hydrogen. While this conclusion is of greatest importance in simplifying calculations of isotope effects, it also means that the simplified discussion of hydrogen vibrations presented above is more appropriate than might appear from an inspection of observed vibration frequencies.

The effect of zero-point energy changes upon equilibria of isotopic molecules is found by summing contributions from different molecular vibrations. As an indication of their relative importance, approximate contributions to kinetic or equilibrium isotope effects from vibrations commonly involved in organic reactions are listed in Table 2.

Table 2

Zero-point energy contributions to isotope effects at 25°C for different vibrations[a]

	ω_H/cm^{-1}	$\exp\{hc(\omega_H - \omega_D)/2kT\}$
C—H	2900	7.0
C—H (C\/\H bend)	1250	2.3
O—H	3600	11.0
O—H (C\/\H bend)	1350	2.5
$\overset{+}{N}$—H	2700	6.0
$\overset{+}{N}$—H (C\/\H bend)	1400	2.5
S—H	2550	5.4
S—H (C\/\H bend)	500	1.4
F—H	4139	14.9
Cl—H	2990	7.5
Br—H	2650	6.2
I—H	2310	4.9

[a] Frequencies rounded to $50\,cm^{-1}$ and ω_D taken as $\omega_H/1.38$, except for hydrogen halides where $\bar{\omega}_e$ is reported and zero-point energies are anharmonically corrected.

The effects are temperature-dependent but, because of the prevalence of solution measurements at 25°C, values are shown at this temperature and unless otherwise specified isotope effects at 25°C will be considered throughout the chapter.

There are two important exceptions to the isotopic dependence of the effective mass of hydrogen vibrations described above. One that occurs commonly is when the vibration involves motion of more than one hydrogen, as in H_2 or in the bending vibrations of H_2O and other molecules with two or more hydrogens attached to the same atom. In this case it is easily seen, e.g. from equation 10, that the frequency shift for isotopic substitution by one deuterium is no longer $1/2^{\ddagger}$ but $3^{\ddagger}/2$. This situation is considered further below.

The second case is when the hydrogen is rendered 'divalent' by hydrogen bonding. This situation is of interest in providing an analogy for proton transfer transition states, and in Table 3 vibration

Table 3

Vibration frequencies of the isotopic bifluoride ions[a]

	ω_H/cm^{-1}	ω_D/cm^{-1}	ω_H/ω_D
ω_a	1577	1150	1.37
ω_s	630	630	1.00
ω_b	1210 (2)	893 (2)	1.35

[a] Solid sodium salts; J. J. Rush, L. W. Schroeder and A. J. Melveger, *J. Chem. Phys.*, **56**, 2793 (1972) and references cited.

frequencies are listed for the linear triatomic bifluoride ion HF_2^-, in which the hydrogen bond is believed to be symmetrical [16]. The ion has two bending vibrations and two stretches. The bending vibrations (ω_b) show the usual sensitivity to isotopic substitution indicated by equation 8, and, as would be expected, one stretching vibration is isotopically sensitive while the other is not. However, the stretching vibrations are quite unlike those of monovalent hydrogens, and correspond to the symmetric and asymmetric vibrations (v_s and v_a) known best for CO_2. It is the asymmetric

$$\overset{\leftarrow}{F}\cdots H\cdots\vec{F} \qquad \overset{\leftarrow}{F}\cdots\vec{H}\cdots\overset{\leftarrow}{F}$$

$$v_s \qquad\qquad v_a$$

stretch that is isotopically sensitive and in this vibration the hydrogen is fully in motion and the effective mass for the vibration $\cong m_H$. The isotopic insensitivity of the symmetric stretch arises from the balance between the equal bond force constants to hydrogen, as a result of which there is no motion of the atom in this vibration.

For unsymmetrical hydrogen bonds, the bond force constants are unequal and the hydrogen is in motion in the 'symmetric' stretch, which now shows some sensitivity to isotopic substitution. At the same time the isotopic sensitivity of the 'asymmetric' stretch is diminished. As is shown below through application of the Product

Rule (p. 214), the isotopic sensitivities of the two vibrations are not independent but, in effect, the change in effective mass normally confined to a single vibration is now spread between two; *i.e.*:

$$\nu_s^H \nu_a^H / \nu_s^D \nu_a^D \cong 2^{\frac{1}{2}} \qquad (12)$$

In the limit of a weak and highly unsymmetrical hydrogen bond, isotopic sensitivity is again confined to a single vibration, and characteristically one vibration approximates that of the monovalent hydrogen while the other is of low frequency and isotopically insensitive [17]. This behaviour has been discussed in detail for triatomic

$$\overrightarrow{F}-\overrightarrow{H}\cdots\overleftarrow{B} \qquad \overleftarrow{F}-\overrightarrow{H}\cdots\overleftarrow{B}$$
$$\nu^H/\nu^D \sim 1 \qquad \nu^H/\nu^D \sim 2^{\frac{1}{2}}$$

transition states by Albery [18], who also points out that if the 'symmetric-asymmetric' classification of vibrations is applied in unsymmetrical cases the 'symmetries' of the vibrations interchange when one bond becomes sufficiently weaker than the other.

Because of the transition-state analogy it is useful to consider not only the effective masses of the stretching vibrations of HF_2^- but also the force constants. The asymmetric and symmetric stretching modes are normal vibrations in which the atoms move in phase, and they are characterized by potential-energy curves and force constants in the same way as the vibration of a diatomic molecule. However, the displacement coordinates (Δa and Δs) now involve simultaneous extension and contraction of both bonds, and to represent them and the potential–energy changes accompanying the displacement it is necessary to consider a potential–energy surface in which the (linear) configurations of HF_2^- are plotted as a function of its bond co-ordinates r_1 and r_2 and the corresponding potential energies are added as contours [3a, 14b, 19].

Such a surface is shown schematically in Figure 2. It has an energy minimum at the equilibrium structure of HF_2^-, with bond lengths r_1^0 and r_2^0, and sections through the surface parallel to one bond axis when the other bond distance is large correspond to the Morse-like curves for dissociation of HF. In terms of the coordinates r_1 and r_2 the displacement coordinates for the normal vibrations are given by straight lines at right angles to each other through the energy minimum. On account of the symmetry of HF_2^-, the lines have slopes $+1$ and -1, since the bond displacements Δr_1 and Δr_2 are equal and $\Delta r_1/\Delta r_2 = +1$ and -1 for the symmetric and asymmetric vibrations respectively. The potential–energy curves corresponding

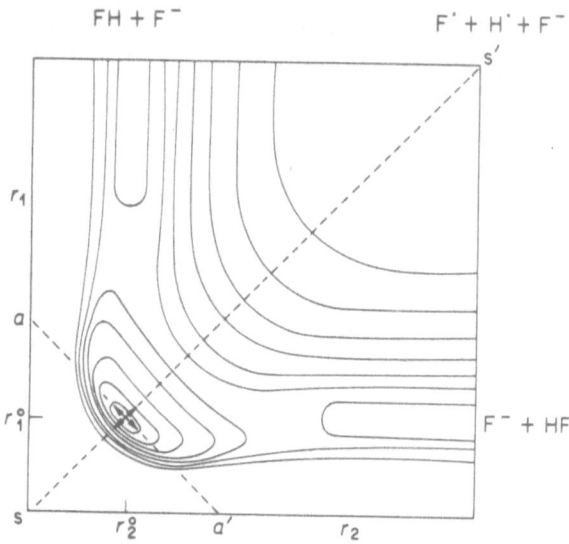

Figure 2. Schematic potential energy surface for a symmetrical hydrogen bond (HF_2^-).

to the displacements in the vibrations are given by sections through the surface along these lines, and the harmonic force constants by the curvatures at their minima:

$$f_a = (\partial^2 V / \partial \Delta a^2)_{r_1^0, r_2^0} \qquad f_s = (\partial^2 V / \partial \Delta s^2)_{r_1^0, r_2^0} \qquad (13)$$

$$\text{(a)} \qquad\qquad\qquad \text{(b)}$$

Together with the effective masses for the vibrations m_a and m_s, these force constants determine the frequencies and zero point energies of the vibrations in the usual way:

$$v_a = (1/2\pi)(f_a/m_a)^{\frac{1}{2}} \qquad v_s = (1/2\pi)(f_s/m_s)^{\frac{1}{2}} \qquad (14)$$

$$\text{(a)} \qquad\qquad\qquad \text{(b)}$$

The force constants f_a and f_s are normal-coordinate force constants and usually one wishes to relate these to the bond force constants f_1 and f_2. These force constants are given by

$$f_1 = (\partial^2 V / \partial \Delta r_1^2)_{r_1^0, r_2^0} \qquad f_2 = (\partial^2 V / \partial \Delta r_2^2)_{r_1^0, r_2^0} \qquad (15)$$

$$\text{(a)} \qquad\qquad\qquad \text{(b)}$$

and measure the potential energy curvatures through the energy minimum parallel to the bond axes in Figure 2. In the quadratic region of the surface close to the minimum the potential energy may be expressed in either bond coordinates or normal coordinates:

$$V - V_0 = \tfrac{1}{2}f_a(\Delta a)^2 + \tfrac{1}{2}f_s(\Delta s)^2$$

$$= \tfrac{1}{2}f_1(\Delta r_1)^2 + \tfrac{1}{2}f_2(\Delta r_2)^2 + f_{12}\Delta r_1 \Delta r_2 \tag{16}$$

The expression based on normal coordinates by definition involves no interaction force constant. However, in general, for any other set of coordinates an interaction force constant f_{12} must be included to take account of the effect of displacement in one coordinate upon the restoring force in the other. For bond coordinates we have

$$f_{12} = (\partial^2 V/\partial \Delta r_1 \partial \Delta r_2)_{r_1^0, r_2^0} \tag{17}$$

The two sets of force constants are interrelated by recognizing that $(\Delta s)^2$ and $(\Delta a)^2 = (\Delta r_1)^2 + (\Delta r_2)^2$, where Δr_1 and Δr_2 refer to bond displacements in the appropriate coordinate, and that the relationships between Δr_1 and Δr_2 in the two cases gives $\Delta r_1/\Delta r_2 = 1$ when $\Delta a = 0$ and $\Delta r_1/\Delta r_2 = -1$ when $\Delta s = 0$. Substitution in the expressions for $V - V_0$ yields:

$$f_s = \tfrac{1}{2}f_1 + \tfrac{1}{2}f_2 + f_{12} \tag{18}$$

$$f_a = \tfrac{1}{2}f_1 + \tfrac{1}{2}f_2 - f_{12} \tag{19}$$

In the more general case of an unsymmetrical linear triatomic molecule, the slopes of the dashed lines in Figure 2 are no longer ± 1 but $+c$ and $-1/c$ with $0 < c < 1$. Combining the modified expressions for $V - V_0$ now gives:

$$f_a = \frac{f_1 + c^2 f_2 - 2c f_{12}}{1 + c^2} \tag{20}$$

$$f_s = \frac{c^2 f_1 + f_2 + 2c f_{12}}{1 + c^2} \tag{21}$$

and the value of c is determined by the condition that the normal coordinates are so located as to involve no interaction terms [3b] (equation 22),

$$\frac{f_1 - f_2}{f_{12}} = c - \frac{1}{c} \tag{22}$$

These relationships between isotopic zero-point energy changes, effective mass, internal and normal coordinate force constants, and potential–energy surface are helpful in considering isotope effects in hydrogen-transfer reactions; they are dealt with in greater detail by Johnston [3]. In conclusion it should be noted that Figure 2 is not a realistic surface for HF_2^- but is simply a schematic adaptation of that shown by Herzberg [14b] for CO_2.

8.2.2 Rotational and translational partition functions

The contributions of translational and rotational partition functions of reactants or products to an equilibrium isotope effect K_H/K_D take the form:

$$\text{MMI} = \left(\frac{M_H}{M_D}\right)^{\frac{3}{2}}\left(\frac{I_H}{I_D}\right)^{\frac{1}{2}} \tag{23}$$

where M_H and M_D are the molecular masses of the isotopic molecules and I_H and I_D are moments of inertia. Strictly speaking (23) applies only when the principal moments of the molecules are equal, and more generally $I_H^{\frac{3}{2}}$ and $I_D^{\frac{3}{2}}$ are replaced by products of the roots of three unequal principal moments. The notation MMI is a convenient abbreviation [20] signifying 'masses and moments of inertia'. More usually it indicates combined contributions from reactants and products.

It is easy to recognise when these partition functions affect K_H/K_D. For appreciable isotope effects the translational contribution is of importance principally for hydrogen molecules:

$$(M_{D_2}/M_{H_2})^{\frac{3}{2}} = 2 \times 2^{\frac{1}{2}}; \qquad (M_{HD}/M_{H_2})^{\frac{3}{2}} = 3 \times 3^{\frac{1}{2}}/2 \times 2^{\frac{1}{2}} \tag{24}$$

while the rotational contribution is significant only when hydrogen exerts an appreciable moment around the centre of mass, i.e. when the molecule contains a single heavy atom or when the heavy atoms have a linear arrangement.

Isotopically sensitive rotations

Values of MMI and $(M_D/M_H)^{\frac{3}{2}}$ for a number of molecules [21] are listed in Table 4. Except in the case of H_2 it can be seen that the rotational term is dominant, and that both factors increase in importance as more than one hydrogen is substituted.* The rotational factor for CH_3Br should also represent the contribution of free internal rotation of a methyl group. For a typical organic molecule in which the

* Even for hydrogen the classical partition function ratio is a good approximation to the quantum value at 25°C [22].

212

Table 4

MMI contributions to isotope effects

	MMI^{-1}	$(M_D/M_H)^{\frac{3}{2}}$		MMI^{-1}	$(M_D/M_H)^{\frac{3}{2}}$
D_2/H_2	5.66	2.82	CD_4/CH_4	3.95	1.48
DCl/HCl	2.02	1.02	CH_2DBr/CH_3Br	1.26	1.01
HDO/H_2O	3.09	1.17	CD_3Br/CH_3Br	1.85	1.03
CH_3D/CH_4	1.48	1.10	$CHDBr_2/CH_2Br_2$[a]	1.10	1.00$_5$

[a] Included as an example of insensitivity to isotopic substitution.

tertiary or secondary hydrogens are isotopically substituted, it is clear that the MMI terms will be very small.

The Product Rule. Within the harmonic approximation, an important relation between the isotopic sensitivities of the classical translational and rotational partition functions and the vibration frequencies of a molecule is provided by the Product Rule [14, 23–25]. For isotopic substitution of a single hydrogen the rule may be stated as follows:

$$\text{MMI} \times \prod_{3n-6} \frac{v_D}{v_H} = \left(\frac{m_H}{m_D}\right)^{\frac{3}{2}} \tag{25}$$

In calculations of isotope effects the rule has been used to evaluate isotopic ratios of moments of inertia from a knowledge of the molecular vibration frequencies [24, 25], and it was used in this way in calculating the MMI contributions in Table 4. It is also useful in providing insight into the qualitative factors affecting isotope effects.

When the factor MMI is unity it can be seen that, in the usual case that motion of hydrogen is independent of the rest of the molecule, the rule implies that a total of three molecular vibrations will show an isotopic sensitivity of $1/2^{\frac{1}{2}}$, as was suggested should be the case in the discussion of zero-point energies above. On the other hand, as the moments of inertia or molecular masses become sensitive to substitution, the number or isotopic sensitivity of hydrogen vibrations will decrease. This is true of the molecules listed in Table 4. In the extreme case of the diatomic hydrogen halides, MMI$^{-1} \cong 2$ and there is a single hydrogen vibration. In the intermediate case of H_2O and D_2O, MMI$^{-1} \cong 2 \times 2^{\frac{1}{2}}$ and instead of there being six vibrations between two hydrogens there are three.

The Product Rule may be applied independently in three cartesian coordinates, with the product of MMI and frequency ratios contributing a factor $(m_D/m_H)^{\frac{1}{2}}$ to each. This is illustrated by the HF_2^-

ion, in which the isotopically sensitive bending and stretching vibrations involve orthogonal displacements of the hydrogen. The small

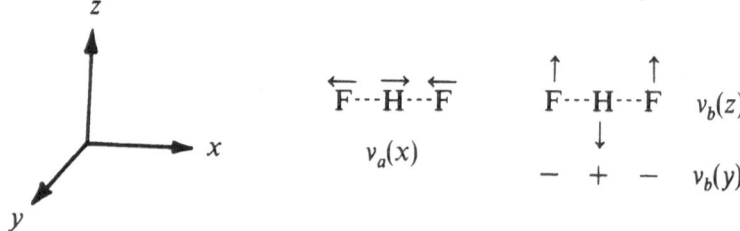

factor $(M_H/M_D)^{\frac{3}{2}}$ is the only contribution to MMI and each frequency shifts by nearly $1/2^{\frac{1}{2}}$ on deuteriation (Table 3). The normal distribution of hydrogen motion in molecules between one stretch and either two bends or two rotations (or both) is indeed a qualitative manifestation of this implication of the Product Rule. As already noted, for a linear hydrogen bond A—H···B, the combined isotopic shifts in the symmetric and asymmetric stretching vibrations are constrained by the Product Rule, i.e. $v_a^H v_s^H / v_a^D v_s^D \cong 2^{\frac{1}{2}}$.

8.2.3 The vibrational partition function

The last factor contributing to equilibrium isotope effects is the ratio of vibrational partition functions, which can be written

$$\text{EXC} = \prod_{3n-6} \{[1 - \exp(-u_H/kT)]/[1 - \exp(-u_D/kT)]\} \quad (26)$$

where $u = hv/kT$. As indicated by the abbreviation EXC this takes account of thermal excitation of low-frequency vibrations and departs from unity when $hv \to kT$. At 25°C, $kT = 207\,\text{cm}^{-1}$, and for $\omega_H = 600\,\text{cm}^{-1}$ and $\omega_D = 435\,\text{cm}^{-1}$ the contribution to EXC is 1.08. From Tables 1–3 it is clear that the great majority of hydrogen vibrations are of such high frequency that their contribution is unity.

When $hv \ll kT$, the partition function ratio reduces to ω_H/ω_D and achieves its maximum value. Since for a vibration of normal isotopic sensitivity $\omega_H/\omega_D \cong 2^{\frac{1}{2}}$, it is apparent that the contribution to the isotope effect from a thermally excited hydrogen vibration is the same as that from a free rotation. In practice weak hydrogen 'vibrations' of a molecule normally correspond to hindered rotations arising from intermolecular interactions, and most commonly occur in solution. They can be important in proton-transfer reactions when, as is quite often the case, these are characterized by an acidic reactant in a hydrogen-bonding solvent. Extreme examples are provided by H_2O and H_3O^+, which in an aqueous lattice [26] achieve 'libration' frequencies of about $700\,\text{cm}^{-1}$; but in other instances lower frequencies occur. There is no difficulty in recognizing the possible intervention of such modes, since they correspond to free rotations of the molecule in the gas phase. They are not expected to be of importance for hydrogen bound to carbon. Occasionally, hindered internal rotations may also need to be considered.

8.2.4 *The magnitudes of equilibrium isotope effects*

Before summing up the relative contributions from different factors it is important to appreciate the magnitudes of equilibrium isotope effects. Since equilibria involve zero-point energy differences and partition function ratios between reactants and products, it is economical to refer values to a common reactant, and for this purpose liquid water has most frequently been used. Values of equilibrium constants ϕ for reactions of the type

$$X-H + \tfrac{1}{2}D_2O \rightleftharpoons X-D + \tfrac{1}{2}H_2O \tag{27}$$

are recorded for a variety of hydrogens in Table 5. Since ϕ measures the distribution of D relative to H between $X-H$ and $O-H$ bonds, it is known as a fractionation factor [27, 28]. Isotope effects upon equilibria are given by the ratios of fractionation factors of isotopically-substituted hydrogens in the products and reactants, *i.e.*:

$$K_D/K_H = \prod \phi_P/\phi_R \tag{28}$$

where as usual the subscripts R and P denote reactants and products.

From Table 5 it is apparent that when a single hydrogen atom is involved, isotope effects upon equilibria are often not far from unity, and for the common organic equilibria between $C-H$ and $O-H$ or $N-H$ bonds fall in the range $K_H/K_D = 0.7-1.5$. One consequence of this is that large changes in zero-point energy (Table 2) are extensively cancelled between reactants and products, and in favourable cases the much smaller partition function ratios may play a role. The two factors may be separated experimentally by measuring K_H/K_D as a function of temperature and plotting $\log(K_H/K_D)$ versus $1/T$. Since the classical partition functions are independent of temperature, while zero-point energy increases exponentially with $1/T$, this fails only in cases rare in the temperature ranges normally studied, where the spacing of energy levels for hydrogen motion is approximately equal to kT.

The temperature-dependence and different contributions to equilibrium and other small isotope effects have been discussed exhaustively by Stern [29]. Here however the isotopic properties of stable molecules are principally of interest in relation to kinetic isotope effects, and it is

Table 5

Fractionation factors for various hydrogen atoms

$$\phi = \frac{[XD_n]^{1/n}[H_2O]^{\frac{1}{2}}}{[XH_n]^{1/n}[D_2O]^{\frac{1}{2}}}$$

	ϕ	Reference		ϕ	Reference
[a]R_3CH	0.97	[1]	H_2O	1.00	
NO_2CH_3	0.91	[2]	[e]ROH	1.04	[7]
[b]$N_2{=}C\begin{smallmatrix}H\\\\\\R\end{smallmatrix}$	0.89	[3]	H_3O^+	0.69	[8]
[c]$Ar{-}H$	0.85	[4]	CH_3CO_2H	0.96	[9]
$H{-}C{\equiv}C{-}H$	0.74	[5]	NH_3	1.04	[6, 12]
NH_4^+	1.08	[6, 11]	[f]RNH_2	~1.30	[7, 12]
[d]R_3NH^+	1.23	[7, 11]	[g]R_2NH	~1.30	[7, 12]
			[h]RSH	0.40	[10]

Compounds:

[a]nitropropane; [b]ethyl diazoacetate; [c]trimethoxybenzene; [d]N,N-dimethyl-*p*-nitroanilinium ion; [e]tri(*p*-methoxyphenyl)carbinol; [f]*p*-nitroaniline; [g]N-methyl-*p*-nitroaniline; [h]ethanethiol (pure liquid).

References:

1. D. M. Goodall and F. A. Long, *J. Am. Chem. Soc.*, **90**, 238 (1968).
2. O. Reitz, *Z. Phys. Chem.*, **176A**, 363 (1936).
3. W. J. Albery and M. H. Davies, *Trans. Faraday Soc.*, **65**, 1066 (1969).
4. A. J. Kresge and Y. Chiang, *J. Chem. Phys.*, **49**, 1439 (1968).
5. Calculated from $C_2H_2 + D_2O \rightleftharpoons C_2D_2 + H_2O$, $K = 0.473$ at 25°C; J. W. Pyper and F. A. Long, *J. Chem. Phys.*, **41**, 1890 (1964) using $p_{H_2O}/p_{D_2O} = 1.152$ (p = vapour pressure); W. M. Jones, *J. Chem. Phys.*, **48**, 207 (1968) and references cited.
6. P. Salomaa, L. L. Schaleger and F. A. Long, *J. Phys. Chem.*, **68**, 410 (1964).
7. V. Gold and C. Tomlinson, *J. Chem. Soc. B*, 1707 (1971).
8. P. Salomaa and V. Aalto, *Acta Chem. Scand.*, **20**, 2035 (1966) and references cited,
9. V. Gold and B. M. Lowe, *J. Chem. Soc. A*, 1936 (1968).
10. F. W. Hobden, E. F. Johnston, H. P. Weldon and C. L. Wilson, *J. Chem. Soc.*, 61 (1939).
11. Values of 0.97 and 0.92 for N—H$^+$ and N—H bonds respectively are quoted by R. L. Schowen, *Prog. Phys. Org. Chem.*, **9**, 275 (1972) from F. J. Karol, PhD Thesis, Massachusetts Institute of Technology (1962).

sufficient to note the general magnitude of the effects and what it implies regarding the relative importance of different contributing factors. This provides a useful contrast to kinetic behaviour and, as will be seen, has a bearing on the interpretation of kinetic effects.

8.3 *Kinetic isotope effects and transition states*

8.3.1 *Zero-point energies*

Within the framework of transition-state theory, kinetic isotope effects may be expressed as differences and ratios of isotopic zero-point energies and partition functions between reactants and transition state, i.e.:

$$k_H/k_D = \text{MMI} \times \text{EXC} \times \text{TUN} \times \exp\{(\Delta\epsilon_R - \Delta\epsilon^{\ddagger})/2kT\} \quad (29)$$

where MMI and EXC now include reactant and transition-state partition functions, and $\Delta\epsilon^{\ddagger}$ is the zero-point energy difference between transition states containing H and D. A new factor for reaction rates, not contributing to equilibria, is the 'correction' for quantum-mechanical tunnelling through the energy barrier to reaction, denoted TUN. However, because kinetic isotope effects are more simply introduced without considering tunnelling, this factor will not be discussed until later.

For primary isotope effects where hydrogen is transferred between two heavy atoms, it is clear that in the transition state the hydrogen must be close to the centre of mass, and that (with a rare exception noted on p. 255) isotopic substitution should lead to effectively no change in molecular mass or moments of inertia. This means that there is no need to revise the discussion of contributions to isotope effects from partition functions already given for stable molecules. However, an important difference between partition-function contributions to rates and equilibria is that at 25°C primary kinetic isotope effects are normally not close to unity but fall in the range $k_H/k_D = 2$ to 10, with even larger values occasionally occurring. As is clear from Table 4, and usually simply from inspection of the reactant molecules, these values are much larger than normal MMI or EXC values, and it follows that in the absence of important tunnelling the dominant factor in the isotope effects must be zero-point energy changes [30].

There are two qualitative manifestations of the importance of zero-point energy changes in primary isotope effects. The first is the wide applicability of Swain's relationship [12] between deuterium and tritium isotope effects: $k_H/k_T = (k_H/k_D)^{1.442}$. Although derived above (p. 205) for equilibria between diatomic molecules, its extension to primary kinetic effects has been confirmed by numerous calculations [13, 26, 31, 32], and its experimental verification [13, 33] is illustrated in Figure 3. The second factor is the common observation of an Arrhenius temperature-dependence,

$$k_H/k_D = (A_H/A_D)\exp[-(E_a^H - E_a^D)/kT] \quad (30)$$

with A_H/A_D close to unity. While the temperature-dependence of isotope effects has more often been studied carefully in the gas phase

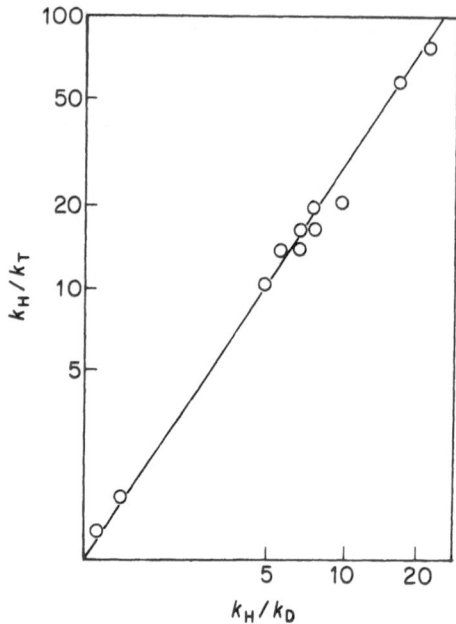

Figure 3. The relationship between deuterium and tritium isotope effects; a log–log plot of values of k_H/k_T against k_H/k_D. The slope of the line is 1.44.

than in solution (see *e.g.* Reference 34), this behaviour has also been established in solution [35].

These results should, of course, be put in perspective. Swain's relationship is not very sensitive to the presence of factors other than zero-point energy [13, 31, 32], and deviations from linear Arrhenius dependences and values of $A_H/A_D < 1$ are found, and normally are taken as indicating the presence of tunnelling [2, 36]. Nonetheless a preliminary interpretation of primary isotope effects will be based solely on zero-point energy changes, and the extent to which this needs to be qualified by further consideration of tunnelling or partition-function contributions will be examined later.

8.3.2 *The Westheimer effect*

The broad characteristics of primary isotope effects that became recognizable as experimental measurements accumulated were that values were generally greater than equilibrium or secondary kinetic effects and that they showed marked variations in magnitude from reaction to reaction, usually within the range 2 to 10. While it was soon recognized that the larger values roughly corresponded to isotopic zero-point energy differences in the stretching vibrations of hydrogens commonly forming the reactant [37] (Table 2), there was no satisfactory explanation of why this was so, or of why smaller values were often observed, until Westheimer showed in 1961 that the behaviour could be related to the structure of the transition state, and in particular to variations in the degree of proton transfer in the transition states of different reactions [1, 24, 38].

Westheimer pointed out that, if the transition state is approximated by a linear 3-centre model $[A \cdots H \cdots B]^\ddagger$, in the case that the force-constants of the partial bonds to the reacting hydrogen are equal, the symmetrical stretching vibration involves no motion of the hydrogen atom, and isotopic substitution leads to no change in zero-point energy. This parallels the behaviour of a symmetrical hydrogen bond such as that in HF_2^-. However, the hydrogen-transfer transition state differs from a hydrogen bond in that the isotopically-sensitive asymmetric stretch, instead of involving motion under contraint of a restoring force about a potential energy minimum, now corresponds to motion along the reaction coordinate and across the potential energy maximum of the barrier to reaction. Clearly this motion possesses no zero-point energy and again there can be no zero-point energy change upon isotopic substitution. It follows that, if zero-point energy changes in the two bending vibrations of the transition state approximately cancel those of the (polyatomic) reactant, the full zero-point energy change in the stretching vibration of the reactant will be felt in k_H/k_D. For proton transfer from carbon this would give an isotope effect of about seven (Table 2), and, if there were a weakening of bending vibrations or an appreciable contribution from tunnelling, larger values presumably could occur.

The distinction between the stretching modes of a symmetrical transition state and a hydrogen bond is seen by comparing Figure 2 with the familiar potential surface for a 3-centre reaction, shown in Figure 4. In Figure 4 the normal coordinates of the transition state are shown as crossed arrows through the saddle-point of the surface. The potential-energy curves for displacement along the coordinates, obtained from sections through the surface, together with the energy profile along the reaction path, are shown below the figure. These curves yield force constants $d^2V/d\Delta r^2$ at the saddle point and 'frequencies', often denoted $v_L^\ddagger$ and $v_R^\ddagger$ for the transition state's normal 'vibrations'. The frequencies are of course notional. Along the reaction coordinate the potential energy passes through a maximum, and the curvature and force constant f_a are negative. Since $v_L^\ddagger = (f_a/m_a^\ddagger)^{\frac{1}{2}}/2\pi$ (where $m_a^\ddagger$ is the effective mass of the vibration), $v_L^\ddagger$ is imaginary, and although the curvature orthogonal to the reaction coordinate is

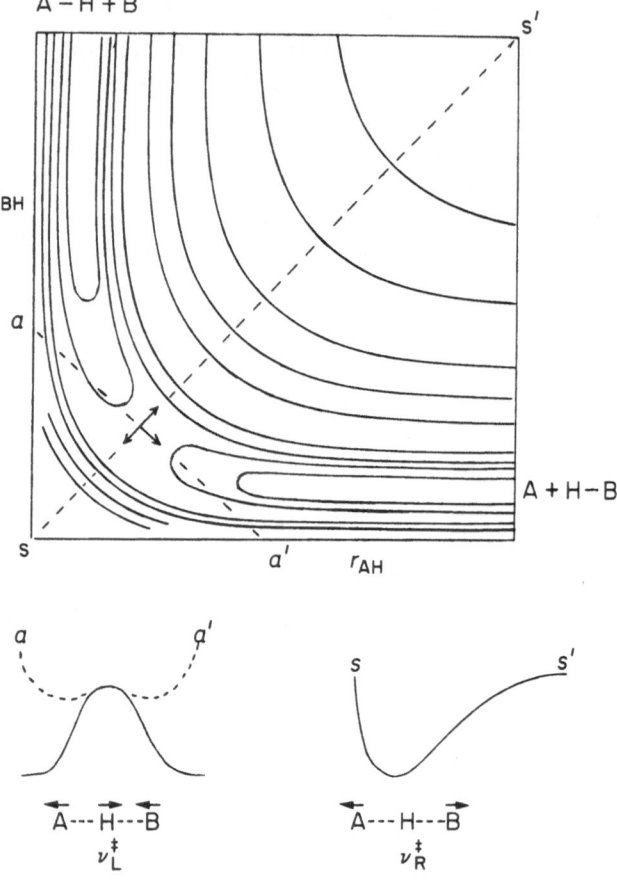

Figure 4. Potential energy surface and transition state stretching modes for the reaction
A—H + B → A + H—B.

positive so that $\nu_R^{\ddagger}$ ls real, the lifetime of the transition state is too short for it to undergo vibrations. Nonetheless $\nu_R^{\ddagger}$ measures the magnitude of the zero-point energy $\frac{1}{2}h\nu_R^{\ddagger}$ which when added to the classical barrier height determines the minimum energy necessary for reaction, while $\nu_L^{\ddagger}$ determines the corresponding quantum correction to motion along the reaction coordinate, which leads to tunnelling.

As presented so far, Westheimer's argument has dealt only with a symmetrical transition state. In transition states where the bond force constants to hydrogen are unequal, the hydrogen will move in the 'real' stretching vibration and isotopic substitution will lead to a zero-point energy change offsetting that of the stretching vibration of the reactant. As the asymmetry of the transition state increases, the degree of cancellation will increase and the value of k_H/k_D will fall. Thus Westheimer's treatment handsomely explains both the overall magnitudes and the range of values observed for primary isotope effects, and shows how the very different behaviour from equilibrium effects may be understood in terms of the special characteristics of the transition state.

If the analogy between proton-transfer transition states and hydrogen bonds is extended, it is possible to speculate as to the limiting forms of the transition state's stretching modes and the corresponding values of k_H/k_D. In the limit of a reactant-like transition state it would be expected that the magnitude and isotopic sensitivity of the real vibration $v_R^{\ddagger}$ should approach that of the stretching vibration of the reactant. This is illustrated on the schematic surface of Figure 5 where the energy contours of the reaction channel are

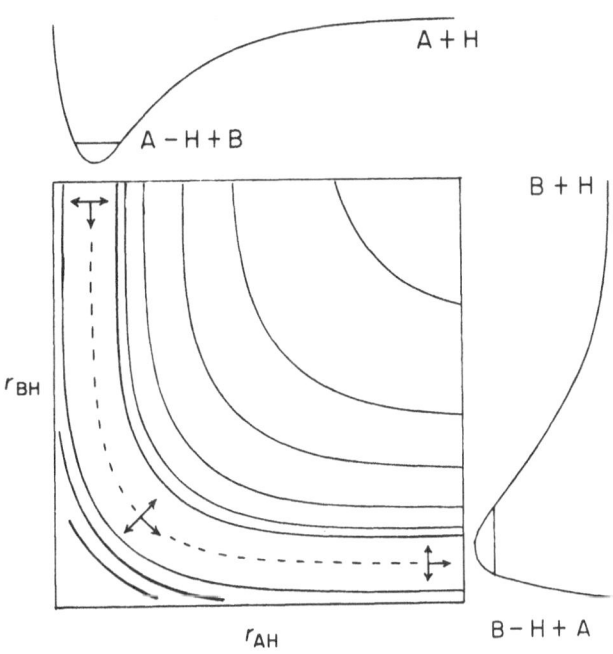

Figure 5. The dependence of normal coordinates on transition state structure.

omitted and the vibrational coordinates of the transition state are shown for various locations of the saddle point. For the reaction coordinate the reactant-like limit is most simply interpreted as a translational approach of reactants. This is consistent for example with the treatment of stretching coordinates in Marcus's adiabatic rate theory [39].

$$\overleftarrow{A}-\overrightarrow{H}\cdots B \qquad \overrightarrow{A}-\overrightarrow{H}\cdots \overleftarrow{B}$$
$$v_R^{\ddagger} \qquad\qquad v_L^{\ddagger}$$

Reactant-like transition-state modes

By the above argument the limiting value of k_H/k_D in a reactant-like transition state should be unity. *Mutatis mutandis*, the limiting value in a product-like transition state should be the equilibrium isotope effect K_H/K_D [40]. For proton transfers in solution, reactions are normally confined to hydrogen transfers between carbon and oxygen or nitrogen atoms, and K_H/K_D will be in the range 0.7 to 1.5 (Table 5) and thus also close to unity. It follows that, for a series of reactions in which the transition state passes from reactant-like to product-like, Westheimer's treatment leads to the expectation that the isotope effect will pass through a maximum [41].

This implication of the Westheimer effect has the important corollary that variations in isotope effect may be used as an empirical guide to changes in transition-state structure. Coming at a time when the importance of the transition state to practical interpretations of reactivity had become widely appreciated, it is not surprising that Westheimer's paper stimulated an experimental search for examples of isotope maxima. Before attempting a more critical evaluation of the effect it is appropriate to consider the results of these investigations.

8.3.3 *Experimental evidence for isotope maxima*

In so far as Westheimer's treatment correlates isotope effects with changes in force constants and the structure of the transition state, it cannot be tested without some experimental measure of these properties, and usually it has been assumed that, within a family of related reactions, the structure of the transition state varies smoothly with the rate constants and equilibrium constants of the reactions, with reactant-like transition states associated with reactive substrates and exothermic reactions. This assumption, which derives from observations of rate-equilibrium and reactivity-selectivity correlations [2, 3, 42], as well as calculations of semiempirical potential energy surfaces [43], is generally known (not quite accurately) as Hammond's Postulate [44]. It should be noted that while the postulate probably applies more generally to proton transfers than to other reactions, recent considerations of its scope and limitations [45, 46], based on extensive experimental experience, strongly suggest that departures from a smooth dependence of isotope effects upon rates or equilibria are as likely to represent a breakdown in the correlation of transition-state structure with the rate or equilibrium constant of a reaction as with k_H/k_D.

There are several advantages in studying isotope effects in proton-transfer reactions. Often the pK_a's of both reactant and product are known and it is possible to check whether the correlation between rates and equilibria themselves, which is a necessary condition for operation of the Hammond postulate, is maintained. Also because proton-transfer reactions are ionic, substituents produce large reactivity changes, and this is important if k_H/k_D is not a particularly sensitive function of reactivity [47]. On the other hand substituents have little effect on the zero-point energy of bonds to the substituted hydrogen, so that variations in k_H/k_D reflect nearly exclusively changes in transition-state structure; and indeed even where hydrogens are bonded to different atoms Table 5 shows that the differences remain small.

Correlations of isotope effects with rate and equilibrium constants. Although systematic variations in k_H/k_D with reactivity were known earlier [48–50], the first example of a maximum, for hydrogen-isotope exchange in aromatic substrates [47, 51], was not reported until 1965. This example was quickly followed by others, however, and a year later a particularly extensive investigation of the ionization of nitroalkanes and ketones to their carbanions, using a variety of oxygen and nitrogen bases ranging in strength from water to the hydroxide ion, was reported by Bell and Goodall [52]. Bell and Goodall's results have since been added to by several authors, [53–55], and the combined measurements, now totalling 74, are shown as a plot of k_H/k_D versus ΔpK, the difference in pK_a's of reactant and product in Figure 6. The range of substrates and bases included in the figure is illustrated below:

Substrates

$(EtOOC)_3CH$

$CH_3COCH_2SO_3^-$
$NO_2CH_2CO_2Et$
$PhCH_2NO_2$

Bases

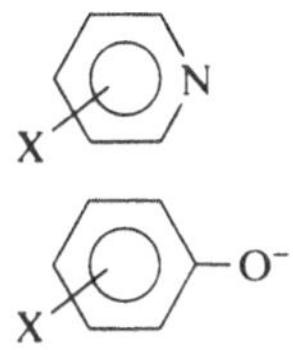

RCO_2^-
HPO_4^-, R_3N
H_2O, OH^-

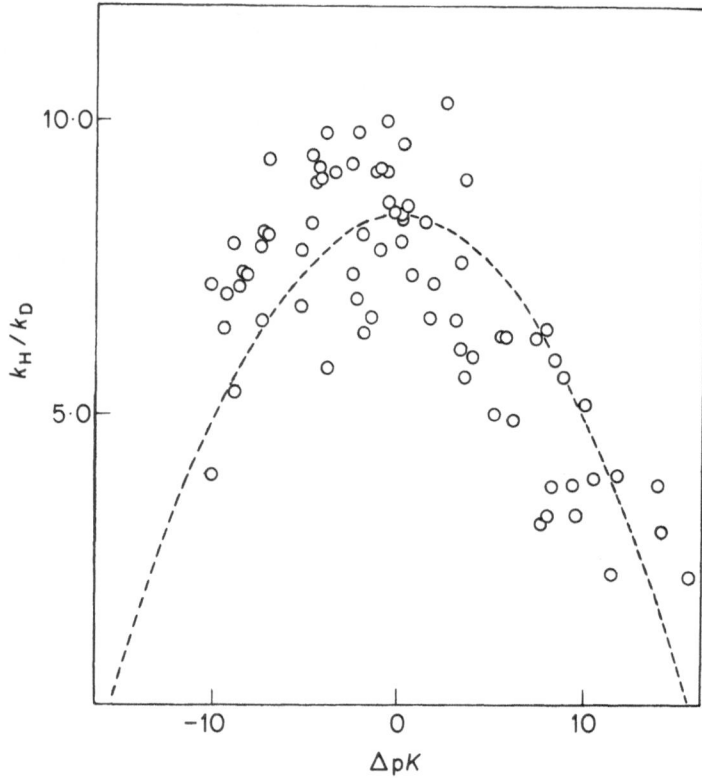

Figure 6. k_H/k_D as a function of ΔpK for the ionization of ketones and nitroalkanes.

Two points are at once apparent in Figure 6. One is that the plot shows considerable scatter, and the other that although a maximum can clearly be seen, close to $\Delta pK = 0$, there is a paucity of measurements for strongly exothermic reactions (large negative ΔpK). The second point is a reflection of an experimental difficulty which could in principle be remedied by determining the equilibrium isotope effects (K_H/K_D) from measurements of the fractionation factors of reactants and products. Combination of k_H/k_D and K_H/K_D would give the isotope effect in the reverse reaction and since K_H/K_D should normally be close to unity (*cf.* p. 215 and Table 5) the left-hand side of the figure will nearly mirror the right. This solution to the problem may be made independent of the equilibrium isotope effect by measuring the fractionation factor ϕ for the reactant and deriving a fractionation factor $\phi^{\ddagger}$ for the transition state from the relationship:

$$k_H/k_D = \phi/\phi^{\ddagger}$$

By the principle of microscopic reversibility, a plot of $\phi^{\ddagger}$ against ΔpK

would give an equivalent point for positive and negative ΔpK's. In so far as the Westheimer effect is a consequence of the behaviour of the transition state, strictly speaking a systematic variation of k_H/k_D without a maximum, though a less dramatic observation experimentally, is sufficient evidence for its operation, provided that the reactant fractionation factor remains approximately constant.

The scatter in Figure 6 has several causes. Many measurements include secondary isotope effects, and although these have been corrected for [54], the necessary use of a constant value for the correction certainly introduces an error [53, 55]. While all the isotopic reactants have $C-H$ bonds, it is possible that there are small changes in zero-point energy (strictly, fractionation factor) with the nature and number of substituents. Measurement of transition-state fractionation factors would obviate this problem, but the wide range of substrate and base structures and especially perhaps the differences in basic atoms and charge types could lead to non-systematic variations in bending vibrations, tunnelling, and the value of ΔpK at which the force constants to hydrogen become equal directly affecting the transition state.

Smoother plots of k_H/k_D versus ΔpK showing maxima are obtained for more homogeneous families of reactions, for example a single substrate with a series of bases. This was found by Bell and Barnes [55] in the ionization of ethyl nitroacetate and by Zollinger and coworkers [56] in an example of general base-catalysed diazo-coupling. Smooth plots are also obtained when the base remains the same but its effective pK_a is varied through a change in acidity function [57–61], for example by increasing the concentration of dimethylsulphoxide in dimethyl-sulphoxide-water mixtures containing hydroxide ions. Isotope maxima observed in this way have been reported for the racemization of optically active menthone [57], elimination of β-phenethyldimethyl-sulphonium ion [58], and ionizations of nitroethane [59] and phenyl-methylacetophenone [60]. However, the accessible range of ΔpK values has usually been too small for large variations in k_H/k_D to be observed.

For reaction of a series of organic substrates with the common acid H_3O^+, Kresge and coworkers [62] have shown that while an excellent correlation of $k_{H_3O^+}/k_{D_3O^+}$ with reactivity exists for a family of seventeen substituted vinyl ethers (equation 31),

$$CH_2{=}CHOEt + H_3O^+ \rightarrow CH_3{-}\overset{+}{C}HOEt + H_2O \rightarrow \text{Products}$$

$$(31)$$

the inclusion of aromatic, acetylenic, or even less closely related vinylic substrates leads to strong dispersion. Although $k_{H_3O^+}/k_{D_3O^+}$ includes a secondary as well as a primary isotope effect, it is not expected that this would affect the correlation. In the ionization of nitroalkanes and ketones, Bell and Barnes [55] and Bordwell and Boyle [54] have similarly noted that the correlations are improved when results are confined to structurally closely-related substrates.

A class of reactions in which the variation of k_H/k_D with ΔpK has been extensively investigated is electrophilic aromatic substitution, in which the rate-determining step is proton loss from a phenonium-ion intermediate. This includes diazo-coupling [56], nitrosation [63] and aromatic hydrogen exchange [47, 51, 64, 65]. Aromatic hydrogen

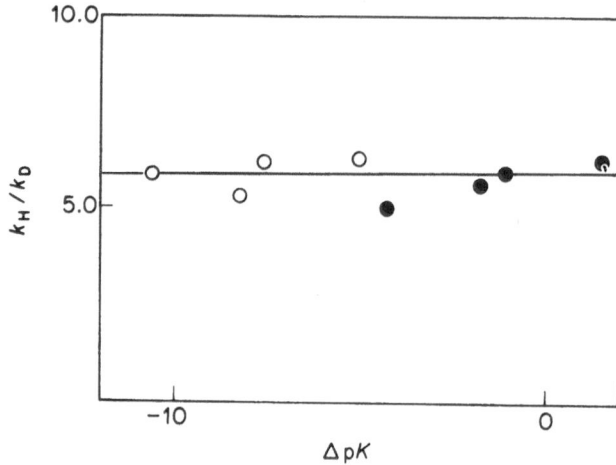

(32)

exchange has been cited as an example of an isotope maximum, but in a study of a series of substituted indoles (equation 32) Challis and Millar [65] have made the interesting discovery that for wide variations in ΔpK and a range of reactivity of 10^7 there is practically no change in isotope effect. This is illustrated in Figure 7 in which, although

Figure 7. Isotope effects for acid-catalysed hydrogen exchange of indoles at 25°C; (●) H_3O^+, (○) CH_2CO_2H.

results for the indoles alone are included, the correlation shown appears to apply to other aromatic substrates in the same ΔpK range [65]. Challis and Millar point out that a similar constancy of

226

k_H/k_D possibly occurs in the ionization of nitroalkanes for ΔpK values close to zero, as indeed had earlier been noted by Bordwell and Boyle [54].

An invariance of k_H/k_D for very large ranges of reactivity and equilibrium constant is clearly not implicit in Westheimer's treatment. Further consideration of this behaviour is postponed, but an additional difficulty with the ionization of ketones and nitroalkanes should be emphasized. Although a broad correlation exists between k_H/k_D and ΔpK, there is essentially no correlation between ΔpK and reactivity spanning the full range of substrates considered [52, 66]. This represents a departure from the Hammond postulate, and indicates that either the rates or the equilibria, or both, fail to reflect the extent of proton transfer in the transition state. Almost certainly this is a contributing factor to the scatter in Figure 6, but it may also be responsible for more systematic departures from 'normal' behaviour.

Correlations of isotope effects with Brønsted exponents. A measure of transition-state structure in proton-transfer reactions which combines rate amd equilibrium measurements, and can be given a more explicit interpretation than either, is the exponent of the Brønsted relation for general acid or base catalysis [2, 67]. This is the slope of a log–log plot of rate constants against equilibrium constants, α or $\beta = d \log k / d \log K$, for reaction of a series of acids or bases with a single substrate. It compares the effect of (small) changes in substituent in the acid or base upon the free energies of activation and reaction, and in so far as the main effect of substituents is in stabilizing ionic charges it is most simply interpreted as measuring the fractional charge on the base in the transition state relative to that in its fully ionized or protonated form, as is illustrated for carboxylic acids and their anions in scheme 2.

$$[RCO_2^{z-} \cdots H \cdots S^{z+}]^{\ddagger}$$

$$k_A \nearrow \qquad \qquad \nwarrow k_B$$

$$S + RCO_2H \; \underset{}{\overset{K_a/K_{SH^+}}{\rightleftharpoons}} \; SH^+ + RCO_2^-$$

$$\alpha = \frac{d \log k_A}{d \log K_a} \qquad \beta = -\frac{d \log k_B}{d \log K_a} = 1 - \alpha$$

Scheme 2

Provided that charge development and the progress of bond-making and bond-breaking remain in step, α and β measure the extent of proton transfer in the transition state.

Values of α or β may be determined in different ways. Since proton transfer in the transition state is expected to vary with the strength of the Brønsted acid or base, α and β should in general represent tangents to a curved $\log k$–$\log K$ plot. For small variations in pK_a systematic deviations from linearity are usually too small to be observed and only an average value of α or β can be obtained, but for larger changes the curvature may be sufficient to evaluate a quadratic term in the dependence (equation 33):

$$\log k = \alpha_0 \log K_a + b(\log K_a)^2 + c \qquad (33)$$

A value of α can then be derived for any acid of known pK_a, as may be seen by differentiation.

$$\alpha = \alpha_0 + 2b \log K_a$$

When Brønsted exponents can be determined in this way, measurements of k_H/k_D for a series of acids or bases allow the variation of k_H/k_D with α or β to be examined. Data suitable for such an analysis have been obtained by Kreevoy and Sea-wha Oh for the hydrolysis of the diazoacetate anion (equation 34) [68].

$$N_2CHCO_2^- + HA \rightarrow {}^+N_2CH_2CO_2^- + A^- \rightarrow \text{Products} \qquad (34)$$

In practice, however, Kreevoy and Sea-wha Oh examined a specific interpretation of equation 33 developed by Marcus.

Marcus showed that if within a $\log k - \log K$ relationship variations in the rate constant solely reflect changes in the equilibrium constant, then $\log k$, or in Marcus's formulation the free energy of activation $\Delta G^{\ddagger}$, may be expressed in terms of a free energy of the proton-transfer step of the reaction $\Delta G_R^{0'}$, and an 'intrinsic' energy barrier $\frac{1}{4}\lambda$ to proton-transfer for reaction of a substrate and acid for which $\Delta G_R^{0'} = 0$:

$$\Delta G^{\ddagger} = w^r + \tfrac{1}{4}\lambda + \tfrac{1}{2}\Delta G_R^{0'} + (\Delta G_R^{0'})^2/4\lambda \qquad (35)$$

It is important that in Marcus's expression, λ and $\Delta G_R^{0'}$ in fact refer not to the overall reaction but to the proton-transfer step occurring within the encounter complex formed between the acid and substrate; w^r is the energy of formation of this complex [69, 70]. The terms of the expression may be evaluated from the parameters of a quadratic fit to the variation of $\Delta G^{\ddagger}$ with ΔG_R^0, the directly-measurable overall free energy of reaction (i.e. from a free-energy analogue of equation 33), by using a structurally-related family of acids or bases and assuming that within the family w^r remains constant and ΔG_R^0 and $\Delta G_R^{0'}$ differ by a constant amount [68].

In Marcus's treatment the variation of isotope effects is represented by supposing that isotopic reactants differ only in their intrinsic energy barriers λ_H and λ_D. Replacement of $\Delta G^{\ddagger}$ in equation 35 by $-(RT \ln \{k/(kT/h)\}$ then gives:

$$\log k_H/k_D = \frac{(\lambda_D - \lambda_H)}{4(2.303RT)} \left\{ 1 - \frac{(\Delta G_R^{0'})^2}{\lambda_H \lambda_D} \right\} \qquad (36)$$

Kreevoy and Sea-wha Oh showed that this expression provides a satisfactory description of the dependence of k_H/k_D upon $\Delta G_R^{0'}$ for the hydrolysis of the diazoacetate anion catalysed by seven tertiary amines, with a value of $(\lambda_D - \lambda_H)$ chosen to optimize the correlation.

When $\Delta G_R^{0'}$ may be approximated by $\Delta G_R^0 = 2.303RT\Delta pK$, equation 36 may be adapted to show the dependence of k_H/k_D on ΔpK, and, if the relation between $(\lambda_H - \lambda_D)$ and the maximum isotope effect in the reaction series is recognized, i.e.:

$$(\lambda_H - \lambda_D)/4 = -RT \ln (k_H/k_D)_{max} \qquad (37)$$

equation (36) becomes:

$$\log (k_H/k_D) = \{1 - m(\Delta pK)^2\} \log (k_H/k_D)_{max} \qquad (38)$$

where $m = (2.303RT)^2/\lambda_H \lambda_D$. In Figure 6 the dotted line drawn through the points is obtained from this relation, with $(k_H/k_D)_{max} = 8.5$, $m = 0.004$, and $(\lambda_H/\lambda_D)^{\ddagger}/4 = 5.4 \text{ kcal mol}^{-1}$.

Marcus's treatment also leads to the dependence of k_H/k_D upon the Brønsted exponent. Differentiation of equation 35 gives:

$$\alpha = \partial \Delta G^{\ddagger}/\partial \Delta G_R^0 = \tfrac{1}{2}[1 + (\Delta G_R^{0'}/2\lambda)] \qquad (39)$$

and neglecting the effect upon α of the small difference between λ_H and λ_D, substitution in 36 gives equation 40:

$$\log (k_H/k_D) = \{1 - (2\alpha - 1)^2\} \log (k_H/k_D)_{max} \qquad (40)$$

The qualitative form of the dependence indicated by equation 40, with a maximum isotope effect at $\alpha = 0.5$, was earlier noted by Lewis and Funderburk [72] as being implied when both isotopic substrates obey Brønsted relationships.

All the expressions obtained from Marcus's treatment are based upon a quadratic approximation for the variation of $\Delta G^{\ddagger}$ and $\log k$, and they cannot be expected to apply for values of ΔpK or ΔG_R^0 far removed from zero [69]. This is the reason for the prediction of inverse isotope effects for sufficiently large ΔpK values in equation 38 and Figure 6, although this 'anomaly' may be removed by taking account of the onset of rate-determining diffusion steps at extreme ΔpK values [70]. The interpretation given for the considerable scatter evident for smaller values of ΔpK is that there is either a variation in intrinsic barriers ($\lambda/4$) between different substrates, or a discrepancy between ΔG_R^0 and $\Delta G_R^{0'}$.

An interesting conclusion reached by Kreevoy and Sea-wha Oh regarding hydrolysis of the diazoacetate anion is that the curvature of the Brønsted plot is so high as to indicate an almost negligible barrier to proton transfer ($\frac{1}{4}\lambda = 1.4$ kcal mol^{-1} for tertiary ammonium ions) and that the bulk of the activation energy derives from formation of the encounter complex [68]. Similarly small barriers have been found by Albery for protonation of other diazo compounds [73], and Kreevoy points out that the values appear to be too small to be consistent with an interpretation of the normal range of isotope effects observed ($k_H/k_D = 1.5$–11) in terms of the large zero-point energy changes envisaged by Westheimer.*

There are some obvious limitations to Marcus's treatment. It represents a quantitative formulation of Hammond's postulate and will not apply where this fails. Also, except where a homogeneous family of acids or bases are used, curvature in a Brønsted plot may be hard to separate from the natural dispersion of the points, and where a homogeneous family is used, the curvature may be too small to be detected within the range of pK_a values available. This point is illustrated by comparison of the Brønsted plot and the dependence of k_H/k_D upon base strength in the ionization of ethyl nitroacetate in Figures 8(a) and (b). The variation in k_H/k_D would seem to imply considerable changes in transition-state structure, and yet for a normal assortment of bases the Brønsted plot is to all intents and purposes linear.

A way of measuring changes in Brønsted exponents that avoids these difficulties was demonstrated some years ago by Bell [74]. A series of substrates is studied using a common set of structurally-related Brønsted acids or bases. Provided that the acids or bases cover a small pK_a range, a linear plot should be obtained, with the slope for each substrate corresponding to α or β for reaction with an acid or base in the middle of the pK_a range. Kemp and Casey [75] in a variation on this procedure have determined relative Brønsted exponents by plotting rate constants for different substrates against

---

* See Reference 73 for an alternative view.

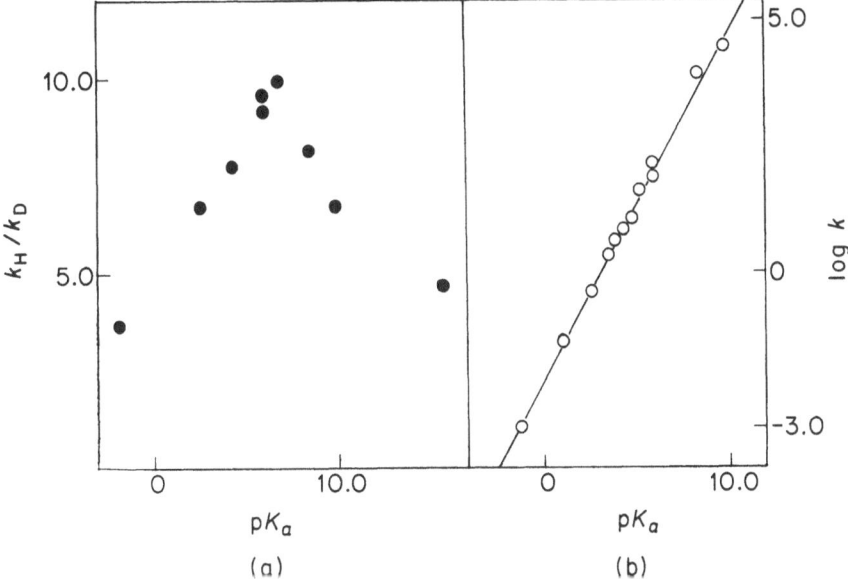

Figure 8. Plots of (a) k_H/k_D, and (b) log k versus pK_a for the ionization of ethyl nitroacetate with a variety of bases, including water, carboxylate and phenoxide ions, and substituted pyridines.

each other for all common bases used. An unsatisfactory aspect of this analysis is that no restriction is placed upon the range of pK_a values, but a comparison of values of β obtained by the two methods showed that as long as rate constants for H_2O and OH^- are omitted differences are usually rather small [75].

Brønsted exponents for the ionization of a series of nitroalkanes and ketones, measured mainly by Bell and his collaborators [48, 55], are shown in Table 6. Although a common set of bases was not used in all cases, the bases were confined to carboxylate anions. The values are compared with isotope effects determined for ionization of the substrates with H_2O as the base [48, 55], and it is apparent that k_H/k_D increases steadily with β, with only ethyl acetoacetate and acetylcyclohexanone out of line, and that the correlation with β is better than that with either the reactivity or pK_a of the substrate. The isotope effects are rather small because water is a weak base and in most cases the transition states should be strongly asymmetric in structure. Moreover because β refers to the much stronger carboxylate anions the absence of an isotope maximum at $\alpha = 0.5$ is not surprising [48]. As would be expected, available measurements for carboxylate anions give large values of k_H/k_D, but the results are too fragmentary to permit any further conclusion.

Table 6

Isotope effects and Brønsted exponents for the ionization of ketones and nitroalkanes[a]

	$10^5 k_{H_2O}$ sec^{-1}	pK_a[b]	k_H/k_D[c]		β
			H_2O	$ClCH_2CO_2^-$	
$CH_2(CO_2Et)_2$	2.45	13.3	1.7		0.79
$CH_3COCH_2SO_3^-$	0.20	13.6	2.2	2.3	0.75[d]
$CHBr(CO_2Et)_2$	21.5		2.7		0.73
CH_3NO_2	0.0065	10.2	2.9	3.3	0.67
$O_2NCH_2CO_2Et$	1570	5.8	3.1	5.7	0.66[e]
	230	10.5	3.4		0.64
$CH_3COCHCH_3CO_2Et$	1.14	12.7	3.8	5.2	0.60
	45.5	9.9	4.6		0.60[f]
$CH_3COCH_2CO_2Et$	116	10.0	3.0		0.59
$CH(CO_2Et)_3$	1480	7.8	3.8	5.0	0.57[d]
$(CH_3CO)_2CH_2$	1320	9.0	3.9		0.48
$(CH_3CO)_2CHBr$	3350	7.0	3.9		0.42
$CH_3COCHBrCO_2Et$	1560		4.3		0.42
$(CH_3)_2C=OH^+$	920[g]	0.3[g]	5.0[g]		0.38

[a] Except as indicated results are from the compilation of R. P. Bell and J. E. Crooks, *Proc. Roy. Soc.*, A, **286**, 285 (1965).

[b] pK_a values were taken mainly from R. P. Bell, *The Proton in Chemistry*, 1st edn, Cornell University Press, Table 17, p. 161.

[c] Experimental values were corrected for secondary isotope effects of 1.15 per hydrogen.

[d] Approximately calculated by the author from the data of D. J. Barnes and R. P. Bell, *Proc. Roy. Soc.*, A, **318**, 421 (1970).

[e] R. P. Bell and T. Spencer, *Proc. Roy. Soc.*, A, **251**, 41 (1959).

[f] T. Riley and F. A. Long, *J. Am. Chem. Soc.*, **84**, 522 (1962).

[g] From the data of J. Toullec and J. E. Dubois, *J. Am. Chem. Soc.*, **96**, 3524 (1974); *Tetrahedron*, **29**, 2859 (1973). A value of -7.5 was used for the pK_a of acetone, G. C. Levy, J. D. Cargioli and W. Racela, *J. Am. Chem. Soc.*, **92**, 6238 (1970).

A further method for obtaining Brønsted exponents is to compare kinetic and equilibrium acidities for proton transfers from the isotopic bases H_2O and D_2O [5, 76, 77]. This involves measurement of the secondary isotope effects in reactions of the type

$$D_2\overset{+}{O}-H + S \rightarrow \left[\begin{array}{c} D \\ \diagdown \\ O \cdots H \cdots S \\ \diagup \\ D \end{array} \right]^{\ddagger} \rightarrow D_2O + SH^+$$

and, since the experimental procedures commonly employed also yield primary isotope effects [77], the results may be used directly to investigate the correlation of Brønsted exponent with k_H/k_D. These isotopic Brønsted exponents possess the advantage that their measurement does not perturb the potential energy surface, and hence the structure of the transition state, for the reaction. However, the measurements do require a high experimental precision [76], and although in some cases this has been achieved, the experimental scatter in a general plot of primary versus secondary isotope effects is such that only a rather poorly defined isotope maximum can be discerned [26].

Isotope maxima in radical reactions. Although hydrogen-atom transfer reactions are not of direct concern here, most interpretations of isotope effects strictly apply to hydrogen-atom rather than to proton transfers, and it is important to know whether the two types of reactions show the same experimental behaviour. It is thus of interest that a good example of a maximum has been observed by Pryor and Kneipp [78] for hydrogen abstraction by alkyl and aryl radicals from thiols and thiophenols (equation 41):

$$R^. + R'SH \rightarrow RH + R'S^. \tag{41}$$

Measurements for seventeen reactions showed a generally smooth dependence of k_H/k_D upon the heat of reaction, calculated from the bond dissociation energies of reactants and products, and gave maximum effects for nearly thermoneutral reactions. Qualitatively similar trends have been reported by Lewis in the radical additions of thiols [79] and hydrogen bromide [80] to olefins.

8.3.4 *Additional factors contributing to isotope effects*

It seems fair to conclude that the experimental data bear out the qualitative dependence of k_H/k_D upon large changes in transition state structure and the existence of isotope maxima predicted by Westheimer. This can be accepted even if major features of the behaviour, such as the occurrence of isotope effects that remain almost constant for large changes in rates or equilibria [54, 65], remain unaccounted for. On the other hand it does not necessarily follow from the observation of maxima that the Westheimer effect is their cause. Apart from the difficulties already noted, since 1961 reservations have been expressed by a number of authors [81–84], and Bell has questioned in particular the likelihood that zero-point energy changes in the real stretching vibration of the transition state could be sufficiently large to account for the small isotope effects observed [2, 52, 55, 81, 82]. The remainder of this chapter considers the possibility that factors other than those emphasized by Westheimer may make an important contribution to the isotope effects.

Bending vibrations. One factor only briefly referred to in Westheimer's treatment is the influence of bending vibrations. Westheimer assumed that zero-point energy changes in bending vibrations cancel between the reactants and transition state, but is this assumption justified?

We have seen that for polyatomic molecules and characteristically for organic molecules there are two bending vibrations sensitive to isotopic substitution. There are also two such vibrations in transition states for proton transfer. For the simple case of linear reactants, products, and transition state, the force constants controlling these vibrations are shown below.

$$A\!-\!B\!-\!H + X\!-\!Y \rightarrow [A\!-\!B\cdots H\cdots X\!-\!Y]^{\ddagger} \rightarrow A\!-\!B + H\!-\!X\!-\!Y$$

$$\overset{}{\underset{f_R}{}} \qquad \overset{f_t}{\underset{f_r \quad f_p}{}} \qquad \underset{f_P}{}$$

The reactant and product each have a pair of equal force constants (in plane and out of plane) and there are three such pairs of force constants in the transition state.

It is easy to see how the transition state force constants, f_r, f_p and f_t, change with the degree of proton transfer: f_r decreases from its value in the reactant and falls to zero; f_p increases from zero to its value in the product, and f_t increases from and falls again to zero. Although the magnitude of f_t is not known it probably has its largest value when f_r and f_p are both weak. Provided this is not substantially greater than values for symmetrical hydrogen bonds, it follows that bending frequencies should vary smoothly between their reactant and product values, with perhaps some increase or decrease for symmetrical transition states, in which f_t is dominant. Thus bending vibrations should normally contribute only a minor modulation to the isotope maximum arising from the Westheimer effect, and several calculations show how this can be so [26, 85–87]. Although a more substantial role for bending vibrations has been suggested [83], this seems to have originated in a mistaken assignment of the 'symmetries' and thence (by analogy with weak hydrogen bonds) the relative magnitudes of the real and imaginary stretching vibrations of unsymmetrical transition states [87]. The form of these vibrations has been clarified by Albery [18].

Strictly speaking the separation of bending and stretching vibrations is only possible when reactants and transition states are linear and the vibrations are orthogonal. In practice it is also possible for monovalent hydrogen compounds and for hydrogen bonds or transition states in which the bonds to the hydrogen are linear since in these cases the vibrations are virtually orthogonal. Thus, Westheimer's treatment of stretching vibrations is consistent with the spirit if not the letter of Wolfsberg and Stern's cut-off procedure [15], while consideration of variations of bending force constants derived from non-linear rather than linear reactants will introduce no qualitatively new factors.

If the above views are correct, one might expect bending vibrations to reveal themselves by their absence, and it has been plausibly argued by Kresge and Chiang [88] and by Lewis and Kozuka [80] that low isotope effects observed for proton or hydrogen transfer from hydrogen halides stems from the presence of uncompensated bending vibrations in the transition state. For diatomic reactants bending vibrations can arise only from intermolecular hydrogen bonding and will be weak if present at all. Although two free rotations contribute a factor of about 2 to k_H/k_D this would not compensate for the adverse zero-point energy change (a factor of 5 for bending frequencies of 1200 cm^{-1}).

A further suggestion made by Saunders [89] is that bending vibrations may play a part in small isotope effects for unsymmetrical transition states through a small increase in frequencies relative to their reactant or product values.

Non-linear transition states. The operation of the Westheimer effect depends on the transition state being linear. If it is not, the balance of forces that leads to isotopic insensitivity for the real stretching vibration in a symmetrical transition state cannot occur. An obvious

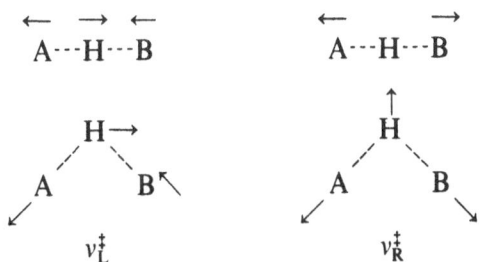

Stretching modes of linear and non-linear transition states

corollary is that for non-linear transition states the real stretching mode will be isotopically sensitive and small isotope effects may be found even in the symmetrical case. A full explanation of the difference between linear and non-linear transition states should take account of bending vibrations, but calculations confirm that for the same force constants k_H/k_D falls as the angle between the bonds to hydrogen decreases [90–92], while for angles of less than 90° k_H/k_D can be brought above a value of 2 to 3 only by substantial tunnelling or by use of what would seem to be unreasonably low bond-stretching force constants [90].

The proposal that low isotope effects may be associated with bent transition states was made before Westheimer's treatment of linear transition states became available [93]. Nonetheless experimental confirmation would clearly provide evidence in its support. In fact there are a number of reactions, most notably 1,2-hyride rearrangements and carbene and nitrene insertions, for which the consistent observation of small isotope effects can reasonably be ascribed to non-linear transition states [90]. On the other hand the mechanisms of the reactions have not always been entirely clear and it may perhaps be questioned whether a definite and general conclusion can yet be drawn.

It should be noted that, although the dependence of isotope effects on transition state geometry derives from changes in zero-point energy, it does not violate the principle that force-constant changes are required for the observation of quantum-mechanical isotope effects [11]. This applies strictly to atomic cartesian force constants and although it is often a good approximation for a general valence force field, for bending vibrations (equation 11), and in the present case even for stretching vibrations, the effective masses of the vibrations depend upon geometry.

The stretching interaction force constant and tunnelling. So far only the effect of the bond force constants f_1 and f_2 upon the stretching vibrations of the transition state has been considered. Normally the transition state also has a large interaction force constant, f_{12}. In the usual linear 3-atom model the magnitude of this constant determines

$$f_1 \quad f_2$$

$$A\text{···}H\text{···}B$$

whether the model represents a hydrogen bond or a transition state. In a transition state, the reaction-coordinate mode is normally represented by the asymmetric stretching vibration, and the condition

236

that motion in this coordinate occurs across an energy maximum is, as we have seen (p. 219), that its force constant f_a should be negative. As is implied in equations 20–22, this is true only if $f_{12} > (f_1 f_2)^{\frac{1}{2}}$.

Through f_a the magnitude of f_{12} has an important influence upon k_H/k_D. When f_a is negative, motion along the reaction coordinate introduces a new contribution to the isotope effect, namely tunnelling. Tunnelling leads to a faster reaction of hydrogen than of its heavier isotopes, and increases in importance with decreasing temperature. As has been pointed out by Bell, within transition-state theory, it corresponds to the quantum correction to the reaction coordinate, analogous in status and formulation to the quantum corrections to stable coordinates that are the origin of zero-point energies.

In practice tunnelling is not confined to the transition state region of a potential surface, and normally a useful correction cannot be based simply on transition state properties. Usually it is evaluated by numerical [22, 94] or exact [2, 3, 95, 96, 97] integration of reaction trajectories along a coordinate extending over the whole reaction path, making use of an assumed or calculated one-dimensional energy barrier. However, since comparisons with exact calculations based on full potential surfaces for simplified model reactions [97] suggest that even these procedures may not be of more than empirical value, there has been some uncertainty as to the quantitative and even qualitative magnitude of tunnelling contributions.

The problem of tunnelling is a large one and has been considered extensively elsewhere [2, 3, 36, 94, 97]. However, because tunnelling was neglected by Westheimer it is important at least to consider whether its presence may modify the interpretation of isotope maxima. Notwithstanding the difficulties of a full interpretation, it is clear that the degree of tunnelling is greatest for the motion of small masses and for high thin energy barriers. These characteristics can be expected to correlate with the following properties of the transition state: (a) curvature of the potential surface along the reaction coordinate; (b) sensitivity of motion along the reaction coordinate to the isotopic mass of hydrogen; and (c) the activation energy of the reaction in the thermodynamically favoured direction [13]. When this is so it is possible to consider the qualitative dependence of tunnelling upon transition-state structure without introducing any new factors into the discussion.

The contribution of (a) and (b) is conveniently expressed in terms of the reaction coordinate frequency $v_L^{\ddagger}$ which, as already noted, is equal to $(f_a/m_a^{\ddagger})^{\frac{1}{2}}/2\pi$ where $m_a^{\ddagger}$ is the reduced mass for motion along the reaction coordinate, and the fact that f_a is negative makes $v_L^{\ddagger}$ imaginary.

Evidently, a large isotopically-sensitive value for $v_L^{\ddagger}$ will be associated with a large degree of tunnelling. In Westheimer's treatment [1] it was assumed that f_a and $v_L^{\ddagger}$ are zero for all transition states, corresponding to a flat-topped energy barrier and no tunnelling correction. A more reasonable behaviour for $v_L^{\ddagger}$ may be deduced from the relationship between transition-state structure and activation energy implied by Hammond's postulate [44]. As shown in Scheme 3, a flat-topped

$[A-H\cdots B]^{\ddagger}$ $[A\cdots H\cdots B]^{\ddagger}$ $[A\cdots H-B]^{\ddagger}$

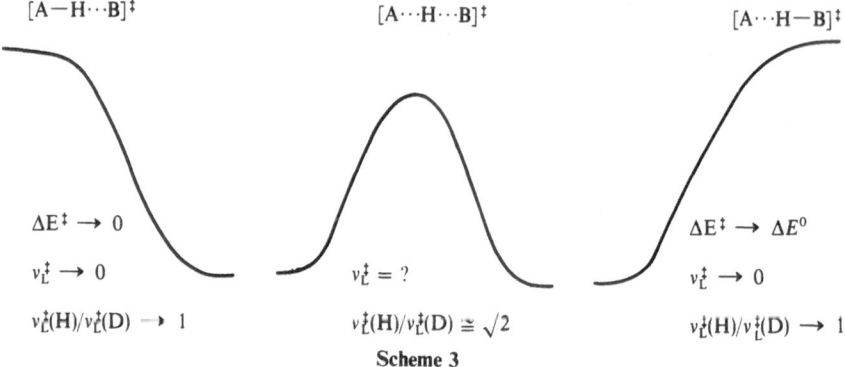

$\Delta E^{\ddagger} \rightarrow 0$ $v_L^{\ddagger} = ?$ $\Delta E^{\ddagger} \rightarrow \Delta E^0$

$v_L^{\ddagger} \rightarrow 0$ $v_L^{\ddagger} \rightarrow 0$

$v_L^{\ddagger}(H)/v_L^{\ddagger}(D) \rightarrow 1$ $v_L^{\ddagger}(H)/v_L^{\ddagger}(D) \cong \sqrt{2}$ $v_L^{\ddagger}(H)/v_L^{\ddagger}(D) \rightarrow 1$

Scheme 3

barrier is expected only for highly exothermic or endothermic reactions, in which the structure and energy of the transition state approach those of the reactants or products. In terms of the criterion of barrier curvature, therefore, a minimum tunnelling correction is expected for extreme reactant-like or product-like transition states, and presumably a maximum correction for a symmetrical transition state. Interestingly, both other criteria show the same transition-state dependence. The isotopic sensitivity of $v_L^{\ddagger}$ is related to that of the real stretching vibration $v_R^{\ddagger}$ through the product rule:

$$\frac{v_L^{\ddagger}(H)v_R^{\ddagger}(H)}{v_L^{\ddagger}(D)v_R^{\ddagger}(D)} \cong 2^{\frac{1}{2}}$$

and the minimum sensitivity of $v_R^{\ddagger}$ for the symmetrical case implies a maximum sensitivity for $v_L^{\ddagger}$. Similarly the maximum activation energy effective for tunnelling is probably associated with a thermoneutral reaction and symmetrical transition state.

Thus each of the factors (a), (b) and (c) suggests that the tunnelling contribution to k_H/k_D will show the same qualitative variation as that from zero-point energy, with a maximum value for a symmetrical transition state. Provided that the correction is not large it will represent only a minor accentuation of the Westheimer effect. A number of calculations have modified Westheimer's model to include behaviour of this type [26, 85, 87]. Typical maximum values of $\omega_L^{\ddagger}(H)$ used have been $700i$ and $1000i$ cm^{-1}, for which at 25 C, Bell's tunnel correction $\Gamma = (h\nu_L^\ddagger/2kT)/\sin(h\nu_L^\ddagger/2kT)$ for a parabolic barrier [2, 95] of curvature $4\pi^2\nu_L^{\ddagger 2}$ yields corrections of $\Gamma_H/\Gamma_D = 1.3$ and 2.0 respectively if an isotopic sensitivity of 1.38 is assumed for $\nu_L^\ddagger$. Judging by calculations of potential energy surfaces [3], these values for $\omega_L^\ddagger$ are on the small side. However, a one-dimensional parabolic barrier gives a large tunnel correction and by most estimates is usually too large [3, 96, 97]. A priori therefore it is not easy to decide whether tunnelling is likely to be of major importance or not.

Experimentally, the principal criteria that have been used to detect tunnelling are (i) unusually large isotope effects, (ii) values of A_H/A_D smaller than estimated on the basis of classical partition functions (for polyatomic reactants in solution, normally unity); (iii) curvature of an Arrhenius plot of $\log k_H$ versus $1/T$. All of these criteria have been reliably observed [2, 36] (though good examples of (iii) are rare), and usually very large isotope effects have been associated with small values of A_H/A_D [13, 98], although it may not be necessary that this should be so [99, 100]. Unfortunately, the criteria are crude and there is no sensitive means of separating tunnelling from zero-point energy changes. Perhaps the best assessment that can be made is that while there is no doubt as to the importance of tunnelling at low temperatures [2, 36], and at higher temperatures a number of dramatically large isotope effects ($k_H/k_D = 15\text{--}25$ at $25°C$) that seem hardly explicable in other terms have been observed in individual cases [2, 13, 36, 50, 72, 101, 102], usually for approximately thermoneutral reactions, the incidence of substantial tunnelling appears to be sufficiently random to suggest an exceptional rather than a general effect.

Of particular interest is Lewis and Funderburk's observation of an isotope effect of 24 for reaction of the sterically hindered base 2,6-dimethylpyridine with 2-nitropropane, while pyridine bases of the same strength but without the 2,6-substituents give much smaller effects [72]. The possibility of sterically inducing tunnelling in this way has been extended to other reactions and the finding, admittedly in a limited number of cases, that the effect is either fairly large [52, 72, 103, 104] or completely absent [55, 56, 75, 105] is again consistent with its occurrence only in specially favourable circumstances. It seems fair to conclude that at present the experimental data offer no compelling reason for invoking large tunnelling corrections as a general rule. Interestingly, in cases where tunnelling probably is important Caldin and Tomalin have suggested that there is qualitative evidence that the effect overlies a normal Westheimer dependence of zero-point energy contributions [106].

The interaction force constant f_{12} is not only important in determining $\omega_L^{\ddagger}$ and the extent of tunnelling; it also affects isotopic zero-point energy changes. In the cases considered above, f_{12} is only slightly larger than $(f_1 f_2)^{\frac{1}{2}}$. Even when $\omega_L^{\ddagger}$ has its maximum value of $1000i$ cm^{-1}, provided that f_1 and f_2 are not themselves very small $f_{12} \not> 1.2(f_1 f_2)^{\frac{1}{2}}$. These values lead to little change from the behaviour found by Westheimer with $f_{12} = (f_1 f_2)^{\frac{1}{2}}$ in all cases [18]. However, it has been pointed out by Wolfsberg and Willi that for much larger values of f_{12} the isotopic insensitivity of the real vibration $\omega_R^{\ddagger}$ applies not only to symmetrical transition states but extends even to very unsymmetrical ones, giving a maximum zero-point energy contribution to k_H/k_D practically independent of the structure of the transition state [84, 107]. The value of $f_{12} = 3.5(f_1 f_2)^{\frac{1}{2}}$ when $f_1 = 10f_2$ obtained from an electrostatic calculation [108] has been cited by Bell [81] as being consistent with this behaviour, and the effects of various combinations of f_1, f_2 and f_{12} have been compared graphically by Albery [18].

Large values for f_{12} in unsymmetrical transition states have several consequences. Probably the most important is that Westheimer's explanation of small isotope effects would no longer apply. There should also be extensive tunnelling, due to the large values of $v_L^{\ddagger}$, but this is perhaps less certain. A further point of interest is that the reaction coordinate is strongly distorted from the minimum reaction path. Because of the instability of unbound H$^+$, conservation of bonding should lead to a strongly curved reaction path [38] for proton transfers as in Figure 4. When the effective mass for motion along the reaction coordinate $m_L^{\ddagger} \sim m_H$ (as is implied by the isotopic insensitivity of $v_R^{\ddagger}$), the coordinate is at an angle of 45° to the bond axes and parallel to the line aa' in Figure 4. Thus it is directed along the reaction path only in a symmetrical transition state. In unsymmetrical cases the deviation of the reaction coordinate from the reaction path leads to difficulties with the application of transition-state theory [109, 110], including the calculation of isotope effects [110]. Also, even though $v_L^{\ddagger}$ is large it underestimates the sharpness of the energy barrier along the true reaction path.

It seems unlikely that extreme behaviour of this kind would occur. However, values of f_{12} might be large enough to 'broaden' the isotope maximum contributed by the zero-point energy. Provided that tunnelling was relatively unimportant (despite the high value and isotopic sensitivity of $v_L^{\ddagger}$), this could be responsible for the large values of k_H/k_D insensitive to wide changes in ΔpK and reactivity noted by Challis [65]. A variation on this behaviour suggested by Bell is that the structure of the transition state may be symmetrical irrespective of the activation energy and energy of reaction, and that differences of isotope effect simply reflect differences in tunnelling for various barrier heights. A model simulating this behaviour has been described by Bell, Sachs and Tranter [82], and Challis and Millar [65] have noted that if tunnelling is again unimportant it too would account for their observations.

Coupling of hydrogen and heavy-atom motions in the reaction coordinate. Since Westheimer's explanation of small isotope effects would be of little significance for transition states with the large interaction force constants or unchanging structure described above, it is important to consider what other factors could be responsible. One possibility pointed out by Bell [81] is that when proton transfer is concerted with displacement of a heavy atom, as in β-eliminations, there must be electronic (as opposed to mechanical) coupling of the hydrogen and heavy-atom motions in the reaction coordinate. This should diminish the importance of hydrogen motion in the reaction coordinate and correspondingly increase its importance in a stable

$$
\begin{array}{c}
\text{B} \\
\quad \diagdown \\
\qquad \text{H} \\
\qquad \vdots \\
\qquad \overset{\rightarrow \; \leftarrow}{\text{C}=\!=\!\text{C}} \\
\qquad\qquad \diagdown \\
\qquad\qquad\quad \text{X}
\end{array}
$$

coordinate, leading to an isotopic zero-point energy change decreasing the isotope effect. Since even in simple proton transfers the charge delocalization that normally occurs has the character of an intramolecular elimination (equation 42), this effect could be of general application [81].

$$
\text{B} + \text{H}-\text{C}-\text{C}=\!\!=\!\!\text{O} \;\rightarrow\; \text{BH}^+ + \;\diagdown\!\text{C}=\text{C}-\text{O}^- \tag{42}
$$

Bell's idea has been tested by Katz and Saunders [86] who have confirmed that in transition-state models for β-eliminations the introduction of suitably strong interaction force constants does lead to the expected effect. It is perhaps surprising therefore that experimentally there is no clear sign of the influence of coupling upon k_H/k_D in concerted eliminations. Extensive measurements show large values (6–10) to be common [6]; an isotope maximum with changes in base strength has been observed [58]; and in general there is an apparently normal correlation of k_H/k_D with the degree of hydrogen transfer in the transition state [7, 111]. Katz and Saunders note that this is understandable if motion of the heavy atom in the reaction coordinate is most important when there is little C—H bond-breaking and the Westheimer effect leads to low isotope effects anyway. It is unlikely that the similarity of concerted and 'non-concerted' proton transfers reflects a pervasive influence of coupling, because small isotope effects are observed for hydrogen transfer between single atoms in the gas phase where no coupling can occur (p. 253 below).

An interesting class of concerted reactions that do appear to give consistently small isotope effects are those involving proton transfer between oxygen or nitrogen atoms. These occur especially in general acid and general base catalysed addition-elimination reactions at a carbonyl group [112], and in so far as they involve solvent isotope effects [5, 113] they are beyond the scope of this chapter. However, it is worth noting that the explanation offered for the smallness of these effects is that there is a (mechanical) *uncoupling* of hydrogen and heavy atom motion and that only the heavy atom moves in the reaction coordinate [5, 45].

8.3.5 *Calculations based on potential-energy surfaces*

It is useful to consider to what extent qualitative interpretations of isotope effects are backed up by calculations. The most fundamental calculations using transition-state theory are those based on a calculated potential-energy surface for the reaction. We have seen how transition-state force constants measure the curvature of a surface at its saddle point. Numerical or analytical evaluation of the appropriate derivatives $d^2V/d\Delta r^2$ for a calculated surface permits calculation of force constants and thence of isotope effects. Moreover the surface also yields an activation energy and energy of reaction, and there is no need for additional assumptions about the relation between these properties and force constants before making comparisons with experiment. This advantage over less fundamental treatments where force constants are assigned directly has been emphasized by Bell [2, 82].

Unfortunately realistic *ab initio* quantum calculations of potential surfaces are possible only for the very simplest reactions. The reaction of $H + H_2$ has now been intensively investigated [114], but comparable calculations for other experimentally accessible cases, such as the reactions of hydrogen with fluorine, are only just becoming available [115, 116]. While these surfaces are of great importance, they are not suitable for the systematic investigation of the variations in isotope effects from reaction to reaction that are of interest in connection with the Westheimer effect, and for this purpose it is necessary to use simpler empirical or 'semi-empirical' methods of calculation.

The most widely used of such methods are the so-called London-Eyring-Polanyi-Sato (LEPS) method based on London's valence bond treatment of H_3, and the empirical Bond-Energy-Bond-Order (BEBO) method of Johnston [3, 19, 117, 118]. The calculations apply strictly only to triatomic systems, and make use of the dissociation energies or Morse potential curves for the diatomic reactants and products together with an assumed repulsive potential between the non-bonded atoms as input parameters. The BEBO method does not give a complete surface but yields sufficient information for calculation of transition-state properties. The validity of the calculations must be judged by comparisons with the results of good *ab initio* calculations where available, and comparison of predicted with experimental measurements of activation energies, isotope effects, and in principle, where reaction trajectories have been computed, with energy partitioning between vibrational, rotational, and translational modes of the products [117–119]. Considering the great difficulty of exact calculations, the methods have been remarkably successful, and comparisons between them usually give good agreement. Nonetheless the occasional occurrence of capricious results [116, 118, 120] is symptomatic of their essentially empirical nature, and it would probably be dangerous to draw more than qualitative conclusions from their application.

The methods apply to radical reactions in the gas phase rather than to proton transfers in solution, but this is true of most interpretations of isotope effects. Experimentally, hydrogen radical and proton tranfers show the same qualitative characteristics with respect to magnitude, susceptibility to tunnelling and dependence on reactivity, and so far interpretations of results from solution have been too crude to distinguish them from gas-phase behaviour. Nevertheless, the reservation should be borne in mind.

Johnston has carried out calculations using both BEBO and LEPS methods for a series of triatomic reaction models involving hydrogen transfer between carbon radicals for (classical) energies of reaction ΔE_c ranging from 0 to ± 33 kcal mol^{-1}. Calculated transition-state frequencies and force constants obtained by the BEBO method are listed, together with isotope effects, as a function of the energy of reaction and of the activation energy ($\Delta E_c^{\ddagger}$) in Table 7. The isotope effects are calculated from zero-point energy changes for the stretching

Table 7

BEBO calculations of transition-state properties and isotope effects for hydrogen transfer between carbon radicals[a]

[b]ΔE_c/kcal mol^{-1}	0.00	-9.10	-16.70	-22.10	-33.30
[b]$\Delta E_c^{\ddagger}$/kcal mol^{-1}	12.80	8.40	5.40	3.60	1.60
$10^5 f_1$/md Å^{-1}	0.71	1.75	2.73	3.46	4.25
$10^5 f_2$/md Å^{-1}	0.71	0.11	-0.19	-0.15	-0.13
$10^5 f_{12}$/md Å^{-1}	1.79	1.64	1.38	1.05	0.59
$\omega_t^{\ddagger}$/cm^{-1}	1950i	1520i	940i	400i	200i
$\omega_r^{\ddagger}$/cm^{-1}	420	480	820	1570	2320
[c]ω_b/cm^{-1}	740	710	660	580	450
[d]H/D($\omega_t^{\ddagger}$)	1.40	1.33	1.26	1.06	1.02
[d]H/D($\omega_r^{\ddagger}$)	1.00	1.05	1.11	1.32	1.38
ω_H(reactant)/cm^{-1}	2960	2820	2700	2600	2400
[e]k_H/k_D(25°C)	7.30	6.30	5.00	2.30	1.10

[a] Results from H. S. Johnston *Gas Phase Reaction Rate Theory*, Ronald Press, New York (1966); Table 13-2, p. 242. The notation is that used throughout the chapter.
[b] Classical energy of activation or reaction.
[c] Bending frequencies.
[d] Isotopic sensitivity of vibration frequency.
[e] Calculated from zero-point energy changes in stretching vibrations only. Values of k_H/k_D for the reverse reaction may be obtained from the equilibrium isotope effects, $K_H/K_D = 1.0$, 0.91, 0.83, 0.79, 0.62 (across the table) calculated from Johnston's product frequencies (ω_H is constant at 2960 cm^{-1}).

vibrations only, and no correction is made for tunnelling. Bending frequencies are not included, since although calculated they are not appropriate to polyatomic reactants. Exothermic reactions only are shown, because isotope effects in the reverse (endothermic) direction, calculable from the equilibrium values in footnote (e), are nearly the same.

The table illustrates many of the points considered qualitatively above. There is a maximum isotope effect for a symmetrical transition state and the expected dependence of the force constants and vibration frequencies on activation energy and energy of reaction is found. Negative bond force constants have not been considered previously but they reflect encroachment of the energy barrier into the reactant or product valleys of the potential surface where the reaction path is nearly parallel to one of the bond axes (*cf.* Figure 5). For a crude comparison with experiment, if it is assumed that $\Delta E_c = \Delta G^0$ the most exothermic reaction with $\Delta E_c = 33$ kcal mol^{-1} corresponds to a ΔpK of 24.

It is noteworthy that the bond force constants for a symmetrical transition state are weak, and that the interaction force constants are sufficiently large ($f_{12}/(f_1 f_2)^{\frac{1}{2}} \sim 2.5$ for $\Delta E_c = 0$) to give large values for the reaction-coordinate frequency $v_L^{\ddagger}$, and to lead to rather small changes in k_H/k_D in the range $\Delta E_c = -17$ to $+17\,\mathrm{kcal\,mol^{-1}}$. The constancy of values of k_H/k_D, which is slightly underestimated because of a variation in reactant frequencies which is unrealistic for C—H bonds*, is more marked if the calculations are made with the LEPS method. The behaviour thus broadly reflects that envisaged by Wolfsberg and Willi [84], and, as already noted, could explain the relatively constant values of k_H/k_D observed experimentally by Challis [65] and Bordwell [54] for proton transfers with values of ΔpK in the range -10 to $+10$.

On the other hand generalization from this behaviour would be rash. Comparison of calculations for the H + H$_2$ hydrogen–exchange reactions with the best available *ab initio* results suggest that both BEBO and LEPS overestimate their barrier curvatures ($v_L^{\ddagger}$), as is shown in Table 8. Moreover, the relatively high *ab initio* value of

Table 8

Stretching force constants and vibration frequencies for the H$_3$ transition state

	Ab initio[a]	BEBO[b]	LEPS[b]
$f_1 = f_2/\mathrm{md\,\mathring{A}^{-1}}$	1.02	1.32	0.96
$f_{12}/\mathrm{md\,\mathring{A}^{-1}}$	1.47	1.89	1.68
$\omega_L^{\ddagger}/\mathrm{cm^{-1}}$	1510i	1700i	1920i
$\omega_R^{\ddagger}/\mathrm{cm^{-1}}$	2050	2330	2110

[a] Calculated by G. W. Koeppl, *J. Chem. Phys.*, **59**, 3425 (1973) for the H$_3$ surface due to B. Liu, *J. Chem. Phys.*, **58**, 1925 (1973). For comparison $f_1 = 0.98$ and $f_{12} = 1.45$ on Shavitt's 'scaled SSMK' surface; I. Shavitt, *J. Chem. Phys.*, **49**, 4048 (1968).

[b] From H. S. Johnston, *Gas Phase Reaction Rate Theory*, Ronald Press, New York (1966); Table 10–7, p. 189.

$1510i\,\mathrm{cm^{-1}}$ for H$_3$ partly reflects a low effective mass. For the same force constants, hydrogen transfers between atoms of mass 12 would give $\omega_L^{\ddagger} = 1180i\,\mathrm{cm^{-1}}$. Although it does not strictly follow, one may tentatively infer that BEBO and LEPS correspondingly exaggerate the 'broadness' of their isotope maxima.

In view of the high values of $\omega_L^{\ddagger}$ and the substantial barrier curvatures they imply, even in the case of the modified *ab initio* value, it may be questioned whether tunnelling corrections can be neglected. However, in what is probably the best available theoretical estimate,

* Probably this derives from an assumed proportionality between force constants and bond dissociation energies. It is not a necessary feature of the BEBO method.

from numerical integration over a two-dimensional surface for H_3, the reaction probability calculated for systems with less energy than required to surmount the classical activation barrier was closely equivalent to that obtained from a one-dimensional parabolic barrier with a curvature corresponding to $\omega_L^{\ddagger} = 922\,\text{cm}^{-1}$ [97], i.e. about 60 per cent of the 'true' value for the surface. This would correspond to $\omega_L^{\ddagger} \sim 700i\,\text{cm}^{-1}$ for H-transfer between heavy atoms which, as we have seen, even for an effective mass $m_L^{\ddagger} = m_H$ leads to a contribution of only 1.3 to k_H/k_D at 25°C.

One further point illustrated in Table 7 is that the barrier curvature is a sensitive function of the energy of reaction and that the structure of the transition state, the activation energy, and the energy of reaction show the expected correlation envisaged in Hammond's postulate. This point has been examined in detail with the aid of BEBO and LEPS surfaces by Johnston [3, 121] and by Mok and Polanyi [43], and the fact that in highly exothermic reactions the saddle point moves into the reactant valley to give a reactant-like transition state is confirmed by SCF calculations including configuration interaction for the hydrogen transfer reaction [115] $H_2 + F \rightarrow HF + H$ and the proton transfer [122] $HeH^+ + H_2 \rightarrow He + H_3^+$. Experimental support for the relationship is also beginning to come from comparisons of correlations between energy distributions in reactants and products obtained from molecular beam and chemiluminescence measurements with predictions based on trajectory calculations for computed surfaces [119].

8.3.6 Anti-Hammond substituent effects

It is now necessary to consider to what extent deviations from smooth correlations of k_H/k_D with $\Delta G^{\ddagger}$ or ΔG^0 in proton transfer reactions reflect departures from the Hammond postulate. The Hammond postulate may be expected to apply to reasonably closely related three-centre reactions of the type considered above, and in particular to their simplified representations by empirical potential-energy surfaces. It does not follow that it applies generally to more complex reactions, and it is now well established that for concerted ionic reactions failure of bond-making and bond-breaking to remain properly in step can lead in the transition state to charge–localizations at atoms that bear no charge in the reactants or products, with the consequence that substituents at these atoms show no correlation between their effects on reaction rates and equilibria [123–126]. This behaviour has been best characterized in E2 β-eliminations, where for example an imbalance of π-bond-making and β-C—H bond-breaking can lead to anionic charge at the β-carbon in the transition state [111, 123].

$$B + \; {}^{H}\diagdown \diagdown_{X} \; \rightarrow \; \left[\begin{array}{c} B \cdots H \\ {}_{\delta^-} \diagdown \diagdown_{X} \end{array} \right]^{\ddagger} \; \rightarrow \; BH^+ + = + X^-$$

The well-known Hofmann effect of β-alkyl substituents of simultaneously decreasing the rate and increasing the equilibrium constant is believed to be a manifestation of this effect [127].

This point is of interest in the light of Bell's suggestion of an analogy between concerted eliminations and proton transfers to or from carbon atoms subject to extensive conjugation [81]. Thus the Hofmann anomaly for alkyl substituents is also observed in the ionization of nitroalkanes [128], and indeed in these reactions a general lack of correlation between kinetic and equilibrium substituent effects has been amply demonstrated by Bordwell [66, 129]. That the behaviour reflects the presence of substituent interactions in the transition state differing in character or relative importance from those present in reactants or products has been well recognized [46, 67, 130, 131], and the parallel with elimination reactions seems particularly clear [46]. If Bell's analogy is correct, common sense

$$B + CH_3NO_2 \; \rightarrow \; \left[B \cdots H \cdots \overset{\delta^-}{C}H_2 - \overset{+}{N} \diagup^{O^-}_{\diagdown O} \right]^{\ddagger}$$

$$\rightarrow BH^+ + CH_2 = \overset{+}{N} \diagup^{O^-}_{\diagdown O^-}$$

and detailed considerations of substituent effects in concerted reactions [45, 46, 125, 126] suggest that the degree of proton transfer in the transition state need not change in the manner envisaged in the Hammond postulate with either the rate or the equilibrium constant. It follows that in a plot of k_H/k_D against ΔpK substituents at carbon subject to these anomalies may cause deviations from a smooth correlation.

A further feature of substituent effects in concerted proton transfers is the possibility that changes in strength of the catalysing base lead to little variation in the extent of proton transfer in the transition state [46, 125, 126]. This has been suggested as an explanation for the observation of extended linear Brønsted plots [46, 126], and it would be of interest to know if it could also lead to constant primary isotope effects. Most examples studied pertain to elimination or addition at a carbonyl group [46, 126] where proton transfer occurs between oxygen atoms. However, Kemp and Casey [75] have measured both Brønsted coefficients and isotope effects for eliminations of substituted benzisoxazoles, in which proton transfer occurs from carbon. In this case it was found that Brønsted exponents were substantially insensitive to substituents in either the base or the substrate, and for

the unsubstituted- and 5-nitro-benzisoxazoles reaction with H_2O, OH^- and tertiary amines gave isotope effects of between 4 and 5 for a reactivity range of 10^8, while for 5,7-dinitro-benzisoxazole, which possibly has a slightly smaller Brønsted exponent (0.65 instead of 0.75), values of k_H/k_D were between 5 and 6.

These results confirm that constant isotope effects are possible in concerted proton transfers. On the basis of the analogy with 'simple' proton transfers, one may speculate that the nearly constant values in the hydrogen exchange of indoles [65] and ionizations of nitroalkanes and ketones [54] over large ΔpK ranges may have a comparable origin and reflect an unchanging degree of proton transfer in the transition state. Consistent with this is the small variation in Brønsted exponents for base catalysis of ionization of a number of nitroalkanes [129] and possibly also of groups of ketones [75]. Constant Brønsted exponents and isotope effects have also been found for the protonation of diazoalkanes [73] differing in reactivity by a factor of 10^5. Values are shown in Table 10, where α_B and α_i refer to exponents determined from the reaction of carboxylic acids and measurement of secondary isotope effects upon proton transfer from H_3O^+ respectively.

The diazo–alkanes nicely illustrate the complexity of substituent behaviour possible in 'simple' proton transfers. While the structure of the transition state shows itself insensitive to substituents in the diazo substrate, the sharp curvature of Brønsted plots [68] and, probably, the difference between α_B and α_i in Table 9 indicate that it is hypersensitive to substituents in the catalysing acid. Interpretations of this behaviour, and of anomalous substituent effects in general [5, 45, 46, 73, 125, 126, 130, 131], are beyond the scope of this chapter, and the object here is simply to make clear that such effects can play an important part in determining the form of correlations between k_H/k_D and $\log k$ or ΔpK.

On the other hand, to an organic chemist, complex reactivity patterns are often of primary interest and isotope effects are seen principally as a tool to assist in their investigation. From this point of view an encouraging feature of Table 9, noted by Albery and co-workers [73], is the agreement between different criteria in their

Table 9

Brønsted exponents and isotope effects for the protonation of diazoalkanes $R_1R_2CN_2$ at 25°C in aqueous solution

$R_1, R_2 =$	Me, MeCO[a]	Me, CO$_2$Et[a]	H, CO$_2$$^-$ [b]
$^c k_{H_3O^+}/M^{-1}s^{-1}$	0.74	19.2	6.5×10^4
$^c \alpha_B$	0.61	0.59	0.51
$^d \alpha_i$	0.27	0.29	0.30
$^e k_{H_3O^+}/k_{H_2DO^+}$	3.1	3.1	2.9

[a] W. J. Albery, A. N. Campbell-Crawford, and J. S. Curran, *J. Chem. Soc., Perkin II*, 2206 (1972).
[b] M. M. Kreevoy and D. E. Konasewich, *Adv. Chem. Phys.*, **21**, 243 (1971).
[c] Brønsted exponent for reaction of carboxylic acids.
[d] Brønsted exponent from secondary isotope effect upon reaction of H_3O^+.
[e] Primary isotope effect.

indication of an unvarying degree of proton transfer in the transition state from one substrate to another. As has been emphasized by Bordwell, the lack of rate-equilibrium correlations of substituent effects at carbon normally precludes their use as simple criteria of transition-state structure. Contrariwise, the consistent observation of rate-equilibrium correlations for substituents in oxygen and nitrogen acids or bases suggests that they may be so used [67, 130]. The conformity between Brønsted exponents and k_H/k_D apparent in Table 9, as well as in Table 6 and in other studies [73, 75, 130], suggests that both may properly reflect the degree of proton transfer in the transition state.

8.3.7 *Isotope effects in gas phase reactions*

We have now nearly concluded our consideration of isotope effects upon proton transfer in solution, and it is of interest to ask whether the conclusions reached are consistent with measurements in the gas phase. Although not of direct concern here, gas-phase reactions in many ways are better understood than solution reactions and at least in simple cases a more fundamental approach to the calculation of isotope effects is possible.

In practice the emphasis of gas-phase studies has often differed from that of solution measurements. Careful experimental measurements over wide temperature ranges have been made, and have been used to test potential-energy surfaces and methods of calculating rate constants for simple atom-plus-molecule reactions [3, 19]. Not surprisingly, most attention has been given to the $H_2 + H$ hydrogen exchange, the simplest of all hydrogen-transfer reactions. In this case it is possible to feel some confidence in the potential-energy surface for the reaction [114], and considerable effort has been expended in the measurement [22, 132, 133] and calculation [117, 134] of rate constants for its various isotopic modifications. Fairly accurate quantum [97] and 'semi-quantum' [135] scattering calculations have now been made, and they offer a crucial test of simpler and more general methods for evaluating rate constants and isotope effects, notably transition-state theory.

For other hydrogen-transfer reactions, satisfactory *ab initio* potential energy surfaces [115, 116] and approximate quantum trajectory calculations of isotope effects [136] are also available in a few instances, but for the most part experimental measurements have been compared with values calculated on the basis of empirical or semi-empirical potential surfaces, making use of transition-state theory, and discussion has centred on the pros and cons of the surface used and the type of tunnelling correction made [3, 19].

Dependence of k_H/k_D upon the energy of reaction. In general less attention has been paid to the point of principal interest here, the variations of k_H/k_D from one reaction to another. However, this has by no means been entirely overlooked [137–141]. The consistent use of BEBO and LEPS surfaces [3, 19] itself implies that systematic variations are being tested, and for families of reactions with common or comparable isotopic reactants, such as hydrogen molecules or hydrocarbons, the possibility has been considered explicitly [137–141]. These reactions are worth looking at more closely, and isotope effects are listed in Tables 10 and 11 as functions of the observed Arrhenius activation energy E_a and the classical energy of reaction ΔE_c computed from the dissociation and zero-point energies of reactants and products. For ease of comparison, Arrhenius parameters are given and values of k_H/k_D are extrapolated to 25°C,

Table 10

Isotope effects for hydrogen molecule reactions $(X + H_2/D_2)$ in the gas phase[a]

X	Reference	Temp.[b]/°C	A_{H_2}/A_{D_2}[c] 'calc.'	obs.	$E_a^H - E_a^D$ cal mol^{-1}	k_{H_2}/k_{D_2} 25°C	k_{HD}/k_{DH}[d] 25°C	E_a^e kcal mol^{-1}	ΔE_c kcal mol^{-1}
F	1	−114–140	2.8	1.04	370	1.9	1.4	1.6	−31.8
Cl	2	−30–70	2.8	1.44	1130	9.7	1.8	5.5	3.1
Br	3	165–350	2.8	1.32	1250	10.9	3.1	17.6	19.2
I	4	360–527	2.8	1.33	1020	7.4	—	32.8	35.6
O	5	143–695	2.8	1.60	600	~6.2	—	10.2	4.3
H	6	17–417	1.7	1.29	1230[f]	10.3[f]	—	6.7	0
D	7	−106–174	2.4	1.5[b]	770[f]	5.6[f]	—	6.9[f]	0
CF$_3$	8	60–597	2.8	1.30	1050	7.7	3.3	10.7	−1.1
CH$_3$	9	125–445	2.8	0.91	1330	8.6[g]	—	10.9	−4.0

[a] Usually measured competitively. [b] Temperature range applies to k_{H_2}/k_{D_2}. [c] Calculated values from MMI contributions assuming X has infinite mass, except when X = H; for H$_3$ the bond length was taken as 0.92 pm; tunnelling and vibrational excitation are not included. [d] Intramolecular isotope effect. [e] Approximate high temperature limit when log $k - 1/T$ plots are curved. [f] Calculated by the author. [g] There is poor agreement with A_{H_2}/A_{D_2} and $E_a^H - E_a^D$ for the corresponding reaction of CD$_3$; c.f. reference 9.

References to Table 10

1. A. Persky and F. S. Klein, *J. Chem. Phys.*, **59**, 3612, 5578 (1973).
2. A. Persky and F. S. Klein, *J. Chem. Phys.*, **44**, 3617 (1966); Y. Bar Yaakov, A. Persky, and F. S. Klein, *ibid.*, **59**, 2415 (1973).
3. R. B. Timmons and R. E. Weston, Jr., *J. Chem. Phys.*, **41**, 1654 (1964).
4. J. H. Sullivan, *J. Chem. Phys.*, **39**, 3001 (1963).
5. A. A. Westenberg and N. de Hass, *J. Chem. Phys.*, **47**, 92 (1967); **50**, 2512 (1969).
6. K. Quickert and D. J. Le Roy, *J. Chem. Phys.*, **53**, 1325 (1970); W. R. Schultz and D. J. Le Roy, *J. Chem. Phys.*, **42**, 3869 (1966).
7. D. N. Mitchell and D. J. Le Roy, *J. Chem. Phys.*, **58**, 4449 (1973); D. J. Le Roy, B. A. Ridley, and K. A. Quickert, *Disc. Faraday Soc.*, **44**, 92 (1967).
8. C. L. Kibby and R. E. Weston, Jr., *J. Chem. Phys.*, **49**, 4825 (1968).
9. J. S. Shapiro and R. E. Weston, Jr., *J. Chem. Phys.*, **76**, 1669 (1972).

Table 11

Isotope effect for reactions of methane derivatives $RCH_3 + X \rightarrow RCH_2 + XH$

X	Reference	Temp./°C	Deuteriated reactant	A_H/A_D[a] 'calc.'	obs.	$E_a^H - E_a^D$ cal mol^{-1}	k_H/k_D 25°C	E_a kcal mol^{-1}	ΔE_c kcal mol^{-1}
F	1	-114-25	CH_2D_2[b]	1.3	0.81	275	1.3	1.9	-35.8
Cl	2	-23-71	CH_2D_2[b]	1.3	1.09	1300	9.9	3.9	-0.7
Br	3	264-339	CD_4	2.1	1.10	1030	6.3	18.6	15.2
Br	4	120-160	$PhCD_3$	1.9	1.08	1430	12.1	7.2	1.7
Cl	5	70	$PhCD_3$	1.9	—	—	2.1[c]	—	-14.2
Cl	6	109-203	CCl_3D[d]	1.0	0.34[e]	1445[e]	3.9	6.5	-10.0
Cl	6	30-160	C_2D_6	1.9	1.70	270	2.7	1.0	-5.1
CF_3	7	42-437	CD_3H[b]	1.3	0.55[e]	2200[e]	22.5	10.5	-5.1
CD_3	8	260-490	C_2D_6	1.9	(1.0)	1500	12.7	11.5	4.4

[a] Calculated from the MMI contribution of the reactants; methyl rotations are treated as free with an MMI contribution equivalent to that of CH_3Br (cf. p. 213, Table 4, p. 217 and footnote c to Table 10); [b] Intramolecular isotope effect; [c] Calculated from $k_H/k_D = 2.1$ at 70°C taking A_H/A_D as the calculated value; [d] Based on four temperatures only; [e] Calculated by the author.

References to Table 11

1. A. Persky, J. Chem. Phys., **60**, 49 (1974).
2. K. B. Wiberg and E. L. Motell, Tetrahedron, **19**, 2009 (1963).
3. T. Yokota and R. B. Timmons, Intl. J. Chem. Kinetics, **2**, 325 (1970).
4. R. B. Timmons, J. de Guzman, and R. E. Varnerin, J. Am. Chem. Soc., **90**, 5996 (1968).
5. C. Walling and B. Miller, J. Am. Chem. Soc., **79**, 4181 (1957).
6. G. Chiltz, R. Eckling, P. Goldfinger, G. Huybrechts, H. S. Johnston, L. Meyers, and G. Verbake, J. Chem. Phys., **38**, 1053 (1963).
7. T. E. Sharp and H. S. Johnston, J. Chem. Phys., **37**, 1541 (1962).
8. J. R. McNesby, J. Phys. Chem., **64**, 1671 (1960).

although this often fails to do justice to the careful experimental studies made.

Inspection of the tables for signs of isotope maxima comparable to those found in solution at once reveals a preponderance of k_H/k_D values in the range 7 to 10. However, closer examination shows that in most cases the reactions are within 3–4 kcal mol^{-1} of thermoneutrality, so that symmetrical transition states and large isotope effects are not unexpected.

A good example of a homogeneous family of reactions in which the energy varies from strongly exothermic to thermoneutral and strongly endothermic is given by the reactions of H_2 with the halogen atoms [137] F, Cl, Br, and I. For the exothermic reaction of fluorine [142], with $\Delta E_c = -32$ kcal mol^{-1}, $k_H/k_D = 1.9$ which is small as expected. For the reaction [143] of Cl, $\Delta E_c = 3$ kcal mol^{-1} and k_H/k_D increases to 9.7, again as expected. However, for the endothermic reactions of Br and I the isotope effect remains large, changing little from the value for Cl, and it has been suggested [19, 138, 139] that here the behaviour is inconsistent with a correlation of k_H/k_D with reactivity or ΔE_c.

An important difference of reactions involving diatomic hydrides from those in which hydrogen is transferred between carbon and oxygen or nitrogen atoms of polyatomic molecules is that the equilibrium isotope effect may not be close to unity. In assessing the significance of k_H/k_D for a highly endothermic reaction, it is important to consider the magnitude of the equilibrium isotope effect, because it is expected to set a lower limit to the observable kinetic effect. Calculated equilibrium values for the H_2 + halogen reactions at 25°C are shown in Table 12, and it is apparent that for iodine, the most strongly endothermic reaction, the equilibrium effect is sufficiently large for the kinetic value of 7.4 [144] to be consistent with a product-like transition state, as Westheimer's treatment would lead one to expect. Thus in the microscopic reverse reaction k_H/k_D would be 1.8, which is only a little larger than for the H_2 + F reaction. Moreover, since the equilibrium values increase in the order I > Br > Cl, it seems likely that the nearly constant kinetic effects reflect a balance of this factor and that of an increasingly unsymmetrical transition state.

There are further peculiarities in the reactions of hydrogen molecules. In comparing H_2 and D_2, isotope effects are contributed by both hydrogen atoms. The central hydrogen should be characterized by an isotope effect maximum in the normal way (though perhaps a fairly mild one in view of uncompensated bending vibrations in the transition state), but that for the terminal hydrogen probably in-

Table 12

Calculated equilibrium isotope effects[a] for the hydrogen plus halogen reactions $(H_2 + X)$ at 25°C

	X	A_H/A_D	$\exp(-\Delta\varepsilon/2kT)$	K_H/K_D
$D_2 + X \rightarrow XD + D$	F	1.00	1.40	1.40
	Cl	1.00	2.80	2.80
	Br	1.00	3.40	3.40
	I	1.00	4.20	4.20
$HD + X \rightarrow XD + H$	F	1.20	0.27	0.32
	Cl	1.20	0.55	0.65
	Br	1.20	0.65	0.80
	I	1.20	0.82	1.00
$DH + X \rightarrow XH + D$	F	0.87	4.00	3.50
	Cl	0.87	4.00	3.50
	Br	0.87	4.00	3.50
	I	0.87	4.00	3.50

[a] Calculated using classical partition functions and anharmonically corrected frequencies; i.e. zero-point energy $= \frac{1}{2}hc(\bar{\omega}_e - \frac{1}{2}\bar{\omega}_e x_e)$.

creases steadily with the degree of hydrogen transfer in the transition state, approaching the substantial equilibrium value of 3.5. The small value of the intramolecular isotope effect [145] for the reaction of Cl and HD (1.8) probably reflects this behaviour, and the compensating changes in k_H/k_D for the two hydrogens which are expected to accompany increasing degrees of hydrogen transfer in product-like transition states may be another factor leading to constant kinetic effects in the halogen reactions.

Isotope effects for the methane derivatives in Table 11 should be more straightforward, and in a number of comparisons the expected qualitative trends are observed. The rather high value [139] for the endothermic reactions of CH_4 and CD_4 with Br may again reflect a large equilibrium effect ($k_H/k_D \sim 3.0$ is calculated for a single hydrogen [21]) and perhaps also a secondary effect. The value for chlorination of ethane has been noted previously as being out of line [3, 140], and for $CH_4 + CF_3$ the large value probably indicates tunnelling [102]; but on the whole the results appear to be quite consistent with the general behaviour of solution reactions.

Partition function contributions to A_H/A_D. An additional feature of the H_2–D_2 isotope effects of Table 11 is that classical partition-function contributions to A_H/A_D are readily predicted. At not too low temperatures $Q_{D_2}/Q_{H_2} = 4 \times 2^{\frac{1}{2}}$, while for the transition state $H\cdots H\cdots X$, where X is a heavy atom, $Q^{\ddagger}_{D_2X}/Q^{\ddagger}_{H_2X} = 2$. It follows that if there are no unusually low force constants in the transition state $A_H/A_D \cong 2.8$. As may be seen, most measurements in the table are low by a factor of 2. In product-like transition states when X is a single atom (perhaps with $H_2 + Br$ or I for example) it is possible that weak bending vibrations are fully thermally excited and that A_H/A_D more closely approaches its equilibrium value of unity (Table 12), especially when the reaction temperature is high. In other cases, however, this seems less likely and tunnelling may occur. For the $H_2 + Cl$ reaction, the complementary roles of low bending frequencies and of a one-dimensional Eckart tunnelling correction (in fitting calculated to observed isotope effects) have been noted by Stern, Persky and Klein [146].

At this point it seems appropriate to sum up the probable contribution of partition functions to isotope effects in solution. In the transition state for proton transfer between heavy atoms, the only likelihood of thermal excitation of hydrogen vibrations at normal temperatures would seem to be in extreme product-like or reactant-like transition states for proton transfer respectively to or from a monatomic ion. In this case the correlation of transition-state bends with reactant or product rotations may be reflected in very weak bends in the transition state. Normally however the transition state will make no such contributions to A_H/A_D, and contributions of isotopically sensitive partition functions will be confined to the reactants.

The values of A_H/A_D listed in Table 12 were calculated from the MMI contributions of the reactants. The product rule was used and the values are those expected on the basis of the discussion on pp. 213–213. In so far as the reactants are usually methane molecules, or involve a multiply-deuteriated methyl group, the values are a little large (1.3 to 1.9). For solution reactions in which a single hydrogen of a polyatomic molecule is substituted, at ordinary temperatures, the MMI contribution will more commonly be in the range 1.0 to 1.2. Except for abnormally small primary isotope effects, values outside this range should indicate experimental error or tunnelling. For a more detailed discussion reference should be made to papers by Stern and his collaborators [29].

Tunnelling. The incidence of tunnelling in the gas phase appears to mirror that in solution. Despite the prevalence of large isotope effects, only in the case of CF_3 and CH_4 does the magnitude of k_H/k_D alone appear to demand tunnelling [102]. Curvature in Arrhenius plots for hydrogen transfers has been noted in a number of instances [92, 132, 133], but where intensively investigated, as in the

D_2 + H reaction [133], the effect is quite mild, as indeed seems to be expected on the basis of the best computational estimates [97], and a rather minor contribution to k_H/k_D at room temperature or above is implied [148]. A moderate contribution from tunnelling may also be indicated by the low values of A_H/A_D noted above. As in solution, the safest generalization would seem to be that tunnelling contributions are present but save in exceptional circumstances are rather small. Again the best authenticated examples apply to thermoneutral reactions with appreciable energy barriers.

8.4 Conclusions

The observation of primary isotope maxima with changing rate or equilibrium constants within families of proton-transfer reactions is now well established. The explanation of the behaviour offered by Westheimer, that it is a consequence of the dependence of the isotopic sensitivity of zero-point energy in the 'real' stretching vibration of the transition state upon the degree of proton transfer, continues to be the most satisfactory one. Although reservations have been expressed regarding the interpretation of small isotope effects, the alternative possibilities of large transition-state bending vibrations or a coupling of hydrogen and heavy-atom motions in the reaction coordinate have so far received little experimental or theoretical support. The importance of tunnelling has not been fully clarified, but probably it shows qualitatively the same dependence upon transition-state structure as do changes in zero-point energy. Deviations from smooth correlations of isotope effects that appear experimentally seem more often to reflect the complexity of structure-reactivity relationships than a failure of k_H/k_D to respond to changes in transition-state structure.

REFERENCES

[1] F. H. Westheimer, *Chem. Revs.*, **61**, 265 (1961).
[2] R. P. Bell, *The Proton in Chemistry*, 2nd Ed., Chapman and Hall (1973).
[3] H. S. Johnston, *Gas Phase Reaction Rate Theory*, Ronald Press, New York (1966). (a) p. 62, (b) p. 334, (c) Chapter 3.
[4] W. A. Van Hook in *Isotope Effects in Chemical Reactions*, edited by J. C. Collins and N. S. Bowman, ACS Monograph 167, Van Nostrand Reinhold, New York (1971). Earlier reviews are cited in this reference.
[5] R. L. Schowen, *Prog. Phys. Org. Chem.*, **9**, 275 (1972).
[6] A. Fry, *Chem. Soc. Revs.*, **1**, 163 (1972).

[7] E. K. Thornton and E. R. Thornton in *Isotope Effects in Chemical Reactions*, edited by C. J. Collins and N. S. Bowman, Van Nostrand Reinhold, New York (1971).

[8] M. Wolfsberg, *Accts. Chem. Research*, **5**, 225 (1972).

[9] J. Bigeleisen, M. W. Lee, and F. Mandel, *Ann. Rev. Phys. Chem.*, **24**, 407 (1973).

[10] R. M. Barrow, *Molecular Spectroscopy*, McGraw-Hill, New York (1962). This text provides an introduction to the theory of molecular vibrations.

[11] M. J. Stern and M. Wolfsberg, *J. Chem. Phys.*, **45**, 2618 (1966).

[12] C. G. Swain, E. C. Stivers, J. F. Reuwer, Jr., and L. J. Schaad, *J. Amer. Chem. Soc.*, **80**, 5885 (1958).

[13] E. S. Lewis and J. K. Robinson, *J. Amer. Chem. Soc.*, **90**, 4337 (1968).

[14] G. Herzberg, *Infrared and Raman Spectra of Polyatomic Molecules*, Van Nostrand, New York (1945). (a) p. 316, (b) p. 203.

[15] M. J. Stern and M. Wolfsberg, *J. Chem. Phys.*, **45**, 4105 (1966).

[16] Z. Iqbal, *J. Chem. Phys.*, **59**, 6183 (1973); J. J. Rush, L. W. Schroeder, and A. J. Melveger, *J. Chem. Phys.*, **56**, 2793 (1972); see also J. M. Williams and L. F. Schneemeyer, *J. Amer. Chem. Soc.*, **95**, 5780 (1973).

[17] G. C. Pimentel and A. L. McClellan, *The Hydrogen Bond*, W. H. Freeman, New York (1960).

[18] W. J. Albery, *Trans. Faraday Soc.*, **63**, 200 (1967).

[19] R. E. Weston, Jr., *Science*, **158**, 332 (1967).

[20] M. Wolfsberg and M. J. Stern, *Pure Appl. Chem.*, **8**, 225 (1964).

[21] Calculated by the author.

[22] K. A. Quickert and D. J. Le Roy, *J. Chem. Phys.*, **53**, 1325 (1970).

[23] E. B. Wilson, Jr., J. C. Decius, and P. C. Cross, *Molecular Vibrations*, McGraw-Hill, New York (1955).

[24] L. Melander, *Isotope Effects on Reaction Rates*, Ronald Press, New York (1960).

[25] J. Bigeleisen and M. Wolfsberg, *Adv. Chem. Phys.*, **1**, 15 (1958).

[26] R. A. More O'Ferrall, G. W. Koeppl, and A. J. Kresge, *J. Amer. Chem. Soc.*, **93**, 1, 9 (1071).

[27] A. J. Kresge, *Pure Appl. Chem.*, **8**, 243 (1964).

[28] E. A. Halevi, F. A. Long, and M. A. Paul, *J. Amer. Chem. Soc.*, **83**, 305 (1961).

[29] M. J. Stern, M. E. Schneider, and P. C. Vogel, *J. Chem. Phys.*, **55**, 4286 (1971); P. C. Vogel and M. J. Stern, *J. Chem. Phys.*, **54**, 779 (1971).

[30] M. E. Schneider and M. J. Stern, *J. Amer. Chem. Soc.*, **94**, 1517 (1972).

[31] M. J. Stern and P. C. Vogel, *J. Amer. Chem. Soc.*, **93**, 4664 (1971); M. J. Stern and R. E. Weston, Jr., *J. Chem. Phys.*, **60**, 2815 (1974).

[32] J. Bigeleisen, *Tritium in the Physical and Biological Sciences*, IAEA, Vienna, Vol. 1 (1962), p. 161.

[33] J. R. Jones, *Trans. Faraday Soc.*, **65**, 2138, 2430 (1969); J. Hine, J. C. Kaufmann, and M. S. Cholod, *J. Amer. Chem. Soc.*, **94**, 4590 (1972).

[34] H. Kwart and M. C. Latimore, *J. Amer. Chem. Soc.*, **93**, 3770 (1971); K. W. Egger, *Int. J. Chem. Kinetics*, **1**, 459 (1969); W. R. Roth and J. König, *Annalen*, **699**, 24 (1966).

[35] R. P. Bell and J. K. Thomas, *J. Chem. Soc.*, 1573 (1939); R. P. Bell and A. Norris, *ibid.*, 118, 854 (1941); E. F. Caldin and J. C. Trickett, *Trans. Faraday Soc.*, **49**, 772 (1953); E. F. Caldin and G. Long, *Proc. Roy. Soc.*, *A*, **228**, 263 (1955).

[36] E. F. Caldin, *Chem. Revs.*, **69**, 135 (1969).

[37] K. B. Wiberg, *Chem. Revs.*, **55**, 713 (1955).

[38] H. S. Johnston, *Adv. Chem. Phys.*, **3**, 131 (1961).

[39] R. A. Marcus, *J. Chem. Ed.*, **45**, 356 (1968) and refs. cited.

[40] J. Bigeleisen, *Pure Appl. Chem.*, **8**, 217 (1964).

[41] E. R. Thornton, *J. Org. Chem.*, **27**, 1943 (1962).

[42] J. E. Leffler and E. Grunwald, *Rates and Equilibria of Organic Reactions*, Wiley, New York (1963).

[43] M. H. Mok and J. C. Polanyi, *J. Chem. Phys.*, **51**, 1451 (1969).

[44] G. S. Hammond, *J. Amer. Chem. Soc.*, **77**, 334 (1955).

[45] E. R. Thornton, *J. Amer. Chem. Soc.*, **89**, 2915 (1967).

[46] J. E. Critchlow, *Faraday Trans.*, *I*, **68**, 1774 (1972).

[47] A. J. Kresge, *Disc. Faraday Soc.*, **39**, 48 (1965).

[48] R. P. Bell and J. E. Crooks, *Proc. Roy. Soc.*, *A*, **286**, 285 (1965).

[49] R. P. Bell, *The Proton in Chemistry*, First Edition, Cornell University Press, New York (1959), p. 201; K. B. Wiberg and L. H. Slaugh, *J. Am. Chem. Soc.*, **80**, 3033 (1958).

[50] R. Stewart and D. G. Lee, *Can. J. Chem.*, **42**, 439 (1964).

[51] J. L. Longridge and F. A. Long, *J. Amer. Chem. Soc.*, **89**, 1292 (1967).

[52] R. P. Bell and D. M. Goodall, *Proc. Roy. Soc.*, *A*, **294**, 273 (1966).

[53] J. E. Dixon and T. C. Bruice, *J. Amer. Chem. Soc.*, **92**, 905 (1970).

[54] F. G. Bordwell and W. J. Boyle, Jr., *J. Amer. Chem. Soc.*, **93**, 512 (1971).

[55] D. J. Barnes and R. P. Bell, *Proc. Roy. Soc.*, *A*, **318**, 421 (1970).

[56] S. B. Hanna, C. Jermini, and H. Zollinger, *Tetrahedron Lett.*, 4415 (1969).

[57] R. P. Bell and B. G. Cox, *J. Chem. Soc.*, *B*, 194 (1970).

[58] A. F. Cockerill, *J. Chem. Soc.*, *B*, 964 (1967).

[59] R. P. Bell and B. G. Cox, *J. Chem. Soc.*, *B*, 783 (1971).

[60] D. W. Earls, J. R. Jones, and T. G. Rumney, *Faraday Trans. I*, **68**, 925 (1972).

[61] L. Melander and N.-Å. Bergman, *Acta Chem. Scand.*, **25**, 2264 (1971).

[62] A. J. Kresge, D. S. Sagatys, and H. L. Chen, *J. Amer. Chem. Soc.*, **90**, 4174 (1968).

[63] B. C. Challis, R. J. Higgins, and A. J. Lawson, *J. Chem. Soc. Chem. Comm.*, 1223 (1970).

[64] A. Campbell Ling and F. H. Kendall, *J. Chem. Soc.*, *B*, 445 (1967); B. C. Challis and E. M. Millar, *J. Chem. Soc.*, *Perkin II*, 1116 (1972).

[65] B. C. Challis and E. M. Millar, *J. Chem. Soc., Perkin II*, 1618 (1972).

[66] F. G. Bordwell, W. J. Boyle, Jr., and K. C. Yee, *J. Amer. Chem. Soc.*, **92**, 5926 (1970).

[67] A. J. Kresge, *Chem. Soc. Revs.*, **2**, 475 (1973).

[68] M. M. Kreevoy and Sea-wha Oh, *J. Amer. Chem. Soc.*, **95**, 4805 (1973).

[69] R. A. Marcus, *J. Phys. Chem.*, **72**, 891 (1968); A. O. Cohen and R. A. Marcus, *ibid.*, **72**, 4249 (1968).

[70] J. R. Murdoch, *J. Amer. Chem. Soc.*, **94**, 4410 (1972).

[71] M. M. Kreevoy and D. E. Konasewich, *Adv. Chem. Phys.*, **21**, 243 (1971).

[72] E. S. Lewis and L. H. Funderburk, *J. Amer. Chem. Soc.*, **89**, 2322 (1967).

[73] W. J. Albery, A. N. Campbell-Crawford, and J. S. Curran, *J. Chem. Soc., Perkin II*, 2206 (1972).

[74] R. P. Bell, E. Gelles, and E. Moller, *Proc. Roy. Soc.*, A, **198**, 308 (1949); R. P. Bell and O. Lidwell, *Proc. Roy. Soc.*, A, **176**, 88 (1940).

[75] D. S. Kemp and M. L. Casey, *J. Amer. Chem. Soc.*, **95**, 6670 (1973).

[76] V. Gold, *Adv. Phys. Org. Chem.*, **7**, 259 (1969).

[77] J. M. Williams, Jr. and M. M. Kreevoy, *Adv. Phys. Org. Chem.*, **6**, 63 (1968).

[78] W. A. Pryor and K. G. Kneipp, *J. Amer. Chem. Soc.*, **93**, 5584 (1971).

[79] E. S. Lewis and M. M. Butler, *J. Chem. Soc. Chem. Comm.*, 941 (1971).

[80] E. S. Lewis and S. Kozuka, *J. Amer. Chem. Soc.*, **95**, 282 (1973).

[81] R. P. Bell, *Disc. Faraday Soc.*, **39**, 16 (1965).

[82] R. P. Bell, W. H. Sachs, and R. L. Tranter, *Trans. Faraday Soc.*, **67**, 1995 (1971).

[83] R. F. W. Bader, *Canad. J. Chem.*, **42**, 1822 (1964).

[84] A. V. Willi and M. Wolfsberg, *Chem. and Ind.*, 2097 (1964).

[85] N.-Å. Bergman, W. H. Saunders, Jr., and L. Melander, *Acta Chem. Scand.*, **26**, 1130 (1972).

[86] A. M. Katz and W. H. Saunders, Jr., *J. Amer. Chem. Soc.*, **91**, 4469 (1969).

[87] R. A. More O'Ferrall and J. Kouba, *J. Chem. Soc.*, B, 985 (1967).

[88] A. J. Kresge and Y. Chiang, *J. Amer. Chem. Soc.*, **91**, 1025 (1969).

[89] W. H. Saunders, Jr., *Chem. and Ind.*, 663 (1966).

[90] R. A. More O'Ferrall, *J. Chem. Soc.*, B, 785 (1970) and references cited.

[91] W. H. Saunders, Jr., *J. Chem. Soc. Chem. Comm.*, 850 (1973).

[92] F. S. Klein, A. Persky, and R. E. Weston, Jr., *J. Chem. Phys.*, **41**, 1799 (1964).

[93] M. F. Hawthorne and E. S. Lewis, *J. Amer. Chem. Soc.*, **80**, 4296 (1958); E. S. Lewis and M. C. R. Symons, *Quart. Revs.*, **12**, 230 (1958).

[94] D. G. Truhlar and A. Kupperman, *J. Amer. Chem. Soc.*, **93**, 1840 (1971).

[95] R. P. Bell, *Trans. Faraday Soc.*, **55**, 1 (1959).

[96] H. S. Johnston and D. Rapp, *J. Amer. Chem. Soc.*, **83**, 1 (1961).

[97] D. G. Truhlar and A. Kupperman, *J. Chem. Phys.*, **56**, 2232 (1971); D. G. Truhlar, A. Kupperman, and J. T. Adams, *J. Chem. Phys.*, **59**, 395 (1973).

[98] E. S. Lewis, J. M. Perry, and R. H. Grinstein, *J. Amer. Chem. Soc.*, **92**, 899 (1970).

[99] A. Bromberg, K. A. Muszkat, and A. Warshel, *J. Chem. Phys.*, **52**, 5952 (1970).

[100] M. J. Stern and R. E. Weston, Jr., *J. Chem. Phys.*, **60**, 2808 (1974).

[101] A. Bromberg, K. A. Muszkat, E. Fischer, and F. S. Klein, *J. Chem. Soc., Perkin II*, 588 (1972).

[102] T. E. Sharp and H. S. Johnston, *J. Chem. Phys.*, **37**, 1541 (1962).

[103] E. Grovenstein, Jr., and F. C. Schmalstieg, *J. Amer. Chem. Soc.*, **89**, 5084 (1967).

[104] J. P. Calmon, M. Calmon, and V. Gold, *J. Chem. Soc.*, **B**, 659 (1969).

[105] E. S. Lewis and J. D. Allen, *J. Amer. Chem. Soc.*, **86**, 2022 (1964).

[106] E. F. Caldin and G. Tomalin, *Trans. Faraday Soc.*, **64**, 2823 (1968).

[107] A. V. Willi, *Helv. Chim. Acta*, **54**, 1220 (1971).

[108] R. P. Bell, *Trans. Faraday Soc.*, **57**, 961 (1961).

[109] K. Morokuma and M. Karplus, *J. Chem. Phys.*, **55**, 63 (1971).

[110] G. W. Koeppl, *J. Amer. Chem. Soc.*, **96**, 6539 (1974).

[111] R. A. More O'Ferrall in *The Carbon-Halogen Bond*, edited by S. Patai, Wiley, Interscience (1973), p. 609.

[112] W. P. Jencks, *Catalysis in Chemistry and Enzymology*, McGraw-Hill, New York (1969).

[113] S. L. Johnson, *Adv. Phys. Org. Chem.*, **5**, 237 (1967); P. M. Laughton and R. E. Robertson in *Solute-Solvent Interactions*, edited by J. F. Coetzee and C. D. Ritchie, Marcel Dekker, New York (1969), p. 399.

[114] I. Shavitt, R. M. Stevens, F. L. Minn, and M. Karplus, *J. Chem. Phys.*, **48**, 2700 (1966); I. Shavitt, *ibid.*, **49**, 4048 (1968); B. Liu, *ibid.*, **58**, 1925 (1973).

[115] C. F. Bender, P. K. Pearson, S. V. O'Neil, and H. F. Schaefer III, *J. Chem. Phys.*, **56**, 4626 (1972); C. F. Bender, S. V. O'Neil, P. K. Pearson, and H. F. Schaefer III, *Science*, **176**, 1412 (1972).

[116] S. V. O'Neil, H. F. Schaefer III, and C. F. Bender, *Proc. Nat. Acad. Sci., USA*, **71**, 104 (1974).

[117] K. J. Laidler, *Theories of Chemical Reaction Rates*, McGraw-Hill, New York (1969).

[118] C. A. Parr and D. G. Truhlar, *J. Phys. Chem.*, **75**, 1844 (1971).

[119] J. C. Polanyi, *Accts. Chem. Res.*, **5**, 161 (1972).

[120] F. S. Klein and A. Persky, *J. Chem. Phys.*, **59**, 2775 (1973).

[121] H. Carmichael and H. S. Johnston, *J. Chem. Phys.*, **41**, 1975 (1964).

[122] M. J. Benson and D. R. McLaughlin, *J. Chem. Phys.*, **56**, 1322 (1972).

[123] J. F. Bunnett, *Angew. Chemie Internat. Edn.*, **1**, 225 (1962).

[124] E. C. F. Ko and A. J. Parker, *J. Amer. Chem. Soc.*, **90**, 6447 (1968).

[125] R. A. More O'Ferrall, *J. Chem. Soc.*, *B*, 268 (1970).

[126] W. P. Jencks, *Chem. Revs.*, **72**, 705 (1972).

[127] R. A. Bartsch and J. F. Bunnett, *J. Amer. Chem. Soc.*, **91**, 1376 (1969); W. H. Saunders, Jr., S. R. Fahrenholtz, E. A. Caress, J. P. Lowe, and M. Schreiber, *J. Amer. Chem. Soc.*, **87**, 3401 (1965).

[128] C. K. Ingold, *Structure and Mechanism in Organic Chemistry*, Bell, London (1953).

[129] F. G. Boldwell and W. J. Boyle, Jr., *J. Amer. Chem. Soc.*, **94**, 3907 (1972).

[130] A. J. Kresge, chapter 7 of this volume.

[131] R. A. Marcus, *J. Amer. Chem. Soc.*, **91**, 7224 (1969).

[132] A. A. Westenberg and N. de Haas, *J. Chem. Phys.*, **46**, 490 (1967); **50**, 2512 (1969).

[133] D. N. Mitchell and D. J. Le Roy, *J. Chem. Phys.*, **58**, 3449 (1973) and refs. cited.

[134] M. Karplus, R. N. Porter, and R. D. Sharma, *J. Chem. Phys.*, **34**, 3529 (1965).

[135] M. Karplus and K. T. Tang, *Disc. Faraday Soc.*, **44**, 56 (1968).

[136] A. Persky and M. Baer, *J. Chem. Phys.*, **60**, 133 (1974).

[137] A. Persky, *J. Chem. Phys.*, **59**, 5578 (1973).

[138] R. B. Timmons and R. E. Weston, Jr., *J. Chem. Phys.*, **41**, 1654 (1964).

[139] T. Yokota and R. B. Timmons, *Intl. J. Chem. Kinetics*, **2**, 325 (1970).

[140] G. Chiltz, R. Eckling, P. Goldfinger, G. Huybrechts, H. S. Johnston, L. Meyers, and G. Verbeke, *J. Chem. Phys.*, **38**, 1053 (1963).

[141] K. B. Wiberg, E. L. Motell, *Tetrahedron*, **19**, 2009 (1963); R. B. Timmons, J. de Guzman, and R. E. Varnerin, *J. Amer. Chem. Soc.*, **90**, 5996 (1968); N. L. Arthur and P. Gray, *Trans. Faraday Soc.*, **65**, 434 (1969).

[142] A. Persky, *J. Chem. Phys.*, **59**, 3612 (1973).

[143] A. Persky and F. S. Klein, *J. Chem. Phys.*, **44**, 3617 (1966).

[144] J. H. Sullivan, *J. Chem. Phys.*, **39**, 3001 (1963).

[145] Y. Bar Yaakov, A. Persky, and F. S. Klein, *J. Chem. Phys.*, **59**, 2415 (1973).

[146] M. J. Stern, A. Persky, and F. S. Klein, *J. Chem. Phys.*, **58**, 5697 (1973).

[147] A. A. Westenberg and N. de Haas, *J. Chem. Phys.*, **47**, 4241 (1967).

[148] G. W. Koeppl, *J. Chem. Phys.*, **59**, 3425 (1973).

9

John Albery

SOLVENT ISOTOPE EFFECTS

9.1 Introduction

The substitution of D for H is an elegant probe to use in the elucidation of chemical dynamics, and in particular the solvent isotope effect is a powerful technique for investigating the role of the solvent. This will be the main theme of this chapter. When H_2O is replaced by D_2O it is inevitable that we are dealing with the effects of substitution at many different sites. The effects of these many substitutions are described by fractionation factor theory [1] and, because many sites are involved, more information can be found if measurements are made in equimolar mixtures of H_2O and D_2O as well as in pure H_2O and in pure D_2O. In 1969 Gold reviewed this particular topic [2]. In this chapter the discussion will be extended and a simple technique for analysing the data obtained in H_2O/D_2O mixtures will be presented. It is the author's hope that, in addition to the complete substitution of H_2O with D_2O, greater use will be made of the 50:50 mixture, which we shall call HDO.

9.2 H_2O and D_2O

In their useful review [3] Arnett and McKelvey have tabulated properties of H_2O and D_2O. Many of the solvent characteristics which affect the rate of reactions, for instance the dielectric constant, the dipole moment and the molar volume, are virtually the same. The most important differences for the dynamics of reactions in solution are those connected with the viscosity and self diffusion of the solvent molecules (Table 1). It can be seen that increasing the mass of the

Table 1

Ratios of viscosities (η) and tracer diffusion coefficients (D) at 25°C

Ratios close to 1.0	Ratios close to 1.2
$\dfrac{\eta_{H_2^{18}O}}{\eta_{H_2^{16}O}} = 1.054^a$	$\dfrac{\eta_{D_2^{16}O}}{\eta_{H_2^{16}O}} = 1.232^b$
	$\dfrac{\eta_{D_2^{16}O}}{\eta_{H_2^{18}O}} = 1.169^a$
$\dfrac{D_{HDO} \text{ in } H_2O}{D_{HTO} \text{ in } H_2O} = 1016^c$	$\dfrac{D_{HDO} \text{ in } H_2O}{D_{HDO} \text{ in } D_2O} = 1.195^d$
$\dfrac{D_{HDO} \text{ in } D_2O}{D_{DTO} \text{ in } D_2O} = 1.029^c$	$\dfrac{D_{H_2O} \text{ in } H_2O}{D_{D_2O} \text{ in } D_2O} = 1.228^c$
$\dfrac{(\eta D_{HDO}) \text{ in } H_2O}{(\eta D_{HDO}) \text{ in } D_2O} = 1.030$	

^{a} A. I. Kudish, D. Wolf and F. Steckel, *J.C.S. Faraday I*, **68**, 2041 (1972); ^{b} F. J. Millero, R. Dexter and E. Hoff, *J. Chem. Eng. Data*, **16**, 85 (1971); ^{c} R. Mills, Molecular Motions in Liquids (ed. J. Lascombe), Reidel, Dordrecht Holland, 1974, p. 391; ^{d} L. G. Longsworth, *J. Phys. Chem.*, **64**, 1914 (1960).

solvent molecule by substituting ^{18}O for ^{16}O has a small effect on the viscosity and on diffusion coefficients, whereas increasing it by the same amount by substituting D for H, has a much larger effect. The change for the oxygen isotopes can be rationalized [4] in terms of a simple translational model, while the change for the hydrogen isotopes must be caused by the difference in the rotational motions of H_2O and D_2O. For the isotopically different solutes in the same solvent the changes are small and of a size that might be predicted from a simple translational model [5]. However, a 'large' difference is found for the same solute (HDO) in the different solvents [6]. This difference can be largely accounted for by the different viscosities of the two solvents: the last entry in Table 1 shows that $D\eta$ is approximately constant.

The values of D in Table 1 were obtained by the classical techniques of a Stokes diffusion cell [5] or a Rayleigh interferometer [6]. Nuclear magnetic resonance (n.m.r.) and neutron scattering can also be used,

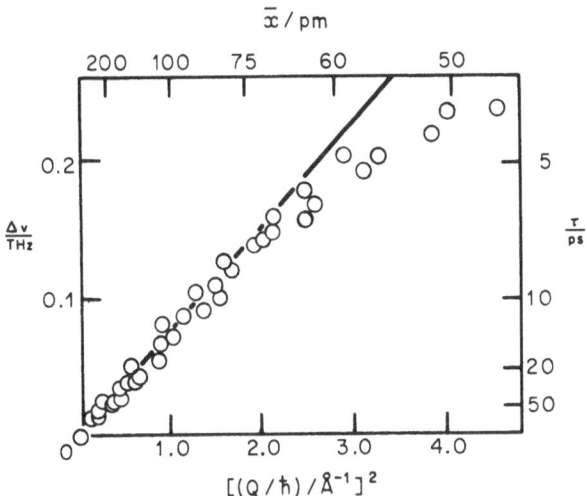

Figure 1. Self diffusion of H_2O at short time scales from neutron inelastic scattering [7]. The line corresponds to equation 1 with $(D/m^2\ Gs^{-1}) = 2.36$.

and although less accurate and more expensive, these techniques allow one to measure the value of D over shorter time scales and distances. Figure 1 shows data obtained from neutron scattering [7], plotted according to equation 1,

$$\Delta v = DQ^2/\hbar^2\pi \qquad (1)$$

where Δv is the line width and Q is the momentum transfer. The axes are also labelled with scales $\bar{x}$ and τ where $\bar{x} = \hbar/Q$ and $\tau = (\Delta v)^{-1}$. These quantities describe the characteristic distance and life time of the interaction of the neutron with the solvent protons. It will be seen that data have been obtained down to $\bar{x} = 50\ pm = 0.5$ Å and $\tau = 4\ ps$. The straight line in Figure 1 corresponds to the value of D measured by Mills using the Stokes cell [5]. Down to $\tau = 10\ ps$ and $\bar{x} = 70\ pm$, this value still holds, and one can conclude that motions on this sort of scale are responsible for macroscopic diffusion. Below this, the points deviate from the straight line and the motion changes from being mainly translational to one containing more rotation. However, there is no discontinuity and the parameter D defined by equation 1 is still of much the same order as the macroscopic value. Similar experiments on D_2O [7] confirm that, as in Table 1, $D_{H_2O}/D_{D_2O} \sim 1.18$. Bragg diffraction from the D_2O solvent prevents such an extensive exploration as for H_2O. However, n.m.r. results on the rotational correlation time do suggest that throughout the region of interest D_{D_2O} is some twenty per cent lower than D_{H_2O}.

265

We now turn to the question of the ideality or otherwise of mixtures of H_2O and D_2O. In the last few years Van Hook and his co-workers [8] have made accurate measurements on the vapour pressure of L_2O solutions. (The symbol L is used to denote either H or D.) When there is only one site on the solvent molecule which can be substituted, as in methanol, then one can simply measure the vapour pressures of MeOH and MeOD over mixtures of MeOL. For water, however, we have three solvent species, H_2O, HDO and D_2O, in equilibrium:

$$H_2O + D_2O \rightleftharpoons 2HOD, \qquad K_{L_2O}$$

If there were no difference in the successive substitution of H by D, then K_{L_2O} would be equal to 4; and at $x = 0.5$ (where x is the atom fraction of D in the solvent) the mole fractions of the isotopic waters would be $x_{H_2O} = 0.25$, $x_{D_2O} = 0.25$ and $x_{HOD} = 0.50$. In fact it is found [9] that $K_{L_2O} = 3.78 \pm 0.03$ at 25°C. This implies that (ignoring statistical factors) it is more difficult to replace H with D in HOH than in HOD; there is a secondary isotope effect on the successive substitutions. This can also be described as a breakdown of the 'rule of the geometric mean' [10], since the rule is derived on the assumption that there are no secondary effects. However, within experimental error, mixtures of H_2O, HOD and D_2O still behave in an ideal fashion, with $K_{L_2O} = 3.8$ [11, 12].

9.3 Fractionation factor theory

To describe isotopic substitution on many sites might be thought to be almost impossible. In the solvent there are three different species (H_2O, HDO and D_2O), for L_3O^+ there are four different species. There will be several isotopic transition states and thus a great variety of possible parallel reactions. However, developing the Gross–Butler theory [13, 14], papers by Gold [15] and Kresge [1] showed how it was possible to describe the variation of reaction rates and equilibria with the isotopic composition of the solvent in terms of fractionation factors.

The fractionation factor ϕ_X describes the equilibrium composition of the $X - L$ site with respect to the isotopic composition of the solvent, according to 2, and

$$\frac{[X - D]}{[X - H]} = \phi_X \frac{[D_2O]^{\ddagger}}{[H_2O]^{\ddagger}} \qquad (2)$$

For the particular case of a 50:50 mixture of H_2O and D_2O, when $[H_2O] = [D_2O]$, ϕ_X would describe whether there was more or less D compared to H on the X site.

Now if $K_{L_2O} = 4$,

$$\frac{[D_2O]^{\ddagger}}{[H_2O]^{\ddagger}} = \frac{x}{1-x} \tag{3}$$

and

$$\frac{[X-D]}{[X-H]} = \phi_X \frac{x}{1-x} \tag{4}$$

or

$$[X-D] = \frac{\phi_X x}{1-x+\phi_X x}[X-L] \left.\begin{array}{l} \\ \\ \\ \\ \\ \end{array}\right\} \tag{5}$$

and

$$[X-H] = \frac{1-x}{1-x+\phi_X x}[X-L]$$

In a chemical reaction $X - D$ and $X - H$ may react at different rates and the total rate of reaction will be given by equation 6.

$$\text{rate} = k_x[X-L] = k_0[X-H] + k_1[X-D] \tag{6}$$

In this equation k_0 describes the rate constant in pure H_2O, k_1 the rate constant in pure D_2O and k_x the rate constant at any value of x. Let

$$\frac{k_1}{k_0} = \frac{\phi_{X,\ddagger}}{\phi_X} \tag{7}$$

the substitution from equations 4 and 5 in equation 6 gives

$$\frac{k_x}{k_0} = \frac{(1-x+\phi_{X,\ddagger}x)}{(1-x+\phi_X x)} \tag{8}$$

This equation describes the effect on the rate constant of isotopic substitution on a single site. The treatment can be extended to many sites, with the simple and elegant result[1]

$$\frac{k_x}{k_0} = \frac{\Pi(1-x+\phi_{\ddagger}x)}{\Pi(1-x+\phi_R x)} \tag{9}$$

The denominator and numerator products each contain a factor for every site that is substituted in the reactants.

There are several particular points to note. For a site with $\phi = 1$, the bracketed factor is equal to unit, and there will be no contribution from that particular site to the isotope effect; this will be true for unchanged bulk solvent. When $\phi_{X,\ddagger} = \phi_X$ the brackets in the numerator and denominator cancel and substitution on that particular site has no effect. In pure D_2O, $x = 1$, and

$$\frac{k_1}{k_0} = \frac{k_{D_2O}}{k_{H_2O}} = \frac{\Pi\phi_\ddagger}{\Pi\phi_R}$$

The measurement of solvent isotope effects in pure D_2O and H_2O only gives a single datum. In principle more information can be found from mixtures.

In his original treatment [1] Kresge identified the various $\phi_\ddagger$ with fractionation factors in the transition state. This is a plausible interpretation which we shall use. However, equations 6 and 7 make no reference to transition state theory, and the final result does not depend upon any hypothesis about equilibrium between the transition state and the reactants.

Returning to equation 9 we can write straightaway the equivalent equation for an equilibrium constant,

$$\frac{K_x}{K_0} = \frac{\Pi(1 - x + \phi_P x)}{\Pi(1 - x + \phi_R x)}$$

where the different ϕ_P describe the fractionation factors in the reaction products.

We now review the approximations and assumptions made to obtain these results. They are three:

(i) the mixtures are ideal;
(ii) $K_{L_2O} = 4$;
(iii) there are no secondary isotope effects on successive substitutions.

As discussed above, for all practical purposes the mixtures are ideal, but $K_{L_2O} = 3.78$ rather than 4. The third approximation is similar to the second but is concerned with the sites around and on the reactants rather than those in the bulk solvent. Since $K_{L_2O} \neq 4$, it is likely that approximation (iii) will also break down. We now examine the effects of the breakdown of approximations (ii) and (iii) on the simple theory.

9.4 Breakdown of the rule of the geometric mean

After it became clear [16] that $K_{L_2O} \neq 4$, Gold [17] examined the effect this would have on the simple theory by carrying out numerical calculations. A different approach was used by Albery and Davies [18] who preserved the simple form of equation 9 and allowed for the breakdown with a small analytical correcting function. The conclusions from both approaches are similar but the analytical method is preferred as being simpler and more general.

In the simple theory we assumed $K_{L_2O} = 4$ to obtain equation 3. If $K_{L_2O} \neq 4$ then in general, equation 3 is not valid, and the value of ϕ will depend on whether one uses equation 2 or equation 4. However, when $x = \frac{1}{2}$, regardless of the value of K_{L_2O},

$$\frac{[D_2O]^{\frac{1}{2}}}{[H_2O]^{\frac{1}{2}}} = \frac{x}{1-x} = 1$$

and

$$\left\{ \frac{[X-D]}{[X-H]} \right\}_{x=\frac{1}{2}} = \phi_X \tag{10}$$

In the following we shall therefore adopt equation 10 as the definition of the fractionation factor. One measures a fractionating factor by analysis of the isotopic content of the reactant or product. The observed value, $(\phi_X)_{obs}$, is

$$(\phi_X)_{obs} = \frac{[X-D]}{[X-H]} \frac{(1-x)}{x}$$

and [18], for $K_{L_2O} = 3.78$,

$$\phi_X = (\phi_X)_{obs}[1 - 0.06(\tfrac{1}{2} - x)] \tag{11}$$

The correcting function is small, being at the most ± 3 per cent, but should be applied when fractionating factors are measured for values of x which are not close to $x = \frac{1}{2}$. Equation 11 implies that D in the HDO molecule is a more active deuterating reagent (after allowing for the statistical factor of 2) than D in D_2O. Hence $(\phi_X)_{obs}$ is larger than ϕ_X for $x < \frac{1}{2}$ and smaller for $x > \frac{1}{2}$.

Equation 11 describes the effect of $K_{L_2O} \neq 4$ on measured fractionation factors. The effect on the rate constant equation can be ignored provided that the values of the fractionation factors used in the equation are those defined by equation 10 for $x = \frac{1}{2}$ [18]. This is therefore a second and even more powerful reason for defining fractionation factors at $x = \frac{1}{2}$. Conversely, fractionating factors measured from analysis of k_x/k_0 or K_x/K_0 are those that apply at $x = \frac{1}{2}$. Thus we can cope with $K_{L_2O} \neq 4$ and preserve the simple form of the equations, provided we also use equations 10 and 11.

Now we turn to the more vexed question of approximation (*iii*) above, the breakdown of the rule of the geometric mean for sites on or around the reactants. A particularly important example is the fractionation on L_3O^+:

$$H_3O^+ + \tfrac{1}{2}D_2O \rightleftharpoons H_2DO^+ + \tfrac{1}{2}H_2O; \qquad 3l_1$$

$$H_2DO^+ + \tfrac{1}{2}D_2O \rightleftharpoons HD_2O^+ + \tfrac{1}{2}H_2O; \qquad l_2$$

$$HD_2O^+ + \tfrac{1}{2}D_2O \rightleftharpoons D_3O^+ + \tfrac{1}{2}H_2O; \qquad \tfrac{1}{3}l_3$$

The factors 3 and $\tfrac{1}{3}$ are statistical factors. If the rule of the geometric mean held and there were no secondary effects, then $l_1 = l_2 = l_3$. However, if successive substitution on L_2O has a secondary effect, it is likely that there will be a similar effect on L_3O^+. We follow the approach of preserving the simple form of the theory.

For a group of n identical sites it can be shown [18] that the correct factors F to be used in equation 9 can be approximated to within a few parts in a thousand by

$$F \simeq (1 - x + \bar{\phi}x)^n[1 - mpx(1 - x)]$$

where,

$$\frac{\phi_{j+1}}{\phi_j} = (1 + p), \quad e.g. \; \frac{l_2}{l_1} = \frac{l_3}{l_2} = 1 + p_{L^+}$$

$$m = \tfrac{1}{2}n(n - 1), \quad e.g. \text{ for } L_3O^+, m = 3$$

and

$$\bar{\phi} = (\Pi\phi_1 \cdots \phi_n)^{1/n}, \quad e.g. \text{ for } L_3O^+, \bar{\phi} = l_2$$

Since p is small and $x(1 - x) \leqslant \tfrac{1}{4}$,

$$\frac{k_x}{k_0} = \frac{\Pi(1 - x + \phi_{\ddagger}x)}{\Pi(1 - x + \phi_R x)}[1 - gx(1 - x)] \qquad (12)$$

where

$$g = [\Sigma mp_{\ddagger} - \Sigma mp_R] \qquad (13)$$

Although $\bar{\phi} = 1$ for the solvent and $(1 - x + \bar{\phi}x)$ is unity throughout, $m = 1$ and $p_{L_2O} = 0.06$. Hence there will be a contribution to the correction function from the water molcule.

In many cases the correction function will be negligible, because, for instance for a neighbouring solvent molecule, $p_{\ddagger} \simeq p_R$ and $x(1 - x) \leqslant \tfrac{1}{4}$.

Neglect of the correction term may not be justifiable for L_3O^+. Unfortunately (see below), it is difficult to determine p_{L^+} for L_3O^+

experimentally; it requires data of very high precision. Measurements by More O'Ferrall, Koeppl and Kresge [19] lead to [20] $p_{L^+} = 0.03$. This value is intermediate between no breakdown ($p_{L^+} = 0.0$) and the amount of breakdown seen in the L_2O molecule ($p_{L_2O} = 0.06$). Using this value we find that $g = -0.03$ for A-S_E2, A-1 and A-2 transition states [20]. Even if at this stage one cannot always go beyond approximation (*iii*), we have at least obtained a simple function to describe the effects of the breakdown of the rule of the geometric mean. It allows us to estimate how uncertain our mechanistic conclusions have to be because of our lack of precise knowledge.

9.5 The medium effect

It has been customary to use fractionation factors to describe the effects of isotopic substitution on particular sites, for instance protons in flight in the transition state, and to describe the effects of solvation in terms of a transfer activity coefficient [2]. However, this separation is not really necessary [21].

In deriving equation 12 we placed no limit on the number of sites, and hence we can include solvent sites involved in solvating the reactants or the transition states. For s such sites (all identical) solution will be described by

$$y_x = (1 - x + \phi_S x)^s$$

Now it is likely that in solvation s will be large and ϕ_S not far removed from unity, since the solvating L_2O molecule will not be very different from a bulk L_2O molecule. Figure 2 shows plots of

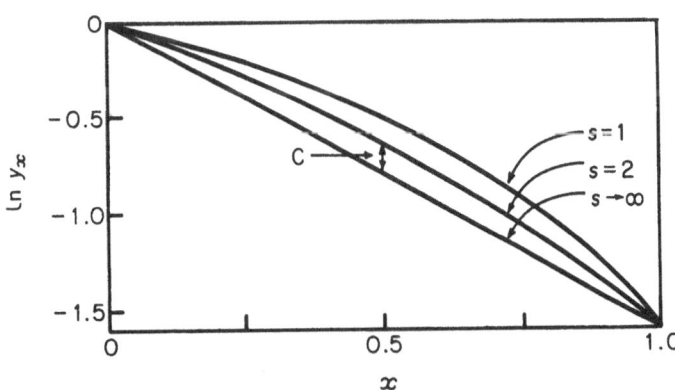

Figure 2. Plots of ln y_x where $y_x = (1 - x + \phi_S x)^s$ and $\phi_S^s = 0.20$ for different values of s. The definition of the curvature parameter C is also shown.

$\ln(y_x)$ against x for different values of s and for a constant value of $y_1 = 0.20$. In the limit, as s becomes large, taking logarithms, followed by expansion, gives

$$\ln y_x = x \ln y_1 = x \ln(\phi_S^s) \qquad (14)$$

For $s > 10$ the limiting expression 14 is valid and the individuality of the different sites disappears. We may then write

$$\Phi_S = \phi_S^s$$

The parameter Φ_S is the same as the transfer activity coefficient between pure H_2O and pure D_2O. The function y_x would then describe the transfer activity coefficient at intermediate values of x. Its logarithm had in fact in the past been assumed to vary linearly with x as required by equation 14. Equation 14 is also confirmed by recent solubility measurements [22].

Different symbols have been used for the transfer activity coefficient; the symbol Φ is suggested to emphasize that fractionation factor theory includes the medium effect and that no special provision need be made for it.

9.6 Curvature in solvent isotope effects

From equation 9, k_x will, in general, be a curved function of x. The extra information about the transition state is contained in this curvature. A simple general method of analysing such curvature, developed from that suggested by Albery and Davies [23], will now be presented. We here concentrate on fractionation in the transition state, but the approach applies also to equilibria and to reactant fractionation.

We first rearrange equation 12 and re-define y_x to give

$$y_x = \frac{k_x}{k_0}[1 + gx(1 - x)]\Pi(1 - x + \phi_R x) = \Pi(1 - x + \phi_{\ddagger}x) \quad (15)$$

The term on the right-hand side contains the information about the transition state. The approach depends on measuring values of k_x at $x = 0$, $x = \frac{1}{2}$ and $x = 1$. The extra information is contained in how far $y_{\ddagger}$, the point at $x = \frac{1}{2}$, deviates from a value interpolated between y_0 and y_1. In our previous work [23] we analysed the curvature in y as a function of x. A great improvement is to analyse the curvature in $\ln(y)$ rather than y. The reason for this, as shown in Figure 2, is that the medium effect contributes no curvature to the variation of $\ln(y)$

with x, *and curvature in this plot is produced only by individual sites.*
It is the existence (or not) of these individual sites and their fractionation factors that is of most interest.

We define curvature parameters C,

$$C = \ln y_{\frac{1}{2}} - \tfrac{1}{2}(\ln y_0 + \ln y_1) = \ln (y_{\frac{1}{2}}/\sqrt{y_1})$$

and γ

$$\gamma = \frac{8C}{(\ln y_1)^2} = \frac{8 \ln (y_{\frac{1}{2}}/\sqrt{y_1})}{(\ln y_1)^2} \tag{16}$$

Now suppose that in the product of fractionation factor brackets
that make up the right-hand side of equation 15 we have a factors in
ϕ_A, b factors in ϕ_B etc. Then

$$\ln y = \sum a \ln (1 - x + \phi_A x)$$

and

$$\ln y_1 = \sum a \ln \phi_A$$

We now define parameters Λ (equation 17) which describe the
contribution of each group of factors to the overall effect $\ln y_1$,
with $\Sigma \Lambda = 1$.

$$\Lambda_A = \frac{a \ln \phi_A}{\ln y_1}, \qquad \Lambda_B = \frac{b \ln \phi_B}{\ln y_1} \text{ etc.} \tag{17}$$

Then

$$\gamma = \sum \frac{\Lambda_A^2}{a}(1 - h_A) \tag{18}$$

where

$$h_A = 1 + \frac{4 \ln (4\phi_A) - 8 \ln (1 + \phi_A)}{(\ln \phi_A)^2} \tag{19}$$

Provided that

$$3.5 > \phi_A > 0.3 \tag{20}$$

h_A is negligible, and we can write, to within 5 per cent,

$$\gamma \simeq \sum \frac{\Lambda_A^2}{a} \tag{21}$$

The definition of γ in equation 16 was chosen so as to achieve this
simple relationship; in particular, $\gamma = 1$ for a linear variation of y
with x and $\gamma = 0$ for a linear variation of $\ln y$ with x. The only site

273

for which the fractionation factor will not fall within the conditions of equation 20, is a site in the transition state corresponding to a proton in flight. We show later how to extend the treatment to include such sites.

From equation 21 it can be seen that, since s is large, the medium effect on the reactants and on the transition state will not contribute anything to γ. Thus in calculating y from equation 15 we need not know the medium effect on the reactants. We can leave it on the right-hand side and include it in a differential medium effect between the transition state and the reactants:

$$\Phi_\Delta = \Phi_\ddagger / \Phi_R \qquad (22)$$

and

$$\Lambda_\Delta = \frac{\ln \Phi_\Delta}{\ln y_1} \qquad (23)$$

The use of a differential medium effect is like using an energy of activation rather than the separate internal energies of the reactant and the transition state. Hence a positive value of γ gives us information about the role of protons that have been singled out from the crowd in the solvent. It also allows us to decide how much of the isotope effect is caused by these special protons and how much by the medium effect.

We can draw the following general conclusions [23] about the curvature of the plot $\ln y$ versus x:

(1) Given n sites in the transition state the curvature, γ, will be a minimum when all n ϕ's are equal.
(2) The curvature (γ) decreases with increasing n to a value of zero corresponding to a linear variation of $\ln y$ with x which means that the isotope effect is entirely caused by a medium effect.
(3) Given an experimental value of γ the number of sites in the transition state must be greater than γ^{-1}.
(4) If $\gamma < 0$, either (a) there is reactant fractionation on individual sites which has not yet been accounted for, or (b) the reaction is taking place by parallel paths.
(5) A model for the reaction with a second parallel transition state containing an equal or greater number of sites than the first will lead to a smaller value of γ than a model with only the first transition state.
(6) A model for the reaction with a second transition state in series (either before or after the first) which contains an equal

or smaller number of sites than the first leads to larger values of γ than the model with only the first transition state, provided that the intermediate is isotopically equilibrated with the solvent.*

These conclusions are general. However, a knowledge of γ and of y_1 can tell us quantitatively about only two groups of factors. Hence we shall be interested in particular in testing models of the transition state containing two groups of factors labelled A and B. Equation 21 then becomes

$$a\gamma = \Lambda_A^2 + \frac{a}{b}(1 - \Lambda_B)^2 \qquad (24)$$

Figure 3 shows plots of $a\gamma$ versus Λ_A for different values of a/b with $|a| < |b|$. If one group of factors is the medium effect then $a/b = 0$ and $a\gamma = \Lambda_A^2$, the parabola passing through $(0,0)$. Also between $\Lambda_A = 0$ and $\Lambda_A = 1$, both Λ_A and Λ_B are positive; that is both groups of factors contribute in the same direction to the overall solvent isotope effect. Outside this central region either Λ_A or Λ_B is negative and this implies that the two groups of factors oppose each other, one being larger than unity and the other less than unity. The sizes of Λ_A and Λ_B describe the relative importance of the groups of factors to the overall solvent isotope effect.

Given an experimental value of γ, one can plot different horizontal lines corresponding to $a\gamma$ for different values of a. Each intersection corresponds to a possible model for the transition state. When Λ_B is close to zero all the curves bunch together, because we cannot find out much detail about a group of factors which do not have much effect on the rate of reaction. Each intersection is associated with a value of a (defined by the horizontal line) and a value of b (defined by the value of a/b associated with a parabola), and values of Λ_A and Λ_B which can also be calculated from equation 25. Determination of a and b decides the model of the transition state. Values of ϕ_A and ϕ_B can be calculated from Λ_A and Λ_B by equation 17.

$$\Lambda_A = \frac{a \pm (ab)^{\frac{1}{2}}[\gamma(a + b) - 1]^{\frac{1}{2}}}{a + b} = 1 - \Lambda_B \qquad (25)$$

The minimum on each parabola in Figure 3 occurs when

$$\Lambda_A = (a/b)\Lambda_B \qquad (26)$$

and corresponds to a model with n equal fractionation factors, where

* I am grateful to Professor Kresge for pointing out the importance of the proviso.

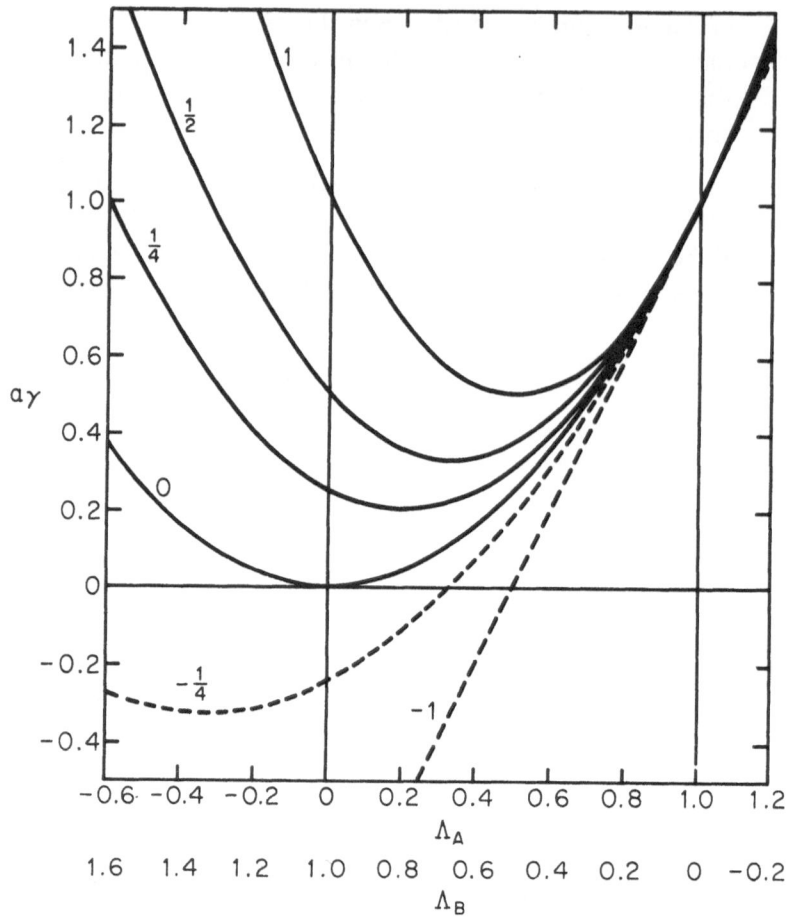

Figure 3. Plot of equation 24 for $a\gamma$ against Λ_A; each curve is labelled with its value of a/b. All transition states have to be in the area between the $a/b = 0$ and $a/b = 1$ parabolas; the broken lines allow for reactant fractionation.

$n = a + b$. Since $a\gamma_{min} = a/(a + b)$, then

$$n \geqslant \gamma^{-1} \tag{27}$$

in accordance with general conclusion (3).

The solution, with n equal factors at the minimum of a parabola, will also be found at $\Lambda_A = 0$ (wherever $a\gamma$ intersects each a/n parabola at a/n) and at $\Lambda_A = 1$ where $a\gamma = 1$ and $a = n$. This last intersection at $\Lambda_A = 1$ is the easiest to use to test for models with only a single group of factors. For any single transition state, no matter what its fractionation, the area of possible solutions is bounded by the $(a/b = 1)$ and the $(a/b = 0)$ parabolas. Negative values of γ can be found if either a or b is negative; this corresponds to fractionation on individual sites in the reactant shown by the dotted lines in Figure 3.

The advantages of this method of analysis are first that all possible transition states containing two groups of factors are considered in a single diagram, second that the calculations are very simple and do not require a computer, third that the experimental error can be shown as an uncertainty in the value of γ and hence of the conclusions, and fourth the medium effect is reasonably innocuous.

Two sources of error can be envisaged in the determination of γ and of the fractionation factors ϕ_A and ϕ_B: first the experimental error in the rate constants and second the uncertainty about the breakdown of the rule of the geometric mean. The errors E in γ and in the fractionation factors are [24]:

$$E_\gamma = \frac{(96E_{\ln k}^2 + 4E_g^2)^{\frac{1}{2}}}{(\ln y_1)^2} \tag{28}$$

$$\frac{E_{\ln \phi_A}}{E_\gamma} = \frac{\frac{1}{2}\ln y_1}{\Lambda_A - (a/b)\Lambda_B} \quad \text{and} \quad \frac{E_{\ln \phi_B}}{E_\gamma} = \frac{\frac{1}{2}(a/b)\ln y_1}{\Lambda_A - (a/b)\Lambda_B}$$

Note that since $(\ln y_1)^2$ occurs in the denominator of equation 28 the approach breaks down when y_1 is close to unity. This is because we cannot hope to find out a lot of detail when the factors are unimportant. (Remember that y_1 is not necessarily the same as k_1/k_0.)

In deriving the simplified equation 21 for γ from equation 18 we had to assume that $\phi > 0.3$. This assumption will be justified for nearly all fractionations except for that in which a proton is being transferred. For this case we cannot ignore the h term in equation 18. With $a = 1$ (since a single proton is being transferred), we obtain

$$\gamma + h\Lambda_A^2 = \Lambda_A^2 + \frac{1}{b}(1 - \Lambda_A)^2 \tag{29}$$

Thus, instead of plotting γ on Figure 3, we should plot the parabola $\gamma + h\Lambda_A^2$. Since h is small, for the region $0 < \Lambda_A < 1$ it will usually be sufficiently accurate to use the linear approximation, 30.

$$\gamma + h\Lambda_A^2 \simeq \gamma + 0.83h\Lambda_A \tag{30}$$

Values of h and $0.83h$ as functions of ϕ_A and $\ln \phi_A$ are obtainable from Table 2.

Test of the analysis

We now test the procedure with an artificial example with $a = 2$ and $b = \infty$. That is the transition state has two protons singled out and a medium effect. We take $\phi_A = 0.80$ and $\Phi_A = 0.90$, for which the

Table 2

Values of h from equation 19

ϕ_A	$-\ln \phi_A$	$100h$	$83h$
0.40	0.92	3.0	3.0
0.35	1.05	4.0	3.5
0.30	1.20	5.5	4.5
0.28	1.27	6.0	5.0
0.26	1.35	7.0	5.5
0.24	1.43	7.5	6.0
0.22	1.51	8.0	7.0
0.20	1.61	9.0	8.0
0.18	1.71	10.0	8.5
0.16	1.83	11.5	9.5
0.14	1.97	13.0	11.0
0.12	2.12	14.5	12.0
0.10	2.30	16.5	14.0

calculated value of γ is 0.325. Figure 4 shows values of $a\gamma$ for $a = 1, 2$ and 3. Possible transition states are then found at each intersection and Table 3 gives values for the fractionation factors for some of the intersections. The transition state we started with (B) is indeed found to be there. We have also calculated the solvent isotope curves for

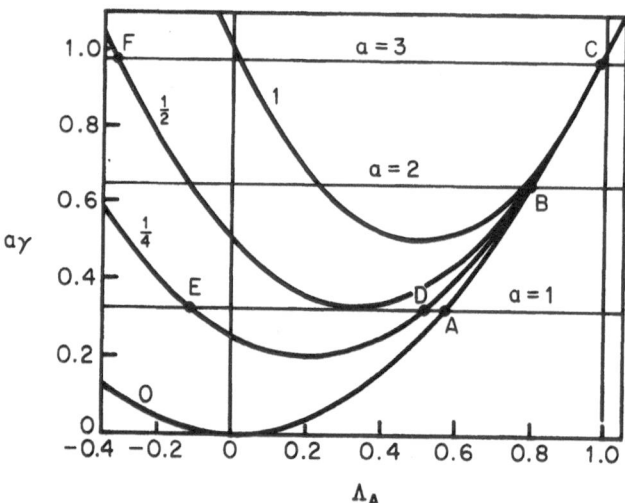

Figure 4. Plot of $a\gamma$ for $\gamma = 0.325$. Parameters for the six transition states A to F given in Table 3.

these six different transition states. The procedure is successful in that all six curves have the same value (to three figures) of $y_{\frac{1}{2}}$ and y_1. What is more all six curves also have the same values (to within

Table 3

Different transition states from Figure 4

Figure 4	A	B	C	D	E	F
a	1	2	3	1	1	3
ϕ_A	0.73	0.80	0.83	0.75	1.07	1.06
b	∞	∞	∞	4	4	6
Φ_A or ϕ_B	0.79	0.90	0.99	0.94	0.86	0.89

2 parts in a thousand) of $y_{\frac{1}{4}}$ and $y_{\frac{3}{4}}$. Despite the different values of a, b and the fractionation factors in Table 3 each model produces virtually the same solvent isotope curve. Experimental results cannot in the foreseeable future discriminate between this the family of curves which have with the same value of γ and the same values of y_x at $x = 0$, $x = \frac{1}{2}$ and $x = 1$. Thus work in mixtures of H_2O and D_2O, compared to work in pure H_2O and pure D_2O, can only provide one more piece of information – the deviation from linearity at $x = \frac{1}{2}$. Even this deviation is small and careful work is required.

Since the deviation from linearity is a maximum at $x = \frac{1}{2}$ and since experimental results are not precise enough for extra information to be obtained by measurements at $x = \frac{1}{4}$, $x = \frac{3}{4}$ etc., it is better to concentrate the measurements at $x = 0$, $x = \frac{1}{2}$ and $x = 1$.

9.7 L_3O^+ and LO^-

For the study of proton transfers in aqueous solution it is vital to know about the fractionation on L_3O^+ and LO^- when they are involved as reactants.

Investigation of the fractionation on L_3O^+ has been carried out by p.m.r. [25–26], vapour pressure studies [27] and measurements of e.m.f. [28]. The data from the three different methods [2] show that there is fractionation on the inner three sites of the L_3O^+ ion and that the value of the corresponding fractionation factor l is 0.69. It is interesting to see if there is any evidence for the breakdown of the rule of the geometric mean. Heinzinger and Weston [27] in their vapour pressure method used solvents with very little D. Thus they measured l_1. Gold [25], on the other hand, using p.m.r. measured his fractionation in 95% D_2O; he therefore measured l_3. Salomaa and Aalto [28] using electrochemical cells with almost pure D_2O measured $l_1 l_2 l_3 = l_2^3$. For our purpose the observed values obtained

279

by the vapour pressure method and n.m.r. must be corrected according to equation 11 to allow for the breakdown of the rule of the geometric mean in L_2O. The value obtained by e.m.f. needs no correction; this is because as discussed above values measured from K_1/K_0 will be those at $x = \frac{1}{2}$. We then obtain the values given in Table 4.

Table 4

Values of l_1, l_2 and l_3

Method	Vapour pressure [32] l_1	e.m.f. [28] l_2	p.m.r. [25] l_3
Observed	0.70	0.69	0.69
Corrected values (equation 11) for l_1 and l_3	0.68	0.69	0.71
Predicted for $p_{L^+} = 0.03$ and $l_2 = 0.69$	0.67	0.69	0.71

We also include the values that would be predicted for the recent value [19, 20] $p_{L^+} = 0.03$. The good agreement may well be fortuitous since the experimental errors are as large as the differences caused by p_{L^+}, but a value of $p_{L^+} = 0.03$ is at least consistent with present experimental data. Batts and Kilford [29] have suggested from an analysis of solvent isotope curves that $l = 0.71$. However they attempted to determine more than two groups of factors from the curves and, as discussed above, (see Table 3) experimental data are not precise enough [30] for this procedure to succeed. Hence for L_3O^+ we will use $l = 0.69$ and $p_{L^+} = 0.03$.

We now turn to OL^-. This problem has taken longer to sort out because the fractionation is more complicated. However, in 1972 Gold and Grist solved the problem [31]. There are three sources of experimental data. First, there are vapour pressure results [32] in strong alkali (7 to 12M). Second, there are results [33] on the auto-protolysis of water. Third, there are recent data [34] on the shift in the p.m.r. spectrum in solutions of MOL where M is Li^+, Na^+ or K^+. We start by writing the autoprotolysis reaction as:

$$L_2O \quad + \quad L_2O \quad \rightleftharpoons \quad L_3O^+ \quad + \quad LO^-$$
$$\bar{\phi} = 1 \qquad \bar{\phi} = 1 \qquad l \qquad \phi_A$$

Then from equation 15, with $p_{L^+} = 0.03$ and $p_{L_2O} = 0.06$,

$$y_x = \frac{K_x}{K_0}[1 - 0.03x(1 - x)](1 - x + lx)^{-3} = \Pi(1 - x + \phi_A x)$$

280

Table 5

Possible solutions for OL^- from Figure 5

Label	γ	ϕ_A	b	ϕ_B or Φ_{OL^-}	Remarks
P	0.43	—	—	—	Impossible single site model
Q	0.43	0.58	3	0.90	Old solution [33]
R	0.43	1.11	3	0.72 ⎫	
S	0.57	1.23	3	0.70 ⎬	New solution [31]
T	0.29	0.94	3	0.76 ⎭	
U	0.43	1.77	∞	0.24 ⎫	Φ_{OL^-} too low
V	0.19	1.46	∞	0.29 ⎭	

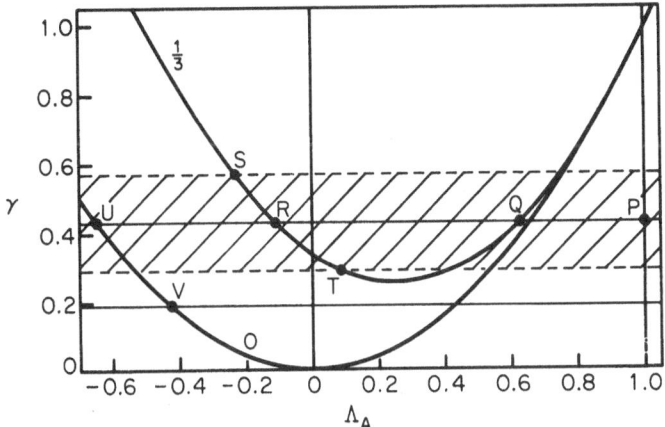

Figure 5. Plot of $\gamma = 0.43 \pm 0.14$ for OL^- with $a = 1$. Parabolas with $a/b = \frac{1}{3}$ and $a/b = 0$. Lower line is $\gamma = 0.19$ calculated with $p_{L^+} = 0.00$ rather than $p_{L^+} = 0.03$. Parameters for models P to V are collected in Table 5.

and

$$\gamma = 0.43 \pm 0.14$$

The γ plot is given in Figure 5 and the results in Table 5. No intersection is found at P ($a = 1$, $\Lambda_A = 1$) which corresponds to the single-site model. The solutions, U and V, on the curve corresponding to the medium effect can be rejected because it is unlikely that the medium effect would be as low as 0.3. This means that no very satisfactory solutions can be found from $\gamma = 0.19$ which is the value calculated for $p_{L^+} = 0.00$; the data are better fitted with $p_{L^+} = 0.03$. Solutions can be found near Q [33] and for some time these values were the accepted ones. However, 5 years later Gold and Grist [31], found a second possible set of solutions at R, S and T. This group of solutions is to be preferred, first because they yield a sensible sequence of fractionation factors, viz.

	L_3O^+	L_2O	LO^-
ϕ	0.69	1.00	1.2 for S
			0.6 for Q

Solution Q gives an unnatural sequence. Second, the fractionation of alkoxide ions in methanol [31, 35] and ethanol [36] gives values for ϕ_B close to 0.7, which again agrees with S rather and Q. The vapour pressure [32] and p.m.r. data [34] are consistent with the $b = 3$ model; in particular the p.m.r. data suggest that ϕ_A is either smaller than 0.5 or greater than 1.2. Giving some weight to this we suggest that $\phi_A = 1.20$ and $\phi_B = 0.70$.

The problem of L_3O^+ and OL^- in L_2O is now largely solved. The advantage of using the new γ procedure is also demonstrated.

9.8 Solutes and salts

The size of the medium effect can be estimated by a study of the relative solubility of 'inert' solutes and salts in H_2O and D_2O. It follows from Section 5 that the solubility, S, in L_2O is given by

$$\frac{S_x}{S_0} = \Phi_S^x \quad \text{and} \quad \frac{S_1}{S_0} = \Phi_S$$

Arnett and McKelvey [3] have collected and reviewed the data on the size of Φ_S. For non-electrolytes which have no exchangeable hydrogens one finds that Φ_S is close to unity. For instance for thirteen different compounds for which accurate data are available (Table 6.2 of reference 3) $\Phi_S = 0.97 \pm 0.09$. Similar results have been found for other organic solutes [37].

Remembering that the rate of a reaction is affected by the differential medium effect and that there will be compensation between the transition state and the reactant, we can see that for neutral species the medium effect will be only a few percent.

For salts, rather than non-electrolytes, we would like to be able to separate the contributions from the two different ions.

$$\Phi_{Salt}^{n_+ + n_-} = \Phi_+^n \Phi_-^n$$

The most promising approach to separating the contributions is that of Salomaa and Aalto [28] who measured the e.m.f.s. of cells without liquid junctions:

$$M(Hg)|MCl(D_2O)|AgCl, Ag - Ag, AgCl|MCl(H_2O)|M(Hg) \quad (I)$$

They showed that Φ_{MCl} was almost identical for $M = Li^+$, Na^+, K^+ and $\frac{1}{2}Cd^{++}$, and suggested that this was because the whole of the effect could be attributed to the anion. This can be rationalized by the different orientation of the solvation water towards cations and anions. Furthermore, making this type of separation, they obtained a value of l_2 (Table 4), which is in agreement with values of l obtained by other methods. We therefore take [28, 38] $\Phi_{Cl^-} = 0.71$ which corresponds to a free energy of transfer of 200 cal mol^{-1} (840 J mol^{-1}). We then obtain the values given in Table 6. These are different from

Table 6

Values of Φ for individual ions at 25°C

L_3O^+	1.00[a]	Ag^+	0.85[e,f]	F^-	1.00[c]	Ac^-	0.90[j]
Li^+	1.00[b]	Tl^+	0.83[g,h]	Cl^-	0.71[a,b]	Pi^-*	0.72[c]
Na^+	0.98[b]	Cd^{2+}	0.94[b]	Br^-	0.62[i]	$CH_3NO_2^-$	0.71[k]
K^+	0.97[b]	$(CH_3)_3S^+$	1.14[d]	ClO_4^-	0.87[c]	Ph_4B^-	0.65[d]
Rb^+	0.97[c,d]	$t\text{-}BuS(CH_3)_2{}^+$	1.15[d]	BrO_3^-	0.82[c]	NPa_2^-†	0.56[d]
Cs^+	0.97[c]	$(C_2H_4)SCH_3{}^+$	1.13[d]	MnO_4^-	0.68[c]	An_4B^-‡	0.55[d]
Cs^+	1.06[d]	$PhCH_2S(CH_3)_2{}^+$	1.14[d]				
		$(n\text{-}Bu)_4N^+$	0.70[d]				

* Pi⁻ is picrate. † NPa₂⁻ is dipicrylamide. ‡ An₄B⁻ is tetra-anisylborate.

[a] P. Salomaa, *Acta Chem. Scand.*, **25**, 365 (1971). [b] P. Salomaa and V. Aalto, *Acta Chem. Scand.*, **20**, 2035 (1966). [c] Jääskeläinen and P. Salomaa, unpublished data. (I am most grateful to Dr. Jääskeläinen and Professor Salomaa for permission to use their data.) [d] B. Hays and C. G. Swain, quoted by E. M. Arnett and D. R. McKelvey [3], p. 373. [e] R. W. Ramette and E. A. Dratz, *J. Phys. Chem.*, **67**, 940 (1963); [f] F. Hein and G. Bähr, *Z. Phys. Chem.* (Leipzig), 270 (1938); [g] P. Salomaa and M. Mattsen, *Acta Chem. Scand.*, **25**, 361 (1971); [h] R. W. Kingerly and V. K. La Mer, *J. Amer. Chem. Soc.*, **63**, 3256 (1941); P. Salomaa, *Acta Chem. Scand.*, **25**, 367 (1971); [j] V. Gold and B. M. Lowe, *J. Chem. Soc.* (A), 1923 (1968); [k] D. M. Goodall and F. A. Long, *J. Amer. Chem. Soc.*, **90**, 238 (1968).

those given by Arnett and McKelvey [3] who relied too heavily on results obtained by Greyson [39] using a cell with a membrane. Although Greyson tried to allow for the effect of the transport of H_2O and D_2O [40], his results, firstly, do not agree with those from the cells without liquid junction, secondly, they are not consistent with solubility measurements [41], and, thirdly, they give a value for l of 0.73 which does not agree with those in Table 4.

Most of the simple cations have factors reasonably close to unity, whereas most of the anions have significantly lower factors. This agrees with the solvation hypothesis on which the separation depends and fits in qualitatively with the values of ϕ_B found for OL^-.

9.9 Weak acid equilibria

The study of weak acid equilibria [2] provides further confirmation of the model for L_3O^+ and also values for the fractionation factors for the sites on the weak acid, LA.

The equilibrium can be written

$$LA + L_2O \rightleftharpoons L_3O^+ + A^-$$
$$\phi_{LA} \qquad\qquad l \qquad \Phi_{A^-}$$

Then, since ϕ_{LA} is close to unity [42] we can write

$$y_x = \frac{K_x}{K_0} = \frac{\Phi_\Delta^x(1 - x + lx)^3}{(1 - x + \phi_{LA}x)} \simeq \left(\frac{\Phi_\Delta}{\phi_{LA}}\right)^x (1 - x + lx)^3$$

We test the model for L_3O^+ by calculating $a\gamma$ from the intersection of the value of Λ_A corresponding to the L_3O^+ fractionation and the parabola corresponding to the medium effect $(a/b = 0)$. Then putting $a = 3$ we have:

$$3\gamma = \Lambda_A^2 = [3 \ln (l)/\ln (K_1/K_0)]^2 \tag{31}$$

Values of γ calculated from this equation are compared with values calculated from the solvent isotope curve in Table 7. Reasonable

Table 7

Comparison of γ for weak acids

Acid	From solvent isotope curve	From equation (31) with $a = 3$ and $l = 0.69$
CH_3CO_2H[a]	0.18 ± 0.09	0.29
$ClCH_2 \cdot CO_2H$[b]	0.44 ± 0.09	0.42
N_3H[b]	0.26 ± 0.13	0.37
![o-nitrophenol] NO_2[c] —OH	0.21 ± 0.18	0.23
O_2N— NO_2[c] —OH	0.40 ± 0.17	0.27
Average	0.30	0.32

[a] V. Gold and B. M. Lowe, J. Chem. Soc. (A), 1923 (1968).
[b] P. Salomaa, L. L. Schaleger, and F. A. Long, J. Amer. Chem. Soc., **86**, 1 (1964).
[c] L. Pentz and E. R. Thornton, J. Amer. Chem. Soc., **89**, 6931 (1967).

agreement is found. What is clear is that the values of γ do cluster around $\frac{1}{3}$ and this is to be expected from equation 27 if the most important fractionation in a weak acid equilibrium is the fractionation on the three sites of L_3O^+. The acetic acid system has been discussed in detail by Gold [2].

Turning to the size of the factor ϕ_{LA}, results for weak acids show [35] that in general $(\phi_{LA}/\Phi_A) \simeq 1.0$

For instance for nitropyridines [44]

$$\ln(\phi_{LA}/\Phi_A) \simeq -0.0005 \, pK_{HA}$$

What seems surprising at first sight is that ϕ_{LA} does not depend on the strength of the acid [43]. The results show that variation in the strengths of the acids cannot be caused by variation in the strength, as measured by ϕ_{LA}, of the O—H or N—H bond; the variation is more likely to be a function of the stability of the anion.

An interesting exception to the general rule of $\phi_{LA} \simeq 1$ is the ionization of LF, studied by Kresge and Chiang [45]. The peculiar observed ratio $(\phi_{LF}/\Phi_A) = 0.67$ is attributed by Kresge to the particularly low value for ϕ_{LF}, which arises from the fact that LF has no bending vibrations.

9.10 Simple proton transfers from weak acids

The simplest type of proton transfer is the single-step process in which the proton is transferred from a weak acid to a base. The same type of process is also the rate-determining step in the $A - S_E^2$ class of reactions [46], for example the acid catalysed decomposition of ethyl diazopropionate [4],

$$HA + CH_3 \cdot CN_2 \cdot CO \cdot OC_2H_5$$

$$\xrightarrow{\text{R.D.S}} A^- + CH_3 \cdot \overset{+}{CHN_2} \cdot CO \cdot OC_2H_5$$

$$\downarrow \text{fast}$$

$$HA + CH_3 \cdot CHOH \cdot CO \cdot OC_2H_5 + N_2$$

The rates of such reactions are usually fairly linear functions of x, corresponding to a value of γ approximately equal to unity. Some results for γ are collected in Table 8 where we have written

$$y_x = (k_x/k_0)(1 - x + \phi_{LA}x) = \Phi_A^x(1 - x + \phi_1 x)$$

and assumed that $\phi_{LA} = 0.98$ [42].

The low value of y_1 in each case shows that there must be at least one proton in flight. Hence $a = 1$. This means that the area of solutions for the transition state has to be close to $\gamma \simeq 1$, $\Lambda_A \simeq 1$. As one might expect, the major part of the overall isotope effect is caused by a

Table 8
Value of γ for LA + S

LA	S	γ	$\gamma + h$	y_1	ϕ_1	Φ_Δ	Reference
CH_3CO_2L	$CH_3NO_2^-$	0.98 ± 0.02	1.09	0.177	0.165	1.08	[66]
CH_3CO_2L	$C_2H_5NO_2^-$	0.97 ± 0.02	1.08	0.166	0.157	1.07	[66]
CH_3CO_2L	$\underset{H}{\overset{CH_3}{NC-C=C-OCH_3}}$	0.83 ± 0.02	0.92	0.189	0.202	0.93	a
CH_3CO_2L	$CH_3 \cdot CN_2 \cdot CO \cdot OC_2H_5$	0.78 ± 0.07	0.85	0.246	0.27	0.90	[47]
$ClCH_2CO_2L*$	$p\text{-}O_2N \cdot C_6H_4 \cdot CH_2N_2$	1.03 ± 0.06	1.08	0.307	0.29	1.05	b

* Reaction carried out in 60 per cent dioxane and 40 per cent L_2O.
a V. Gold and D. C. A. Waterman, *J. Chem. Soc. B*, 839 (1968); b C. Diederich and H. Dahn, *Helv. Chim. Acta.*, **54**, 1950 (1971).

single proton in flight. The fractionation factors are worked out from the intersection of $\gamma + h$ with the medium effect parabola $\gamma = \Lambda_A^2$. The values of Φ_A are all close to unity and not too much should be made of the differences; but it is interesting that for the two nitro compounds, where the charge type of the reaction is preserved, Φ_A is a little greater than unity, while for the next two systems, where charge is created $\Phi_A \simeq 0.9$, which is roughly the value found for the ionization of acetic acid [42].

For certain $A - S_E2$ reactions one can also measure the isotopic fractionation in the $C-L$ bond that is formed during the reaction. After the bond is formed, the subsequent steps of the reaction are rapid and the $C-L$ bond can no longer exchange with the solvent; its isotopic composition is determined at the moment of its conception. For these reactions analysis of the product then allows one to determine the fractionation factor, ϕ_P, for the transition state which formed the $C-L$ bond. For instance for ethyl diazopropionate

$$(\phi_P)_{obs} = \frac{[CH_3 \cdot CDOH \cdot CO \cdot OC_2H_5]}{[CH_3 \cdot CHOH \cdot CO \cdot OC_2H_5]} \frac{(1-x)}{x}$$

(Other authors have labelled this function r^{-1} and κ_D/κ_H; we prefer to emphasize that it is a fractionation factor.)

The analysis of the curvature presented so far strongly suggests the simple transition state:

$$A \cdots \vec{L} \cdots C$$

$$\phi_1$$

If this is the case then ϕ_1 should equal ϕ_P. Where both have been measured this is indeed found to be the case. For instance for the acetic acid catalysed decomposition of ethyl diazopropionate [47], $\phi_1 = 0.27 \pm 0.01$ and $\phi_P = 0.25 \pm 0.01$.

9.11 Simple proton transfers from L_3O^+

Before discussing these results in more detail we now turn to the similar simple proton transfer from L_3O^+. For the reaction

$$L_3O^+ + S \rightarrow L_2O + LS^+$$

we write

$$y_x = [1 - 0.03x(1-x)](1 - x + lx)^3 k_x/k_0$$

287

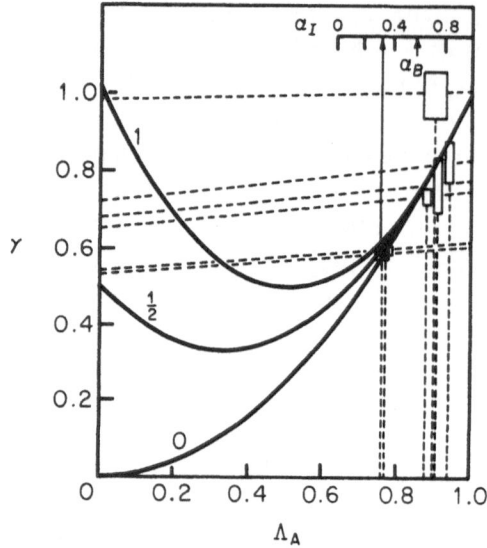

Figure 6. Plots of $(\gamma + 0.83h)$ for six $A - S_E2$ transition states where ϕ_P has been determined. Vertical broken lines are values of Λ_P from equation 32. Each box represents $\pm$ one standard deviation on γ or Λ_P. Solutions for the simple $A - S_E2$ transition state should lie on the parabola with $a/b = \frac{1}{2}$. The α_I scale refers to ethyl diazopropionate only and is defined by equation (33); α_B is the Brønsted coefficient. The compounds in order of increasing γ are $CH_3 \cdot CN_2 \cdot CO \cdot CH_3$,[a] $CH_3 \cdot CN_2 \cdot CO \cdot C_2H_5$,[a] $C_2H_5OCH:CH_2$,[b] $(CH_3)_2C:CH \cdot HgBr$,[c] $CH_2:CH \cdot CH_2HgI$[d] and $(CH_3)_2C:CH_2$.[e]

[a] [20]; [b] A. J. Kresge and Y. Chiang, *J. Chem. Soc. B*, **58** (1967); M. M. Kreevoy and R. Eliason, *J. Phys. Chem.*, **72**, 1313 (1968); [c] M. M. Kreevoy and R. A. Landholm, *Int. J. Chem. Kinetics*, **1**, 157 (1969); [d] M. M. Kreevoy, P. J. Steinwand and W. V. Kayser, *J. Amer. Chem. Soc.*, **88**, 124 (1966); [e] V. Gold and M. A. Kessick, *Disc. Faraday Soc.*, **39**, 84 (1965).

Values of γ for six systems where ϕ_P has been determined as well are plotted in Figure 6. The values of y_1 are small enough to be clear that a proton is in flight in the transition state; hence $a = 1$. We can then check whether ϕ_P is the fractionation factor for the proton in flight. We calculate

$$\Lambda_P = \ln \phi_P / \ln y_1 \qquad (32)$$

Figure 6 shows for each system the area corresponding to one standard deviation on either side of the mean where, from equation 30, $(\gamma + 0.83h \, \Lambda_A)$ meets Λ_P. If ϕ_P is the fractionation factor for the proton in flight then a possible transition state must be found close to the intersection. This is found to be the case for all six reactions. With infinite experimental precision one might hope that the points would delinate clearly one of the parabolas. This is not the case.

The points for the diazo systems suggest that it is unlikely that $b = 1$. That is, the secondary effect cannot be attributed to another single proton with a low fractionation factor. Also, although the

288

parabola describing the medium effect ($a/b = 0$) does fit the data, for the diazo compounds the medium effect would have to be $\Phi_A \simeq 0.6$. This seems too far from unity for a reaction where there is no change in charge type. Hence we conclude that the secondary effect must be caused by a small number of specific sites. The obvious transition state, which fulfils this condition with $b = 2$, is

$$\phi_2\ \begin{array}{c}\text{L} \\ \diagdown \\ \end{array}\qquad \overrightarrow{\ \ } \\ \text{O}\cdots\text{L}\cdots\text{S} \\ \diagup \\ \phi_2\ \text{L}\qquad \phi_1$$

The purpose of the present discussion is to show how far one can deduce the type of transition state, rather than assume it and then merely show that it is consistent with the data.

9.12 Solvent bridges

The question of whether a solvent bridge is involved in an aqueous proton transfer or not has been reviewed [48] and discussed by a number of authors [46, 49, 50] including Grunwald and Eustace in Chapter 5 of this book. The transition states established by the iso-topic method for proton transfer to carbon show that there are no solvent bridges for this type of reaction. As discussed above we can rule out transition states with two protons in flight:

$$-\text{O}\cdots\overrightarrow{\text{L}}\cdots\text{O}\cdots\overrightarrow{\text{L}}\cdots\text{S} \\ | \\ \text{L}$$

More complicated models involving two steps or restricted rotation have been discussed but can similarly be rejected [20].

Further support for the model with no solvent bridge is found from the fact that for these reactions the point for H_3O^+ as a catalyst lies well below the Brønsted plot obtained from the carboxylic acids. Typically the rate is about 20 times slower than expected. It has recently been claimed [51] that this effect can be partly attributed to electrostatic interactions between the charged catalyst and the developing charge in the transition state. However the deviation for H_3O^+ is the largest [51] and another plausible explanation [52] is that it is caused by the desolvation step

$$H_9O_4^+ + S \rightleftharpoons H_2O + (H_2O\cdots H-)_2O^+-H\cdots S$$

This explanation fits in with the model in which there is no solvent bridge.

We can safely conclude that, for the proton transfer reactions to carbon discussed here, there are no solvent bridges and the rate-determining step is the direct proton transfer from the acid to the carbon base. This is because there are no hydrogen bonds between the receiving carbon and the aqueous solvent. These systems therefore contrast with the oxygen and nitrogen bases discussed by Grunwald and Eustace (Chapter 4).

9.13 Degree of proton transfer

Having established the nature of the transition state for the transfer from L_3O^+ as being

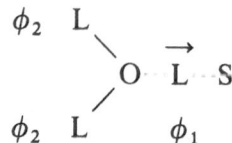

we can make further deductions about the degree of proton transfer from the size of the secondary factor ϕ_2. Kresge [1] suggested that

$$\phi_2 = l^{1-\alpha_I}$$

where α_I measures the degree of proton transfer. Depending on the degree of proton transfer, the secondary factor changes from the value of ϕ appropriate for L_3O^+ to that appropriate for L_2O.

For the transition state

$$y_x = (1 - x + \phi_1 x)(1 - x + \phi_2 x)^2 \Phi_\Delta^x$$

A knowledge of y_1 and γ allows the determination of ϕ_1 and $\phi_2^2\Phi_\Delta$. If one has measured ϕ_P then, assuming $\phi_P = \phi_1$ one can apply the γ treatment to $y_x(1 - x + \phi_1 x)^{-1}$ to see whether the separation of ϕ_2 and Φ_Δ can be achieved. This has been tried [20] but, even with precise data, the separation is not very successful. Hence to obtain ϕ_2 we are forced to assume that $\Phi_\Delta = 1.0$. This is probably not a bad assumption since the species are cations or neutral solutes and there is no change in charge type during the reaction.

Once we assume $\Phi_\Delta = 1$, α_I can be read directly off the γ plot. To do this we calculate

$$(\Lambda_B)_{max} = \frac{2 \ln l}{\ln y_1} \tag{33}$$

then $\alpha_I = 1 - \Lambda_B/(\Lambda_B)_{max}$. This procedure is shown in Figure 6 for ethyl diazopropionate. The transition state 'area' then represents how 'far' the proton is transferred. Another route to finding α_I, without measuring a solvent isotope effect curve, is to find values of the overall solvent isotope effect y_1 and the product fractionation factor ϕ_P. The slope of the Brønsted plot, α_B, has also been interpreted as a measure of the degree of proton transfer. The results in Figure 7

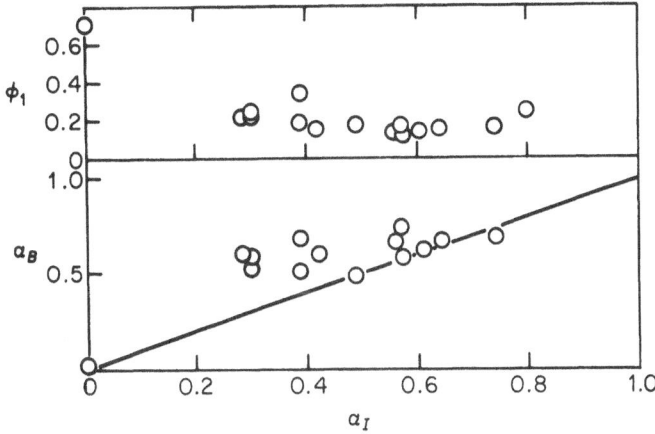

Figure 7. Characteristics of transition states for simple proton transfers to carbon; ϕ_1 is the primary fractionation factor, α_B the Brønsted coefficient for RCOOH, α_I is the degree of proton transfer as measured by the secondary solvent fractionation.

show that in general α_I is not equal to α_B and it has been an interesting question [2, 48, 53] as to whether α_I and α_B should be the same for each reaction or not. The reasons for the change in rate when H is substituted by D and when it is substituted by, for instance, Cl are rather different. To see whether α_I is different from α_B for the *same* acid we have recently measured the secondary effect in the general acid catalysed [53] decomposition of 3-diazobutanone.

The results were $\alpha_{B,RCO_2H} = 0.61 \pm 0.03$, $\alpha_{CL_3CO_2H} = 0.64 \pm 0.10$, and $\alpha_{LCO_2H} = 0.64 \pm 0.07$. They may be contrasted with the value of $\alpha_I = 0.28 \pm 0.02$ for L_3O^+. Thus we can conclude that for the *same* acid α_I is the same as α_B. The use of the secondary isotope effect differentiates the Brønsted plot and allows one to measure a value of α for each acid. This is particularly important for H_3O^+ which, as discussed above, deviates from the Brønsted plot.

From Figure 7, it can also be seen that some reactions are well behaved (as good as Gold) and $\alpha_I \simeq \alpha_B$. For these the degree of proton transfer in the transition state is independent of the strength of the catalyst. For other reactions we find that α_I is significantly smaller than α_B. For instance, inspection of Figure 6 shows that the transition state area for ethyl diazopropionate is well to the left of α_B. The change in α means that the proton is less transferred when H_3O^+ is the catalyst than when RCO_2H is the catalyst. In accordance with the Hammond postulate, the more powerful the catalyst the earlier is the transition state. For proton transfer this change can be expressed quantitatively by the Marcus theory [55].

However, if one plots the rate of the reaction against α_I for the different substrates, rather than the different catalysts, no clear pattern emerges. If we plot ϕ_1 against α_I [19], the points (Figure 7) do show a vague minimum around α_I of ~ 0.6. (A minimum in ϕ_1 corresponds to a maximum in the primary isotope effect $(k_H/k_D)_1$.) The scatter is not much worse than a similar plot made by Bell of k_H/k_D against ΔpK [56]. Two different explanations have been advanced [56, 57] for this type of behaviour. Whichever one is correct, the scatter on the plot for ϕ_1 shows that it is a poor measure of the details of transition state symmetry. Too much happens to the transferring proton for it to be a good probe. The non-transferring protons are only spectators and hence make better witnesses.

9.14 Parallel transition states

General conclusions (5) and (6) in Section 6 are interesting in that one explanation of unusual values of γ may be the existence of parallel or consecutive transition states. We now discuss two acid-catalysed reactions each of which probably occurs through two parallel transition states. The first example is Kankaanperä's work [58] on the L_3O^+-catalysed hydrolysis of the three pyrans:

	A	B	C
	Acetal hydrolysis	Vinyl ether hydrolysis	Both?
k_1/k_0	2.91	0.43	1.25

A gives a normal A-1 (or A-2) curve for the reaction:

292

For B we obtain a normal $A-S_E2$ curve for the reaction:

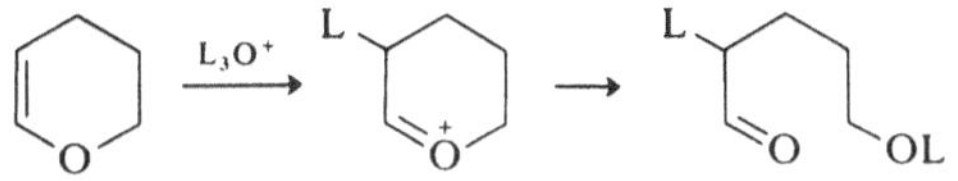

For C the curve is neither one nor the other; it might just be a very unsymmetrical $A-S_E2$ transition state with $\phi_1 \sim 0.45$ and $\phi_2 \sim 0.95$. The value of γ however is only 0.36 ± 0.11. This value gives impossible factors for an $A-S_E2$ transition state ($\phi_1 \sim 0.64$ and $\phi_2 \sim 0.80$) and confirms that the reaction must be proceeding along the two parallel routes:

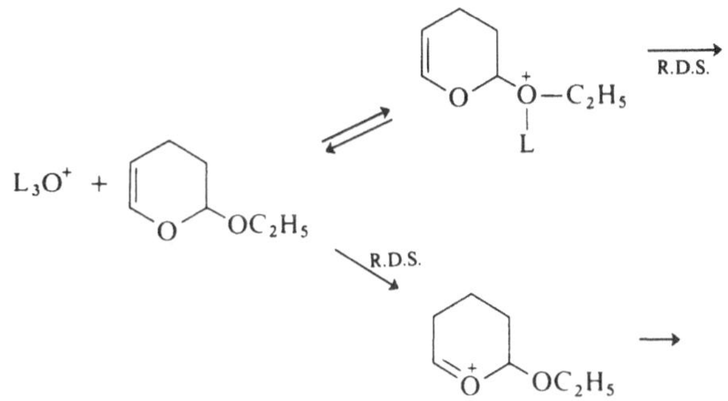

An even more striking example is provided by the work of Pihlaja [59] on the L_3O^+-catalysed decomposition of oxathiolanes:

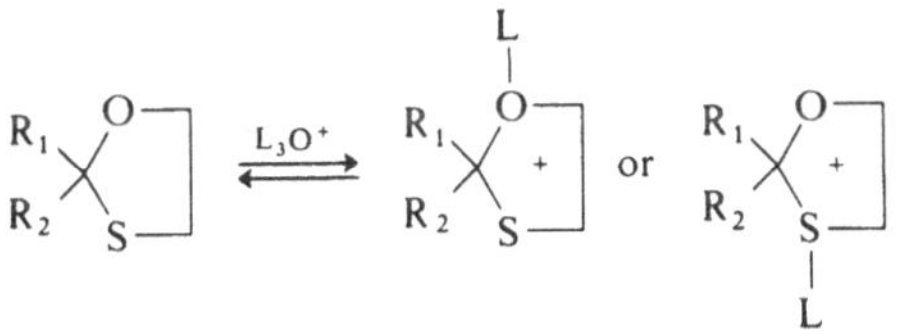

$$\rightarrow R_1R_2C = O + LOCH_2 \cdot CH_2 \cdot SL$$

where R is H or CH_3. Pihlaja recognized that the solvent isotope curves were much too curved and attributed this partly to a medium effect on the transition states. However, if this alone were to blame, then γ should still have a sensible value. In fact the values are:

R_1	R_2			
CH_3	CH_3	$y_1 = 0.52$	and	$\gamma = -0.9 \pm 0.5$
CH_3	H	$y_1 = 0.97$	and	$(\ln y_1)^2 \gamma = -1.2 \pm 0.5$

(Since y_1 is close to unity, γ for the second system is equal to -1683.) The negative values of γ show from general conclusion (4) that the

293

reaction cannot be taking place through a single transition state and must involve parallel routes. They probably involve protonation on the O and on the S atoms.

9.15 The reverse reaction

Simple proton transfers can also be investigated by studying the kinetics of the reverse reaction to those discussed above $i.e.$:

$$L_2O + LS \rightarrow L_3O^+ + S$$

A thorough investigation of this reaction for three cyanocarbon acids has been carried out by Long and his co-workers. Figure 8 shows

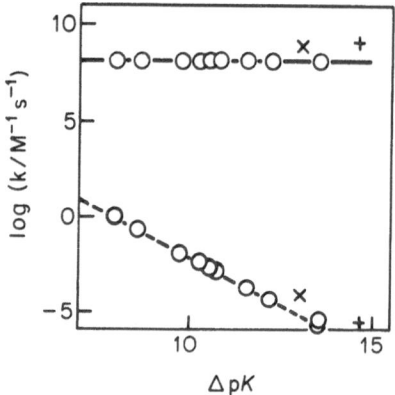

Figure 8. Brønsted plots [60] for proton transfer to (solid line) and from (broken line) cyanocarbon acids with RCO_2H/RCO_2^- shown by circles; H_3O^+/H_2O shown by $\times$ for malonitrile and by $+$ for t-butyl-malonitrile.

the Brønsted behaviour for the three compounds [60]. For all three substrates the points for the carboxylic acids lie on a common line with $\beta = 1$. The converse of this is that the reaction

$$LA + S^- \rightarrow A^- + LS$$

has $\alpha = 0$ with $k \simeq 10^8 \ \mathrm{dm^3 \ mol^{-1} \ s^{-1}}$.

This behaviour is reminiscent of the rates of reactions found by Eigen [61] for proton transfer to and from oxygen and nitrogen bases. Using tritium, Hibbert and Long have measured the ratio of the primary fractionation factors ϕ_1/ϕ_{SL} for the SL site [62] (Table 9). Since ϕ_{SL} is approximately unity we find that the factor for the transferring proton from SL is about 0.69 which is a familiar number. This suggests that in the reaction with L_2O the proton is already fully

294

Table 9

Fractionation factors for proton transfer from cyanocarbon acids[a]

Acid	Base	ϕ_1/ϕ_{SL}	$\Phi_\Delta\phi_2^2$	Φ_Δ^*	Φ_{eq}
$HCL(CN)_2$	L_2O	0.69	0.27	0.57	—
$(CH_3)_3C \cdot CL(CN)_2$	L_2O	0.71	0.28	0.59	0.80
$(CH_3)_3C \cdot CL(CN)_2$	OAc^-	0.68	—	0.89	0.89

[a] Calculated from F. Hibbert and F. A. Long, *J. Amer. Chem. Soc.*, **93**, 2836 (1971).
* From $\phi_2 = 0.69$ and $b = 2$ for L_2O as base.

transferred in the transition state:

$$S^- \cdots L - \overset{+}{\underset{\underset{\phi_1}{|}{L}}{O}} - L \cdots \overset{}{\underset{\underset{\phi_2}{|}{L}}{O}} - L$$

$$\phi_2$$

Furthermore, the secondary fractionation for L_2O is considerably smaller than unity [62]. Even with $\phi_2^2 = (0.69)^2$, $\Phi_\Delta < 0.6$. This value is *smaller* than the equilibrium value and so cannot be intermediate between the reactants and the products [63]. Since we would expect the structure of the solvation to change steadily from reactants through to products, this suggests that there must be a 'dynamic medium effect' which applies to the kinetics of the process and not to the thermodynamics. The diffusion coefficients of H_3O^+ in H_2O and D_3O^+ in D_2O differ by 40 per cent [64] and this difference would help to explain the reason why the reaction is even slower in D_2O than in H_2O. Thus the characteristics of the L_2O reaction fit a model in which the reaction is in some way 'diffusion' controlled. It is interesting that for this reaction H_3O^+ is a more effective catalyst than predicted by the Brønsted plot for RCO_2H whereas for the normal $A-S_E2$ reaction H_3O^+ is less effective. This agrees with the explanation that attributes the slower rate for H_3O^+ for normal $A-S_E2$ reactions to the desolvation step. For the mechanism proposed here there is no desolvation in the transition state and instead H_3O^+ may be more effective because of its greater mobility. This point is further discussed below.

Turning to the reaction with acetate ion, we find that ϕ_1 is again close to 0.69. This was used by Hibbert and Long [62] to argue against the solvent bridge on the grounds that, since L_2O and OAc^- had the same ϕ_1, they had the same type of transition state. But there is a great difference in that for product-like transition states the limiting factors for L_2O and OAc^- are different, being 0.69 and 0.98 respectively. Hence, while for L_2O it is clear that the L^+ has reached its destination and is safely grounded in the transition state, for OAc^- it is equally clear that the L^+ must still be in flight. Hence for acetate (and presumably the other carboxylate bases) there is no solvent bridge and no diffusion control. The reaction is still controlled by the proton transfer step, even though the transition state is very product-like. This conclusion is further reinforced by the observation [62] that the limiting rate constant of $10^8\,dm^3\,mol^{-1}\,s^{-1}$ is two orders of magnitude smaller than the calculated value for diffusion-controlled rates of $10^{10}\,dm^3\,mol^{-1}\,s^{-1}$; for proton transfer to carboxylate anions the observed and calculated rates are in good agreement [61]. Thus, despite the similarity of the Brønsted plot to that predicted by Eigen [61], it is better explained by the Marcus theory of proton transfer [55] rather than by diffusion control. The Marcus theory [55] is a quantitative expression of the Hammond postulate and describes how, as the reaction becomes more thermodynamically favourable, the transition state becomes more reactant-like. The theory predicts that in the limit α tends to zero even though the proton transfer is still rate-determining (see Figure 8). One must be careful to distinguish between Marcus curvature and Eigen curvature [52]. Indeed many of the systems which were originally thought [61] to show diffusion control are probably displaying limiting Marcus behaviour. Hibbert and Long's work [62] with the isotopic probe shows that this is definitely the case for the reaction between the cyano-carbon acids and carboxylate bases. The same general picture has also been found by Hibbert [65] for proton transfer to and from disulphonyl carbon acids.

Another reverse reaction studied in detail by Dahlberg and Long [63] is

$$L_2O + CH_3 \cdot CO \cdot CL(CH_3) \cdot CO \cdot CH_3$$

$$\rightarrow L_3O^+ + CH_3 \cdot C\bar{O}{=}C(CH_3) \cdot CO \cdot CH_3$$

The reaction can be followed by measuring the detritiation (L = T) of the substrate. For this transition state,

$$\phi_2\ L \diagdown$$
$$\overset{\leftarrow}{O\cdots T\cdots S}$$
$$\phi_2\ L \diagup$$

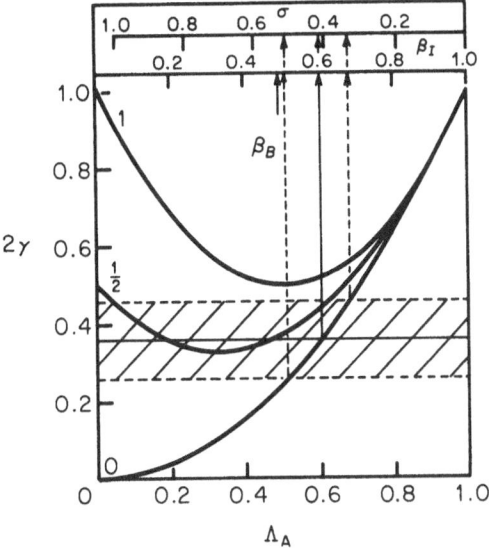

Figure 9. Plot of 2γ for detritiation of methyl acetylactone [63]. The scales for the degree of proton transfer (β_I) and the solvation scale (σ) are defined in equation 34; β_B is the Brønsted coefficient.

there is only specific fractionation on the two secondary sites, and

$$y_x = (1 - x + \phi_2 x)^2 \Phi_\Delta^x$$

Figure 9 shows the γ plot for $a = 2$. The value of Φ_Δ is as low as 0.76 ± 0.04. However, for the overall equilibrium $\Phi_{eq} = 0.50$ and this particularly low value is attributed to the solvation of the enolate ion. In Figure 9 we can now plot two scales which characterize the transition state:

<table>
<tr><td>Proton transfer</td><td>Solvation</td></tr>
</table>

$$
\begin{array}{ll}
\phi_2 = l^{\beta_I} & \Phi_\Delta = \Phi_{eq}^\sigma \\
(\Lambda_A)_{max} = 2 \ln (l)/\ln y_1 & (\Lambda_B)_{max} = \ln \Phi_{eq}/\ln y_1 \\
\beta_I = \Lambda_A/(\Lambda_A)_{max} & \sigma = \Lambda_B/(\Lambda_B)_{max} \\
\quad = 0.60 \pm 0.09 & \quad = 0.40 \pm 0.10
\end{array}
\qquad (34)
$$

Dahlberg and Long [63] tested the hypothesis that $\sigma = \beta_I$, but, since there is no reason why the degree of proton transfer should be the same as the degree of solvation, we prefer to define the new parameter σ. Figure 10 shows the area where the transition state for the L_2O catalysed reaction has to be found as a function of the L^+ transfer and the solvation change.

Dahlberg and Long also studied the detritiation reaction with acetate ion [63]. Here there are no ϕ_2 effects and they found $\Phi_\Delta = 1.00$.

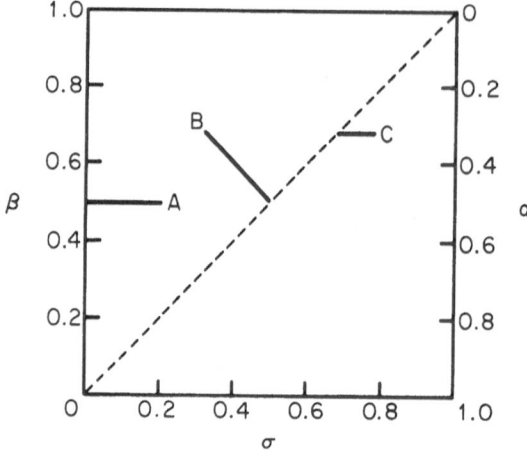

Figure 10. Transition state characteristics for methyl acetyl acetone (A, B) and nitromethane (C) in terms of the degree of proton transfer (α, β) and the solvation scale (σ). The bases are OAc$^-$ (A, C) and L_2O (B).

In view of the $\Phi_\Delta = 0.76$ for L_2O this was rather surprising. However, $\Phi_\Delta = \Phi_\ddagger/\Phi_{OAc^-}$ and, in going to the precursor of the transition state, the solvation around the CO_2^- group which is responsible for Φ_{OAc^-} will be replaced by the substrate [19]. Hence we may find that

$$\Phi_\ddagger \simeq \Phi_{OAc^-} \simeq 0.9$$

Therefore the transition state area for the OAc$^-$ catalysed reaction is shown in Figure 10 for σ calculated with $\Phi_\Delta = 1.0$ and with 0.90. A similar line is also plotted for the nitromethane system [66],

$$CH_2NO_2^- + CH_3CO_2H \rightleftharpoons CH_3NO_2 + CH_3CO_2^-$$

for which $\beta_B = 0.67$, $\Phi_{eq} = 1.27$ and $\Phi_\Delta = 1.08$. The use of the medium effect to characterize transition state solvation is an important new application of solvent isotope effects.

9.16 Concerted proton transfers

We now turn to some reactions where the proton transfer is concerted with other changes taking place in the reaction co-ordinate, and in particular with the making or breaking of carbon oxygen bonds. Many of these reactions involve water as the reactant. The solvent isotope effect is simpler to use when water is either the acid or the base because we do not have to bother about reactant fractionation.

We start with the elegant work of Bell and Critchlow [67] on the hydration and dehydration of carbonyl compounds. Following a suggestion by Eigen [68], the order of the reaction with respect to water in water-dioxan and water-acetonitrile mixtures for the water-catalysed reaction of 1,3-dichloroacetone was shown to be 3 for the hydration reaction and 2 for the dehydration reaction [69]. This suggested a transition state somewhere between **4** and **5**

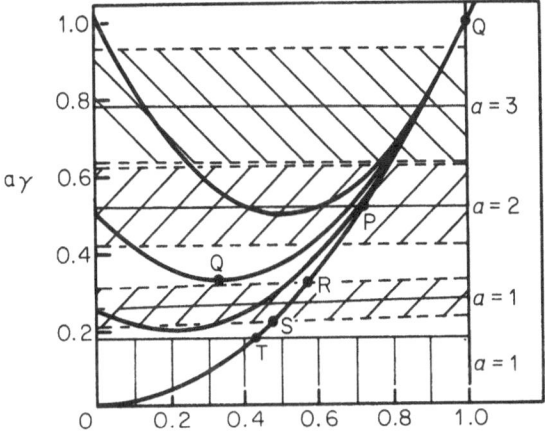

From Bell and Critchlow's data [67], $\gamma = 0.26 \pm 0.05$, and this value is plotted in Figure 11. From general conclusion (3) of Section 6

Figure 11. Plot of $a\gamma$ for hydration of 1,3 dichloroacetone (for $a = 1, 2$ and 3) and for the mutarotation of tetramethylglucose (for $a = 1$, vertical hatching). The characteristics of transition states P to T are given in Table 10.

the number of sites must be greater than γ^{-1} and, since it is unlikely that all the factors are equal [67], the solvent isotope effect shows that there are at least five or six sites contributing to the fractionation. This is what one would expect for the cyclic transition state. An even more striking example is found from the mutarotation of tetramethylglucose [70]. The reaction is similar in type to the hydration-dehydration reaction, and for catalysis by water one finds

$$\gamma = 0.07 \pm 0.11 \quad \text{for} \quad \phi_R = 1.23$$

and

$$\gamma = 0.00 \pm 0.12 \quad \text{for} \quad \phi_R = 1.00$$

The calculations have been carried out for two values of ϕ_R for the exchangeable site on the tetramethylglucose since the measured factor [70] of 1.23 seems a little large; the value of γ, however, is not much affected. These low values of γ again suggest that at least five or six sites must be involved, somewhat as in the transition states **4** and **5** above. It is just possible that, as discussed above, the particularly low value of γ might be caused by parallel reaction paths (general conclusions of Section 6). The mutarotation of glucose is catalysed by both acids and bases and H_2O lies close to the Brønsted plot [71], whether it is plotted as an acid or as a base. However, it is unlikely that any reduction in γ due to parallel paths would be so large as to invalidate the conclusion about the cyclic transition state.

The solvent isotope effect curve does not give us the detailed fractionation, but we can draw conclusions [67] about the size of the most important factor, ϕ_A, that is singled out in the transition state. To see if any pattern emerges we also consider two further reactions which have been studied in sufficient detail, first, the hydrolysis of acetic anhydride [72] and, second, the hydrolysis of dichloroacetyl salicylate anion [73]. Values of $a\gamma$ are plotted in Figure 12. Since the overall isotope effect is small (y_1 low) in each case we need plot $a\gamma$

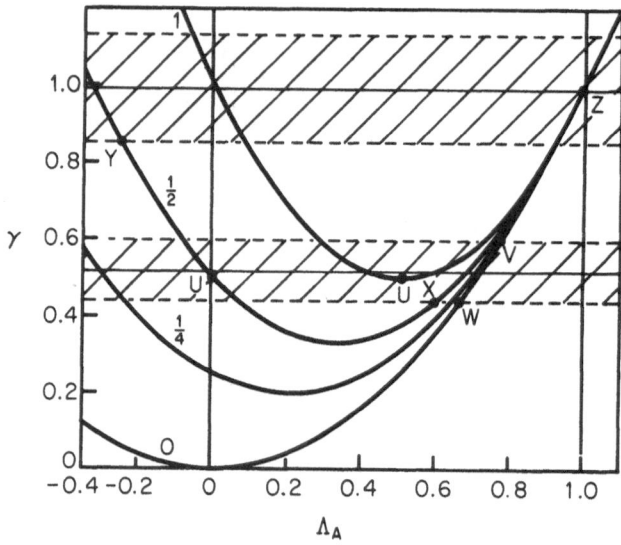

Figure 12. Plots of γ for hydrolysis of acetic anhydride (V to X) and of dichloroacetyl salicylate anion (Y, Z). X is the simple $A - S_E2$ transition state discussed by Batts and Gold [72]. Parameters for other possible models are collected in Table 10.

only for $a = 1$; no new solutions are found for higher values of a. We divide the possible solutions into those where it is likely that only one proton is in flight and those which require more than one proton to be in flight. In deciding whether a proton is in flight or not we take the value of $l = 0.69$ as a limiting value of a grounded proton; it is also interesting that ϕ_{LF} and ϕ_{FLF^-} are about 0.7 [45]. We can also examine models where no protons are in flight, that is all the factors are greater than 0.7. We do this by working out the minimum value of n which would be required for the overall isotope effect:

$$(0.70)^n = y_1$$

Then, since γ is a minimum for n equal factors, it follows that $a\gamma$ must be greater than $n\gamma$. However values of $a\gamma$ greater than 1 can only be caused by fractionation factors greater than 1. Apart from OL^- these are rare, and therefore we can conclude that models with $n\gamma$ greater than 1.5 are extremely unlikely. The results are collected in Table 10. Straightaway we can see that for all but two of the reactions a model with no protons in flight is either unlikely or completely ruled out. Secondly, models with more than one proton in flight are impossible for some of the reactions. The size of the factors to be expected for such a process has been a matter of dispute [67] but, if anything, factors of about 0.6 seem rather large. When one turns to the single-proton column all the reactions have a possible model and, apart from L_2O + tetramethylglucose, all the factors for the proton in flight are very similar. The values for the primary factor are not much affected by the different possible models for the secondary factors given by $\Pi\phi_B$. Thus the 'one-proton' model is the only one that can be common to all the reactions. Incidentally, as Minor and Schowen [73] pointed out, the characteristics of the reaction of the dichloroacetyl salicylate anion are very similar to that of the *acetate* catalysed reaction of the glucose; this is because [74] their reaction is an intramolecular one involving the carboxylate group on the ring. Also we must be a little cautious about the acetic anhydride reaction in view of the curious temperature dependence of k_1/k_0 [75]. However, the most important point to note is that the fractionation factors in Table 10 are in the range 0.4 to 0.5. The fractionation factors for the proton in flight in the $A-S_E2$ reactions were nearly all less than 0.25. These low values are found for a wide range of transition structure and reactivity, from diazoacetate anion to isobutene (see Figure 8). In his paper to the Faraday Discussion in 1965 [76] Bell discussed the inadequacies of the three-centre model for proton transfers and emphasized that other vibrational modes

Table 10

Possible transition states for reactions of S with L_2O etc.

S	Acid or base	y_1	γ	n protons in flight				One proton in flight			No protons in flight	
				Figure	Label	n	ϕ	Label	ϕ_A	$\pi\phi_B$	n	$m\gamma$
$(ClCH_2)_2CO$ [67]	L_2O	0.252	0.26 ± 0.05	11	P	2	0.60	R	0.52	0.50	4	1.0
				11	Q	3	0.63	S	0.46	0.55		
Tetramethyl[a] glucose [70]	LOAc	0.330	0.65 ± 0.08					—	0.41	0.81	3	(1.9)
	L_2O	0.346	0.07 ± 0.11	11	—		—	T	0.65	0.53	3	0.2
	L_3O^+	0.302	0.51 ± 0.02		—	2	0.55	—	0.43	0.71	4	(2.0)
	OAc^-*	0.551	1.7 ± 0.2					—	0.46	1.20	2	(3.4)
$(CH_3CO)_2O$ [72]	L_2O	0.344	0.52 ± 0.08	12	U	2	0.59	V	0.47	0.73	3	(1.6)
								W	0.42	0.82		
[structure: benzene ring with O·CO·CHCl₂ and CO₂⁻ substituents] [73]	L_2O	0.458	0.99 ± 0.14	12	Y	2	0.63	Z	0.46	1.00	2	(2.0)

[a] With $\phi_R = 1.23$; no significant difference is found in ϕ_A with $\phi_R = 1.00$.

involving more atoms than three must also be considered. If the motion of the proton in the reaction coordinate is coupled to the motion of the heavy atoms then the isotopic substitution will have less effect and the fractionation factor will be closer to unity. In all the reactions in Table 10 we are concerned with breaking or making a C—O bond. No such bond breaking or bond making process is involved in the A-S_E2 reactions. Thus the fractionation factor of 0.4 to 0.5 suggests that the proton is in flight but its motion is coupled to other heavy atom motion. Other reactions, such as ester hydrolysis [73, 77] which have $k_1/k_0 \sim 0.5$ no doubt also fall into this category, even though the separation of the factors has not yet been carried out.

However we have to be a little careful in using this criterion. The results presented by Bell [76] for the overall isotope effect, k_1/k_0, for the reverse reaction,

$$SL + L_2O \rightarrow S^- + L_3O^+$$

suggest that it is possible to have $k_1/k_0 \sim 0.5$ for unsymmetrical transition states ($\alpha < 0.3$) which are outside the range of values plotted in Figure 8. Hence the use of ϕ_1 to decide whether the proton transfer is simple or is concerted can apply only to 'symmetrical' transition states. Thus provided $0.3 < \alpha < 0.7$, if

$\phi_1 < 0.3$ the proton transfer is 'simple',
if
$0.4 < \phi_1 < 0.7$ the proton transfer is 'concerted',
and if
$0.7 < \phi_1$ the proton is not transferring in the transition state.

For the hydration/dehydration reaction Critchlow [78] has set up a simple model using bond orders for the possible transition state:

He concludes that, in the transition state for the acid-catalysed reaction, proton A is half transferred, protons B and C have not moved and the C$\cdots$O bond is partially formed. In discussing in a complex reaction whether the different atom transfers are concerted or take place in a series of discrete steps, it is helpful to use contour diagrams depicting the reaction coordinate as a function of two geometrical variables. These diagrams were introduced to describe the role of the

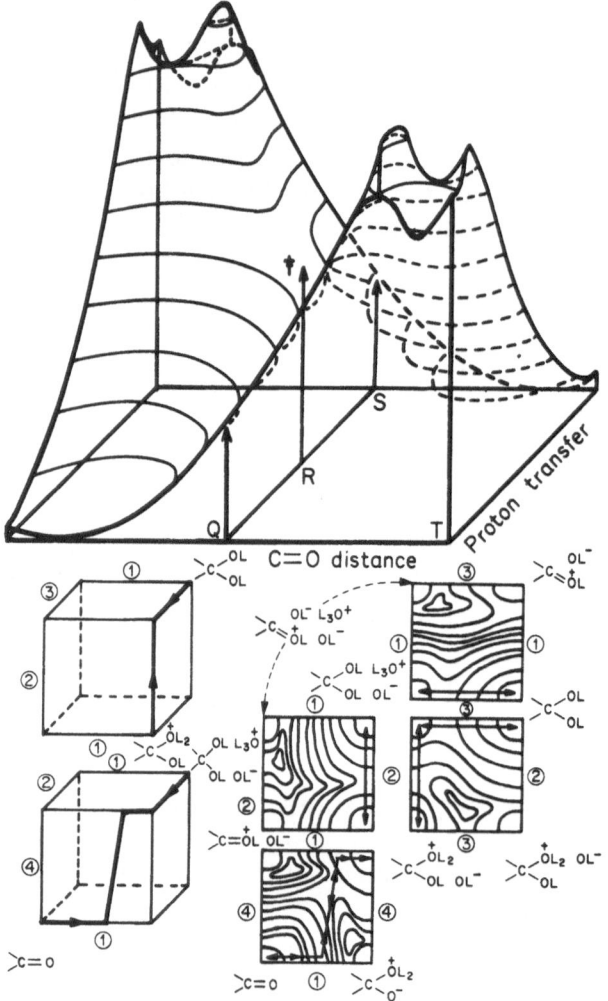

Figure 13. Exaggerated free energy surface showing two volcanoes with contours for a concerted proton transfer. At all values of the C–O distance there is a barrier to proton transfer. The Marcus theory of proton transfer predicts that $\alpha = \frac{1}{2}$ since the saddle occurs at the value of x where the thermodynamics driving the proton transfer is in balance (QRS). The surface has been drawn for a symmetrical case, but this conclusion still holds for an unsymmetrical case. The strip cartoon shows the mechanism for the hydration dehydration reaction of a ketone.

The numbers refer to the geometry: $\begin{array}{c} \text{①} \ \underset{\text{O}}{\overset{\backslash}{\text{O}}}\ \text{②}\ \overset{/}{\text{O}} \\ \text{C} \qquad \text{③} \\ \overset{\backslash\backslash}{\text{O}}\ \text{④}\ \underset{\backslash}{\text{O}} \end{array}$ The different surfaces can be joined together

on the sides of a cube as shown. Note, in the dehydration reaction the first step is the auto-protolysis of water. It is this step which gives rise to the secondary fractionation in Table 10.

solvent [48] and have since been applied to elimination reactions [79] and to the type of reaction we are considering here [80]. Figure 13 shows a schematic diagram for our particular case.

The reason why the reaction does not go round the sides of the diagram (the stepwise route) is that the North-West and South-East corners are too 'high'; the species there are unstable. Furthermore, following Jencks [80], we can see that for the concerted mechanism to operate the proton transfer has to change from being 'uphill' to start with

$$C = O + L - O - R \rightleftharpoons C = \overset{+}{O} - L + {}^-O - R \qquad K_s \ll 1$$

to being 'downhill' after the change along the C—O co-ordinate,

$$K_s \gg 1$$

In Jencks' words 'the pK of the catalyst must be intermediate between the pKs of the initial and final sites'. The actual proton transfer can be treated by the Marcus theory [55] which relates the rate of the proton transfer to the thermodynamic driving force. As the system moves along the x-co-ordinate, the driving force becomes larger. It can be shown [52] that such a surface will have a saddle point or transition state at the point where $K_s = 1$. Thus the movement along the x-co-ordinate (in either direction) brings the 'thermodynamics' for driving the proton transfer into balance. If $K_s \simeq 1$, then $\alpha \simeq \frac{1}{2}$. Changing the catalyst, as one does in making a Brønsted plot, does not then alter the degree of proton transfer. That remains at $\alpha = \frac{1}{2}$. What it does do, is to alter the position along the x-co-ordinate [52] where $K_s = 1$. The stronger catalyst has less far to go. Thus this simple model explains why these concerted reactions give long linear Brønsted plots with $\alpha \simeq \frac{1}{2}$. The Critchlow model is different from the Marcus theory but the same features are found.

In an interesting paper, Swain, Kuhn and Schowen [81] argued that the proton was not 'in flight' in this type of reaction. Counselling against anthropomorphic motivation they argued that the proton in the transition state will be found grounded in its most stable environment. They were against having the proton flying as well as the CO distance closing because that would require more energy. Their argument is plausible but wrong. As can be seen from Figure 13, it is worth finding the extra energy (Q to R) to transfer the proton at Q rather than travelling on to the 'high' corner at T. Kreevoy and Cordes [82] have advanced an ingenious idea in which the proton transfers without even 'flying'. Throughout its journey it is cradled in a valley. The best argument against this picture is the size of the primary factor ϕ_1; as discussed above, it is less than that for $(FLF)^-$. Further arguments against this model are advanced by Jencks [80]. However, in an interesting review, Schowen paints a different picture [83] and there is still plenty of controversy to be enjoyed.

9.17 Comparison with electron transfer

In an electron transfer [84, 85] there is no emission or absorption of radiation and hence, in view of the Franck–Condon principle, the actual electron transfer has to be isoenergetic. This condition is achieved by suitable reorganization of the solvation and ligands. For instance, we can write the symmetrical isotopic reaction as:

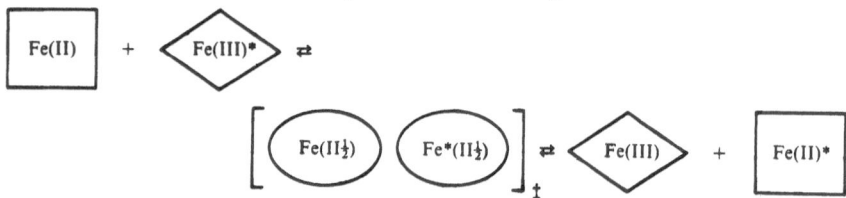

The squares, diamonds and circles represent the solvation and ligand distances appropriate to Fe(II), Fe(III) and the transition state. Figure 14 shows schematically the energetics of the reaction. The free energy of activation is expended entirely on changes along the x-co-ordinate and not on the actual electron transfer. An actual electron transfer takes place in a shorter time than an actual proton transfer and the electron almost certainly tunnels whereas for the proton the tunnelling is usually optional (even in Stirling). Despite these differences there is a similarity between the picture for the electron transfer, and that for the proton transfer. Indeed one might go further and ask, if the solvation terms are so important for electron transfer what part do they play in proton transfer? And here we come to the dog that has not barked [86]*. Right at the beginning we saw that the self-diffusion coefficients and viscosities of H_2O and D_2O were 20 per cent different, and that this difference continued down to

---

*'To the curious incident of the dog in the night-time'. 'The dog did nothing in the night-time'. 'That was the curious incident', said Sherlock Holmes.

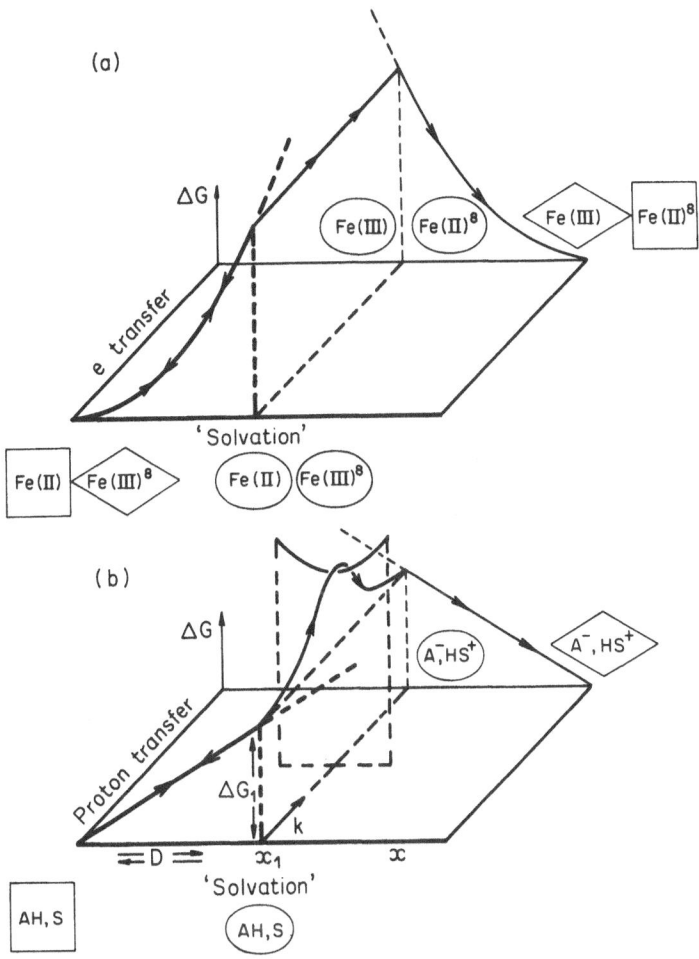

Figure 14. Comparison of free-energy surfaces for a symmetrical electron transfer reaction and for a proton transfer reaction. The simplified proton transfer model is treated in equation 34. The term 'solvation' includes the approach of reactants inside the solvent cage.

very short times ($\tau \sim 10^{-12}$ s). Yet in all our treatment of solvent isotope effects no mention has been made of this difference. Only once did the dog make a tiny growl and that was in the rather special case of the reaction of the cyanocarbon bases with H_3O^+. Yet if in D_2O all the translational movement is 20 per cent slower, one might expect that the energy transfer and the approach of the reactants would be that much slower and that there would be a dynamic medium effect on all kinetic, but not thermodynamic, parameters. There would be a contribution throughout to Φ_Δ of 0.8. Not only does the whole pattern presented here then collapse, but many absurd results are produced.

For instance, in Figure 7 many reactions would have values of α_I greater than α_B even though L_3O^+ is a stronger acid than RCO_2H. Wherever in a Table, Φ_A is now greater than 0.80, then on putting in the new factor, Φ_A would become greater than unity and in many cases as large as 1.25. For many reactions Φ_{eq} is less than unity and it is hard to see any reason for Φ_A being both greater than unity and at the same time not intermediate between unity and Φ_{eq}. Hence we conclude that the 20 per cent viscosity and self-diffusion difference is generally not seen as a dynamic medium effect on the rates of proton transfer.

9.18 Diffusion model

To investigate what this means we set up a very simplified model for the diffusion/solvation process and the actual proton transfer. We are not concerned here with the gross diffusion of the reactants together into the same solvent cage, but rather with the diffusive motion of the solvent and reactants once they are in the cage. The neutron-diffraction data show that the same types of motion are still occurring for $\tau \sim 10^{-12}$ s while the lifetime of the caged pair is of the order of 10^{-10} s. Writing the caged reactants as R and showing as before the changes in the solvating environment, a kineticist might be tempted to write

$$\boxed{R} \underset{k_{-1}}{\overset{k_1}{\rightleftharpoons}} \bigcirc{R} \xrightarrow{k_2}$$

where the diffusion/solvation process is controlled by k_1, k_{-1} and the proton transfer by k_2. But this model must be too crude for the fine detail of motion that we are considering. The neutron scattering data in Figure 1 supports earlier conclusions [87] that the Rabinowitch-Eyring model [88] of diffusion is probably invalid. There is no series of regular little hops. But Eyring did show us that the diffusion process can be treated as a series of kinetic jumps. We now reverse his argument and describe the kinetic process with a diffusion equation to describe motion along the x-co-ordinate:

$$D\frac{\partial^2 c}{\partial x^2} + g'\frac{\partial c}{\partial x} - k(x)c = 0 \tag{35}$$

We have assumed that the free energy increases linearly with x, i.e.

$$\frac{\Delta G}{RT} = g'x$$

The rate constant $k(x)$ describes the chances of proton transfer at any distance x. This equation is similar in form to the differential equation for the diffusion and migration of an unstable ion in an electric field.

Figure 14b illustrates the model. The equation is not only applicable to translational diffusion but can be used to describe any two-step process in which motion along the x-co-ordinate is uncoupled from that along the y-co-ordinate. For electron transfers the Franck–Condon principle means that this separation is virtually complete, leading to movement on the x, y plane which obeys the rules of a rook in chess rather than a bishop. The question of whether in a proton transfer the motion is coupled or uncoupled is less clear. But, if the motion is coupled, then the fractionation factor would be close to unity since there would be heavy atom motion in the reaction co-ordinate. The primary factor is very sensitive to the coupling of the motion. Even factors as close to unity as 0.4 suggest that the L^+ motion is fairly independent. Hence we assume as a reasonable simplification that the processes are uncoupled. Figure 10 may be regarded as a ground plan for Figure 14b with progress along the x-co-ordinate measured by σ.

To obtain analytical solutions instead of taking a complicated expression for $k(x)$ we assume that there is a simple 'window' through which reaction can take place between x_1 and x_2. That is we take

$$k(x) = 0 \quad \text{for } x < x_1$$
$$= k \quad \text{for } x_1 < x < x_2$$
$$= 0 \quad \text{for } x_2 < x$$

Instead of presenting the complete and complicated analytical solution we present three approximate solutions which hold under different conditions. It turns out that the switching off of $k(x)$ at x_2 is unimportant since for reasonable surfaces the system does not climb further along the x-co-ordinate than it need.

The solutions are:

$$\text{I} \quad k_{\text{obs}} = A'x_T k \exp\left(-\Delta G_1/RT\right)$$
$$\text{II} \quad k_{\text{obs}} = A'x_1^{-1}D$$
$$\text{III} \quad k_{\text{obs}} = A'x_T^{-1}D \exp\left(-\Delta G_1/RT\right)$$

where k_{obs} is a first-order rate constant, ΔG_1 is the difference in G between $x = 0$ and $x = x_1$, $x_T = (g')^{-1}$, and A' is the area available for the diffusion process per unit volume and unit concentration.

The distance x_T describes the distance on the x-co-ordinate for which $\Delta G = RT$; it will be large for small gradients and vice versa. The quantity A' arises because the diffusion equation describes a flux per unit area while kineticists measure rates of reaction per unit volume; for instance in solution I the factor $A'x_T$ describes the fraction of the volume in which reaction can take place.

One important feature of the solutions is that they can all be expressed as a first-order rate constant. This means that one can apply equation 35 successively to all the different separate processes from the reactants up to the transition state. For instance in the sequence,

$$\text{translation diffusion} \rightarrow \text{rotational diffusion} \rightarrow \text{atom transfer}$$

the rotational-diffusion and atom-transfer steps can provide a rate constant to put into the differential equation for translational diffusion. Thus, although the equations have been derived for translational diffusion, their form would be the same for rotational diffusion or indeed any smooth process on the x-co-ordinate which altered the separate rate process on the y-co-ordinate. Thus they can be applied to the C—O concerted proton transfer discussed above.

The conditions under which the different solutions hold can be characterized by two dimensionless parameters $\chi_1 = \Delta G_1/RT$, and $\kappa = kx_T^2/D$.

The parameter κ compares the lifetime of a species undergoing a reaction with rate constant k to the time it takes to diffuse the distance x_T where x_T is the distance which increases ΔG by RT.

Figure 15a shows the regions where one finds the different solutions. For solution I, which is found at low values of κ, the rate-determining process is the k-step and the diffusion processes are in equilibrium. For

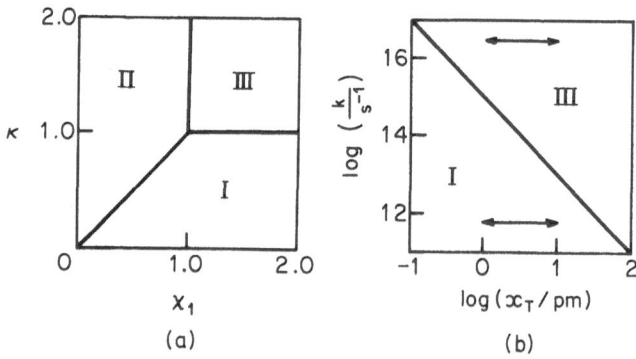

Figure 15. Conditions under which solutions I, II and III are found. Figure 15b shows a plot of equation 36 for $\kappa = 1$ and $D = 10^{-5} \text{ cm}^2 \text{ s}^{-1}$; the arrows indicate the most likely values of x_T.

solutions II and III, on the other hand, which occur at large values of κ, the k-process is so fast that the diffusion process is rate determining. Solution II corresponds to the case where there is hardly any free-energy barrier on the x-co-ordinate; there is no exponential term and the reaction is transport-controlled. This is not the case with the reactions we have been discussing, and hence we are more interested in the solutions with $\chi_1 > 1$. In solution III the fast k-process draws on the population near x_1 which is replenished by diffusion from the reactant reservoir over the thermal distance x_T. We can now summarize in Table 12 the expected behaviour of the different solutions with respect to isotopic substitution. The primary fractionation factor will be low for solution I since these rates depend on k. However, there will be no dynamic medium effect because the rates do not

Table 11

Isotopic behaviour of solutions I to III

	Primary isotope effect	Dynamic medium effect	Examples
I	$\phi_1 < 0.7$	$\Phi_\Delta = 1.0$	Most proton transfers to carbon
II	$\phi_1 \simeq \phi_R > 0.7$	$\Phi_\Delta = 0.8$	Most proton transfers to O and N
III	$\phi_1 \simeq \phi_R > 0.7$	$\Phi_\Delta = 0.8$	Cyanocarbon bases + H_3O^+ Simple electron transfers

depend on D. Exactly the reverse will be found for solutions II and III. The distinction between II and III can be made from the size of the observed rate constant, since for II the rate determining step will be the diffusion-controlled encounter. Once inside the solvent cage, the intimate process we have been discussing will take place much faster ($\tau \sim 10^{-12}$ s) than the break up of the reactant pair which requires a larger displacement ($\tau \sim 10^{-10}$ s). Thus most of the reactions that we have been discussing fall into class I. The reaction between the cyanocarbon bases and H_3O^+ fits nicely into Class III. In this mechanism there is a free-energy barrier to reaction but this barrier is on the x-co-ordinate. The rate-determining process is the 'diffusion' of the system within the solvent cage; the actual proton transfer is fast enough so that it is not rate determining. Electron-transfer reactions also fit into this category (see Figure 14). Those reactions studied by Eigen [61], which are genuine diffusion-controlled reactions ($k \sim 10^{10}$ dm^3 mol^{-1} s^{-1}), will be in class II.

The distinction between class I and class III depends on the size of κ, where

$$\kappa = kx_T^2/D \tag{36}$$

The neutron diffraction data show that $D \sim 10^{-5} \, \text{cm}^2 \, \text{s}^{-1}$ even down to the small time scales. We are not concerned here to separate out the different types of motion that contribute to D. We can treat it merely as an observed parameter that describes the solvent motion and hence the chance for reaction. The value of x_T probably lies between 1 and 10 pm. Figure 15b shows for what values of k one then finds the system in class I or in class III. For electron transfers $k \sim 10^{16} \, \text{s}^{-1}$ and so it is not surprising they are in class III. Class III proton transfers require a value of k close to the maximum that one could expect for a single vibration. As soon as there is any barrier to the transfer of the proton, k will be reduced so that the system shows class I behaviour. This is what is found for most proton transfers. In particular, for the cyanocarbon base the change from H_3O^+ to RCO_2H switches the system from class III to class I. The absence of the dynamic medium effect therefore suggests that for most proton transfers, as opposed to electron transfers, the solvent motion is not predominent in the rate-determining step and is not too closely coupled to the proton transfer. The protons transfer in those reactants that happen to find themselves in the right environment, and we can separate the reaction sequence into a series of steps:

Encounter	Ionic atmosphere	Solvation	Atom transfer
$\rightleftharpoons$	$\rightleftharpoons$	$\rightleftharpoons$	$\rightarrow$

This complete separation is certainly an oversimplification. However, in understanding the kinetics of reactions in solution, oversimplification is essential. What one has tried to show is how the dynamics of these processes can be investigated by using neutrons, either by bombarding the solution with them, or by packing them on to the reactant sites and allowing them to take a trip through the transition state.

It is a pleasure to record my gratitude to Dr John White for stimulating and helpful conversations.

REFERENCES

[1] A. J. Kresge, *Pure Appl. Chem.*, **8**, 243 (1964).
[2] V. Gold, *Adv. Phys. Org. Chem.*, **7**, 259 (1969).

[3] E. M. Arnett and D. R. McKelvey, *Solute-Solvent Interactions* (ed. J. F. Coetzee and C. D. Ritchie), Dekker, New York and London (1969), p. 343.

[4] A. I. Kudish, D. Wolf, and F. Steckel, *J. Chem. Soc. Faraday I*, **68**, 2041 (1972).

[5] R. Mills, *Molecular Motions in Liquids* (ed. J. Lascombe), Reidel, Dordrecht Holland (1974), p. 391.

[6] L. G. Longsworth, *J. Phys. Chem.*, **64**, 1914 (1960).

[7] J. W. White, *Ber. Bunsenges. physik. Chem.*, **75**, 379 (1971).

[8] J. Pupezin, G. Jakli, G. Jancso, and W. A. Van Hook, *J. Phys. Chem.*, **76**, 743 (1972).

[9] W. A. Van Hook, *J. Chem. Soc. Chem. Comm.*, 479 (1972).

[10] J. Bigeleisen, *J. Chem. Phys.*, **23**, 2264 (1955).

[11] W. A. Van Hook, *J. Phys. Chem.*, **72**, 1234 (1968).

[12] W. A. Van Hook, *J. Phys. Chem.*, **76**, 3040 (1972).

[13] P. Gross, H. Steiner, and F. Krauss, *Trans. Faraday Soc.*, **32**, 877 (1936).

[14] J. C. Hornell and J. A. V. Butler, *J. Chem. Soc.*, 1361 (1936).

[15] V. Gold, *Trans. Faraday Soc.*, **56**, 255 (1960).

[16] L. Friedman and V. J. Shiner, *J. Chem. Phys.*, **44**, 4039 (1966).

[17] V. Gold, *Trans. Faraday Soc.*, **64**, 2770 (1968).

[18] W. J. Albery and M. H. Davies, *Trans. Faraday Soc.*, **65**, 1059 (1969).

[19] R. A. More O'Ferrall, G. W. Koeppl, and A. J. Kresge, *J. Amer. Chem. Soc.*, **93**, 9 (1971).

[20] W. J. Albery and A. N. Campbell-Crawford, *J. Chem. Soc. Perkin II*, 2190 (1972).

[21] V. Gold, *Trans. Faraday Soc.*, **64**, 2143 (1968).

[22] P. Salomaa and M. Mattsen, *Acta Chem. Scand.*, **26**, 2137 (1972).

[23] W. J. Albery and M. H. Davies, *J. Chem. Soc. Faraday Trans. I*, 68, 167 (1972).

[24] H. Margenau and G. M. Murphy, *The Mathematics of Physics and Chemistry*, Van Nostrand, Princeton (1956), p. 515.

[25] V. Gold, *Proc. Chem. Soc.*, 141 (1963).

[26] A. J. Kresge and A. L. Allred, *J. Amer. Chem. Soc.*, **85**, 1541 (1963).

[27] K. Heinzinger and R. E. Weston, *J. Phys. Chem.*, **68**, 744 (1964).

[28] P. Salomaa and V. Aalto, *Acta Chem. Scand.*, **20**, 2035 (1966).

[29] B. D. Batts and J. Kilford, *J. Chem. Soc. Faraday I*, **69**, 1033 (1973).

[30] A. J. Kresge and W. J. Albery, *J. Chem. Soc. Chem. Comm.*, **507** (1974).

[31] V. Gold and S. Grist, *J. Chem. Soc. Perkin II*, 89 (1972).

[32] K. Heinzinger and R. E. Weston, *J. Phys. Chem.*, **68**, 2179 (1964).

[33] V. Gold and B. M. Lowe, *J. Chem. Soc. (A)*, 936 (1967).

[34] C. E. Taylor and C. Tomlinson, *J. Chem. Soc. Faraday I*, **70**, 1132 (1974).

[35] R. A. More O'Ferrall, *J. Chem. Soc. Chem. Comm.*, 114 (1969).

[36] P. Beltrame, A. M. Bianchi, and M. G. Cattania, *Gazzeta*, **102**, 456 (1972).

[37] D. B. Dahlberg, *J. Phys. Chem.*, **76**, 2045 (1972).

[38] P. Salomaa, *Acta Chem. Scand.*, **25**, 365 (1971).

[39] J. Greyson, *J. Phys. Chem.*, **71**, 2210 (1967); J. Greyson and H. Snell, *J. Phys. Chem.*, **73**, 3208, 4423 (1969); H. Snell and J. Greyson, *J. Phys. Chem.*, **74**, 2148 (1970).

[40] J. Greyson, *J. Phys. Chem.*, **71**, 259 (1967).

[41] P. Salomaa and M. Mattsen, *Acta Chem. Scand.*, **25**, 361 (1971).

[42] V. Gold and B. M. Lowe, *J. Chem. Soc. (A)*, 1923 (1968).

[43] R. P. Bell, *The Proton in Chemistry*, Chapman & Hall, London (1973), p. 234.

[44] I. R. Bellobono and E. Diani, *J. Chem. Soc. Perkin II*, 1707 (1972).

[45] A. J. Kresge and Y. Chiang, *J. Phys. Chem.*, **77**, 822 (1973).

[46] J. M. Williams and M. M. Kreevoy, *Adv. Phys. Org. Chem.*, **6**, 63 (1968).

[47] W. J. Albery, A. N. Campbell-Crawford, and R. W. Stevenson, *J. Chem. Soc. Perkin II*, 2198 (1972).

[48] W. J. Albery, *Prog. Reaction Kinetics*, **4**, 353 (1967).

[49] F. Hibbert and F. A. Long, *J. Amer. Chem. Soc.*, **93**, 2836 (1971).

[50] E. Grunwald and M. Cocivera, *Disc. Faraday Soc.*, **39**, 105 (1965).

[51] A. J. Kresge and Y. Chiang, *J. Amer. Chem. Soc.*, **95**, 803 (1973).

[52] W. J. Albery, A. N. Campbell-Crawford, and J. S. Curran, *J. Chem. Soc. Perkin II*, 2206 (1972).

[53] W. J. Albery, *Reaction Transition States* (ed. J. Dubois), Gordon and Breach, London (1972), p. 273.

[54] W. J. Albery, J. R. Bridgeland, and J. S. Curran, *J. Chem. Soc. Perkin II*, 2203 (1972).

[55] R. A. Marcus, *J. Phys. Chem.*, **72**, 891 (1968).

[56] R. P. Bell, *The Proton in Chemistry*, Chapman & Hall, London (1973), p. 265.

[57] F. H. Westheimer, *Chem. Rev.*, **61**, 265 (1961).

[58] A. Kankaanperä, *Acta Chem. Scand.*, **23**, 465 (1969).

[59] K. Pihlaja, *J. Amer. Chem. Soc.*, **94**, 3330 (1972).

[60] F. Hibbert, F. A. Long, and E. A. Walters, *J. Amer. Chem. Soc.*, **93**, 2829 (1971).

[61] M. Eigen, *Angew. Chem.*, **75**, 489 (1963).

[62] F. Hibbert and F. A. Long, *J. Amer. Chem. Soc.*, **93**, 2836 (1971).

[63] D. B. Dahlberg and F. A. Long, *J. Amer. Chem. Soc.*, **95**, 3825 (1973).

[64] *Landolt-Bornstein II*, **7**, 278, 279, 280 (1960).

[65] F. Hibbert, *J. Chem. Soc. Perkin II*, 1289 (1973).

[66] D. M. Goodall and F. A. Long, *J. Amer. Chem. Soc.*, **90**, 238 (1968).

[67] R. P. Bell and J. E. Critchlow, *Proc. Roy. Soc. A*, **325**, 35 (1971).

[68] M. Eigen, *Disc. Faraday Soc.*, **39**, 7 (1965).

[69] R. P. Bell, J. F. Millington and J. M. Pink, *Proc. Roy. Soc. A*, **303**, 1 (1968).

[70] H. H. Huang, R. R. Robinson, and F. A. Long, *J. Amer. Chem. Soc.*, **88**, 1866 (1966).

[71] J. N. Brønsted and E. A. Guggenheim, *J. Amer. Chem. Soc.*, **49**, 2554 (1927).

[72] B. D. Batts and V. Gold, *J. Chem. Soc. A*, 984 (1969).

[73] S. S. Minor and R. L. Schowen, *J. Amer. Chem. Soc.*, **95**, 2279 (1973).

[74] A. R. Fersht and A. J. Kirby, *J. Amer. Chem. Soc.*, **90**, 5826 (1968).

[75] R. E. Robertson, B. Rossall and W. A. Redmond, *Can. J. Chem.*, **49**, 3665 (1971).

[76] R. P. Bell, *Disc. Faraday Soc.*, **39**, 16 (1965).

[77] S. L. Johnson, *Adv. Phys. Org. Chem.*, **5**, 275 (1960).

[78] J. E. Critchlow, *J. Chem. Soc. Faraday Trans. I*, **68**, 1774 (1972).

[79] R. A. More O'Ferrall, *J. Chem. Soc. B*, 274 (1970).

[80] W. P. Jencks, *Chem. Rev.*, **6**, 705 (1972).

[81] C. G. Swain, D. A. Kuhn, and R. L. Schowen, *J. Amer. Chem. Soc.*, **87**, 1553 (1965).

[82] E. H. Cordes, *Prog. Phys. Org. Chem.*, **4**, 1 (1967).

[83] R. L. Schowen, *Prog. Phys. Org. Chem.* **9**, 275 (1972).

[84] R. A. Marcus, *J. Phys. Chem.*, **67**, 853 (1963); R. A. Marcus, *J. Chem. Phys.*, **43**, 1261 (1965); R. A. Marcus, *Electrochim. Acta*, **13**, 995 (1968); R. A. Marcus, *Ann. Rev. Phys. Chem.*, **15**, 155 (1964).

[85] V. G. Levich, *Adv. in Electrochem. and Electrochem. Eng.*, **4**, 249 (1966). R. R. Dogonadse, *Reactions of Molecules at Electrodes* (ed. N. S. Hush), Wiley, London (1971), p. 135.

[86] A. Conan Doyle, *The Strand Magazine*, **4**, 645 (1892).

[87] R. M. Noyes, *J. Chem. Phys.*, **23**, 1982 (1955).

[88] E. Rabinowitch, *Trans. Faraday Soc.*, **33**, 1225 (1937). S. Glasstone, K. J. Laidler, and H. Eyring, *The Theory of Rate Processes*, McGraw-Hill, New York (1941), p. 524.

10

Edward S. Lewis

TUNNELLING IN HYDROGEN-TRANSFER REACTIONS

10.1 Meaning of tunnelling

Tunnelling is defined in terms of a barrier separating two stable situations. If the system acquires energy equal to or greater than the height of the barrier, then it can cross the barrier and get from one side to another. If the energy is less than that of the top of the barrier then classically it cannot get from one side to another. According to quantum mechanics, however, there is a perceptible chance that a system with insufficient energy initially at one side of the barrier can be discovered on the other side of the barrier. This process has been called 'tunnelling', by analogy to the only classical way to cheat a geographical energy barrier.

One way to consider quantum mechanical tunnelling is in terms of the uncertainty principle. Consider an object rolling toward a barrier as in Figure 1. As it gets higher its momentum falls, and so therefore does the uncertainty in its momentum. By the uncertainty principle, the uncertainty in its position increases, represented by the short line through the centre of the fast object on the left, and the longer line on the one at the right. At this point the object has enough uncertainty in its position so that it might be on either side, and there is then a significant chance that it is on the other side, without ever having had the energy to reach the top. Furthermore, this simple way of looking at the matter makes it clear that the extent of tunnelling would be greater with a thin barrier, and with an object of low mass (for which the momentum is smaller).

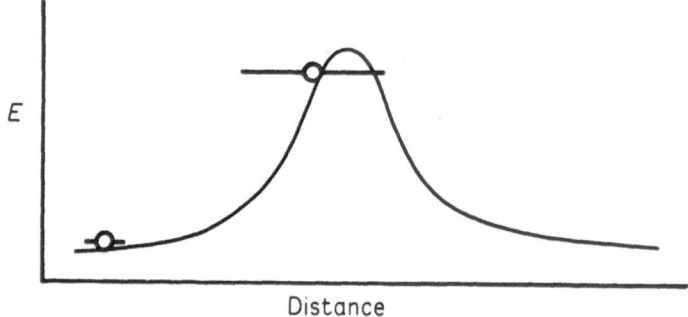

E

Distance

Figure 1. Tunnelling as a consequence of the uncertainty principle. The two objects have the same total energy; the one on the right has less kinetic energy, and hence less momentum and greater positional uncertainty. The positional uncertainty is illustrated by the lengths of the horizontal lines through the objects.

In his book, Bell [1] points out the analogy of tunnelling to total reflection at an air-gap between two plates of glass as the gap becomes as small as the wave length of the light.* Similarly, an energy barrier becomes less effective as the barrier becomes thin compared to the de Broglie wave length of an object, h/mv. A 200-kg lion moving slowly at 1 cm/sec has a wave length of 3×10^{-32} cm and is not likely to penetrate its cage walls,† but a hydrogen atom at the same slow speed has a wave length of 0.04 mm, which is quite significant.

Tunnelling has been considered important for electrons, both in β-decay and in the behaviour of solid-state devices, as well as in nuclear decay to give α-particles, but our concern here will be with the transfer of hydrogen atoms. Before leaving the nuclear processes but in connection with the chemistry to follow, mention should be made of recent studies of the depolarization of muons (half-life $\sim 2 \times 10^{-6}$ sec) in solvents, in which an uncharged intermediate (muonium, $\mu^{+}e$) is produced by electron capture, and shows chemistry similar to that of the hydrogen atom but with far greater rates. Since muonium has only $\frac{1}{9}$ the mass of hydrogen, the possibilities for tunnelling are very great indeed, and this has been postulated but not yet clearly separated from the potentially extremely large zero-point energy effects [2].

* The microwave analogue of this optical experiment is quite simple to demonstrate, with the help of a 3-cm emitter and receiver and two prismatic blocks of paraffin wax.

† This is the basis for the quantum mechanical method of lion hunting, which is to place a firmly closed cage in Africa, then wait for the time that the lion will behave quantum mechanically and tunnel through the cage walls. Although it is theoretically sound, it is not very likely to happen during the time between the first lion and the extinction of the species.

10.2 Properties of the tunnel correction

A complete quantum-mechanical calculation of the passage of a system from one side of a barrier to another is in principle possible, but has not yet been entirely successful for a real chemical reaction for which even the barrier is somewhat uncertain. Some progress has been made, however, but it will not be discussed here. When this calculation becomes more practical, there will be no tunnel correction, since this is an artifact of treatment of the barrier passage as a purely classical process.

With current methods the tunnel correction is a useful consideration, and can often have a significant value. The tunnel correction is usually applied as a multiplicative factor, Q, on the classically calculated rate. It has the following properties. (1) $Q > 1$. (2) Q decreases with increase in temperature. (3) Q is greater for light systems than for heavy ones. (4) Q is greater for a thin barrier than a thick one. These properties are all directly deducible from a consideration of the relation between the barrier thickness and the De Broglie wavelength. A possible alternative treatment of a tunnel correction as an additive term has not been attempted; it apparently has the advantage that it would not need to have as strong a temperature dependence to compensate for the Arrhenius dependence of the classical term.

The tunnel correction is not now a fundamentally defined number; rather it is defined by the equation $Q = k_{obs}/k_s$, where k_{obs} is the observed rate constant for a chemical reaction and k_s is that calculated on the basis of some model which is as good as possible except that it does not allow tunnelling. In this chapter the definition used for k_s is that calculated by absolute reaction rate theory [3], i.e., $k_s = (\kappa RT/Nh)K^{\ddagger}$ where $K^{\ddagger}$ is the equilibrium constant for the formation of the transition state. The factor κ, the transmission coefficient, is also a quantum correction on the barrier passage process, but it is in the other direction, that is $\kappa < 1$. We shall here follow the customary view (though it is not solidly based) that κ is temperature-independent and not markedly less than unity. The term k_s is used following Bell [1]; the s stands for semi-classical, that is quantum mechanics is applied to vibrations and rotations, but translation along the reaction coordinate is treated classically.

There is a close analogy between the tunnel correction and resonance energy. Both represent the difference between a real system and an artificial model. In both we know the direction of this difference ($Q > 1$, resonance energy > 0); in both we know the factors which will make this quantum correction large enough to be worth considering; and in both there is an unsatisfactory model, with the particular choice of model determining the magnitude of the correction.

10.3 Calculation of the tunnel correction

Once a rigorously defined semiclassical model has been chosen, the tunnel correction is in principle calculable, although it is seldom easy to write it explicitly. Since the potential energy surface describing the reaction is multidimensional, the treatment becomes extremely complex, and the calculation of Q to better precision than that of experimental determinations of k_s is rather futile. Since k_s is roughly calculable only for the $H + H_2$ reaction and even more roughly for a few other very simple systems, a major simplification in the calculation of Q has generally been used, namely to reduce the multidimensional surface to a one-dimensional plot of energy against reaction coordinate, and calculate Q for crossing this one-dimensional barrier.

The barrier can be described in a number of ways, of which the most useful are the height above the energy of the reagent or product, the curvature at the top, and the overall analytic form. As in other problems the shape can be approximated by a parabola near the top, and the curvature expressed as the classical frequency v of a particle of mass m in a parabolic well of the same shape as the barrier. It is convenient to define $u = hv/kT$.

The first estimate of the tunnel correction, for this one-dimensional treatment, was that of Wigner [4], who wrote $Q = 1 + u^2/24$. This equation is still believed to give the tunnel correction when the correction is small.

It is necessary to consider the details of the barrier when we wish to estimate larger tunnel corrections. The first approach in this direction was by Bell [1, 5], who approximated the barrier for a symmetric reaction by an inverted parabola with a base width $2a$ and a height E. The tunnel correction for this situation is given explicitly by equation* 1.

$$Q = \frac{\frac{1}{2}u}{\sin \frac{1}{2}u} - u \exp(E/kT)\left(\frac{y}{2\pi - u} - \frac{y^2}{4\pi - u} + \frac{y^3}{6\pi - u} - \ldots\right)$$

$$v = E^{\frac{1}{2}}/\pi a(2m)^{\frac{1}{2}} \tag{1}$$

$$y = \exp(-2\pi E/hv)$$

* This form of the equation is that given in Reference [1], p. 275; it is easier to use than that given in Reference [5].

In this equation the first term alone suffices for high barriers and small values of u, but for large tunnel corrections the later terms are needed. The Wigner correction is the first two terms in the expansion of the first term of equation 1 as a power series in u.

This equation has been used on experimental data to evaluate the parameters E and a. The expression however is limited by the artificiality of its assumptions. The first artificiality is the form of the barrier, which may be satisfactory near the top but is obviously unreal at the point of truncation. It may be smoothed out in various ways; the Eckart barrier, equation 2, and the Gaussian barriers, equation 3, have also been used [6, 7] with quantitatively different but qualitatively

$$V_{(x)} = E/[\cosh^2 (\pi x/l)] \tag{2}$$

$$V_{(x)} = E \exp (-x^2/a^2) \tag{3}$$

similar results. Use of the truncated parabolic barrier is best justified by the fact that the computations are simpler and the correct form of the barrier is unknown; furthermore it is consistent with the harmonic approximation used for all the real transition-state vibrations.

The major criticism of the Bell treatment is that it is treated one-dimensionally, and an alternative treatment which still assumes the linear transition state but allows a multiplicity of tunnelling 'paths' between the reagent valley and the product valley of the potential energy surface, in addition to that along (or under) the reaction coordinate, has been presented by Johnston and Rapp [6a]. Although this is intuitively more realistic than the one-dimensional treatment, the treatment is not shown to be exact, and it still does not consider the possible non-linear arrangements. Another attack on the complete passage over the barrier limited to a linear configuration is that of Russell and Light [6b]. The primary problem, our ignorance of the potential energy surface, is a further limitation on the extensive use of this approach, although Johnston has made a reasonable attack here also.

For the present purposes the Bell truncated-parabola treatment will be adopted when a specific barrier is needed. The conclusion should not be drawn that the barrier is a truncated parabola with such and such dimensions E and a, but rather that the data resemble those calculated for a hypothetical one-dimensional barrier of these dimensions. The statement in this form recognizes the artificiality and inadequacy of the model, yet allows a level of quantitative treatment which now seems adequate.* Presumably, this simple model will ultimately be found inadequate and a better model (one which does not merely introduce another arbitrarily adjustable parameter) will be applicable to all the known cases.

A comment should be made on the magnitude of the tunnel correction when the reaction is irreversible. In Figure 2, which represents

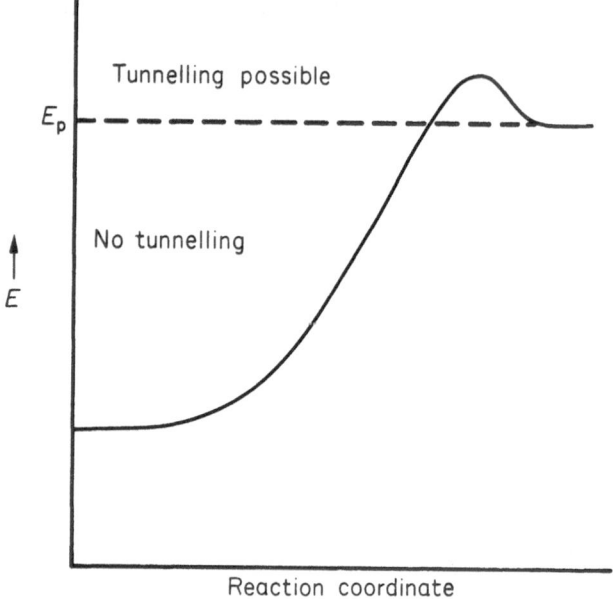

Figure 2. Restriction on tunnel corrections for a highly unsymmetrical barrier. Systems of energy below E_p to the left of the barrier can neither tunnel to the right side, nor are they accessible by tunnelling from the right side.

Vertical axis: E.
Horizontal axis: reaction coordinate.

a highly endothermic reaction, it is clear that no low-energy reagents can become products, and that only those with an energy equal to or greater than the product energy E_p can tunnel. Thus the tunnel correction will be much smaller than for the symmetric case with the same barrier curvature and height.

It is less obvious how the tunnelling in the reverse direction will be influenced by the extreme exothermicity. The consideration of the relation between rates and equilibrium makes this clear. The equilibrium constant K is independent of the path and mechanism of

* References [6] and [7] both claim that the Bell treatment is inadequate for the system under investigation. It is unclear whether the improvement claimed may not be due primarily to the introduction of further parameters.

conversion of reagent to product, and (for any one-step reaction) we can write equation 4, in which the superscripts $+$ and $-$ refer to the forward and reverse processes.

$$K = \frac{k_{obs}{}^{+}}{k_{obs}{}^{-}} = \frac{Q^{+}k_{s}{}^{+}}{Q^{-}k_{s}{}^{-}} \tag{4}$$

In absolute reaction-rate theory, the ratio $k_{s}{}^{+}/k_{s}{}^{-}$ is also the equilibrium constant K, since in this ratio all transition-state properties cancel. Thus we can write $K = (Q^{+}/Q^{-})K$, or $Q^{+} = Q^{-}$. With more rigour, to avoid assumptions about the transmission coefficient, we should write $Q^{+}\kappa^{+} = Q^{-}\kappa^{-}$.

10.4 The experimental estimation of the tunnelling correction in hydrogen-transfer reactions

The applicability of quantum mechanics to chemistry is generally accepted, hence the existence of the tunnel correction is no more open to question than any other consequences of quantum mechanics. The only question is whether it is important enough to be worth consideration. Bell [1] states: 'In any kinetic treatment which is sufficiently refined to take into account the zero-point energy of the transition state it is not justifiable to neglect the tunnel effect'. This follows in part from his treatment which shows the close analogy between the tunnel effect and zero-point energy. There is a long step, however, between the conviction that there is a tunnel correction and the experimental demonstration that a rate is substantially faster than that classically calculated, because of the difficulty of this classical calculation. This section is concerned with that step. It is specifically pointed to the hydrogen transfer reaction (equation 5):

$$A + HB \longrightarrow AH + B \tag{5}$$

There are two ways in which absolute reaction rate theory is now far more accurate than in its prediction of the absolute value of any rate constant. One of these is the prediction of the form of the temperature dependence of the rate, and the other is the value of isotope effects. It is the classically-anomalous behaviour of these two aspects that constitute most of the generally accepted evidence of significant tunnel corrections. Virtually all these methods were anticipated by Bell long before his first experimental example [8].

The temperature-dependence of hydrogen-transfer reactions is given by the equation 6, in which we express $\ln k_s$ by the Arrhenius

$$\ln k_{obs} = \ln Q(T) + \ln k_s$$
$$= \ln Q(T) + \ln A - E_a/RT \qquad (6)$$

expression. (The Arrhenius expression is used instead of the absolute reaction rate expression for simplicity; experimentally they are virtually indistinguishable.)

The slope of the Arrhenius plot will then be given by:

$$d \ln k_{obs}/d(1/T) = d \ln Q/d(1/T) - E_a/R$$

Since Q increases as T falls, the first term will be positive and the absolute value of the slope will be less than E_a/R. Since E_a is only an experimental number determined by a plot of $d \ln k_{obs}$ against $1/T$, we cannot recognize the tunnel correction from this alone. However, the Bell parabolic-barrier calculation leads to the conclusion that $d \ln Q/d(1/T)$ is not a constant; instead it becomes larger at lower temperatures, and thus the Arrhenius plot (if Q is substantial) becomes concave upward. This is Bell's first criterion for tunnelling, and examples in proton-transfer reactions have been explored extensively by Caldin and his group [9]. Curvature of Arrhenius plots can result from a number of experimental difficulties and wide ranges of temperature are necessary to detect it. Caldin has achieved this range by a combination of conventional slow kinetic methods at low temperature with stopped-flow methods at higher temperatures and has established curvature convincingly in several cases.

When the Arrhenius plot is concave upward, a study over a small range will not establish curvature, but the extrapolated A factor will be lower than that extrapolated from higher temperature measurements. If the A factor is theoretically calculable, the discrepancy can indicate a substantial tunnel correction. Again, like the low value of E_a, this is not experimentally very useful.

The availability of three isotopes of hydrogen allows the comparison between the behaviour of any two of these experimentally and the comparison to that calculated by absolute reaction rate theory. Since the isotope effect cancels out the major uncertainty in absolute reaction rate theory in calculations of rate, namely the barrier height, the theory puts some sharp restrictions on the magnitude of the isotope effect and its temperature dependence [10]. However, the semi-classically calculated isotope effect k_s^H/k_s^D is multiplied by the factor Q^H/Q^D, a number greater than unity which increases as the temperature falls. Thus k_{obs}^H/k_{obs}^D can be larger than that calculated classically and will have a stronger temperature dependence. Over a short range of temperature both observed and semiclassical isotope effects should follow equation 7.

$$k^H/k^D = (A^H/A^D)\exp(E_a^D - E_a^H)/RT \tag{7}$$

The semiclassical calculation places limits both on A_H/A_D for which $(2^{\frac{1}{2}} > A_H/A_D > 0.5)$ and on the activation energy difference $E_a^D - E_a^H$, which cannot be greater than the zero-point energy difference between the reagents BH and BD.

When there is a significant tunnel correction, the term $E_a^D - E_a^H$ factor will increase, and may increase beyond that allowed semiclassically. Since the contributions of bending modes have not been found to be of great importance in isotope effects, evidence of tunnelling may be found (with less rigour) from values of $E_a^D - E_a^H$ in excess of the difference between zero-point energy of BH and BD for the stretching mode only.

It was shown above that tunnelling could lead to a reduction in the experimental A-factor (because of curvature in the Arrhenius plot). Since $Q_H > Q_D$, we may conclude that $A^H < A^D$ when the tunnel corrections are significant, and no alternative explanation has been able to explain many experimentally determined examples with $A^H < 0.5A^D$. An early example, and the first example used to demonstrate tunnelling, was due to Bell [8] in some proton transfers from 2-carbethoxycyclopentanone. A number of subsequent examples are to be found in Caldin's review [9] and experimental values of A_H/A_D as low as 0.004 [11] have been reported. This test has been frequently applied to reactions of very large isotope effects, but less often to isotope effects in the more normal range, yet when these are studied it is found that low values of A_H/A_D are quite common even with smaller isotope effects. Thus the isotope effects in the attack of various bases on phenylnitromethane are not unusually large, yet they nearly all give $A_H < A_D$ [12].

In many of these cases equation 1 has been used to fit the experimental data, with E and a used as adjustable parameters, and with $m^D = 2m^H$. These calculations must be taken with a grain of salt since not only is the truncated parabola a rather unrealistic barrier, but the assumption about the mass is an extreme one. We shall return later to this problem. Nevertheless the values of E and a are qualitatively reasonable, and support the contention that the low values of A^H/A^D demonstrate tunnelling.

Several cases have now been collected with isotope effects so large as to be improbable in terms of the semiclassical treatment. These are presented in Table 1. The values of Q_H/Q_D are a rough lower limit assuming $k_s^H/k_s^D \leq \exp[(h/2kT)(v_0^H - v_0^D)]$, where v_0 is the vibrational frequency of the stretching mode of the bond to hydrogen in the reagent. This is often referred to as the 'maximum isotope effect', but this is true only in a one-dimensional approximation. However, in most reasonable treatments [13] the inclusion of bending vibrations and more dimensions does not lead to far different values of k_s^H/k_s^D, so that this rougher estimate is sufficient. The parenthetical Q_H/Q_D entries in Table 1 are those based on some other calculations, in most cases based on the low values of A^H/A^D. Of these, all are higher than the lower limit, except for reaction (2), where the parenthetical

Table 1

Isotope effects showing substantial tunnel corrections

Reaction (temp./°C)	k^H/k^D	A^H/A^D	$(Q_H/Q_D)^a$	Reference
1. MnO_4^- oxidation of $C_6H_5C^*HOHCF_3$ (25°C)	16.0	0.25	1.9	[14]
2. $Me_2C^*HNO_2$ + 2,4,6-lutidine (32°C)	21.4	0.15	2.7 (2.4)	[15, 16]
3. $(p\text{-}Me_2NH_6H_4)_3C^*H$ + chloranil (25°C)	11.7	0.04	1.4 (3.0)	[16, 17]
4. 4a–4b di*hydrophenanthrene + $O_2(-31°C)$	95	—	7.0 (11.8)	[18, 19]
5. Mesityl-S-*H^b + $(C_6H_5)_3C\cdot$ (25°C)	25^b	0.2^b	2.0^b	[20]
6. $CH_3\cdot$ + *$H_3CNC(-163°C)$	>1100	—^{c}	>3.5	[21]
7. $p\text{-}NO_2C_6H_4C^*H_2NO_2$ + tetramethylguanidine in toluene (25°C)	45	0.03	5.4 (6.5)	[22]

^{a} In the calculation of k_s^H/k_s^D, all starting CH bonds were estimated to have $v = 3000 \text{ cm}^{-1}$, and the SH bond was estimated at 2500 cm^{-1}; ^{b} Tritium was used, read k^H/k^T, A^H/A^T, Q_H/Q_T; ^{c} k^D was too small to measure; the Arrhenius plot of k_H was strongly curved.
Other values are given by Bell in *Chem. Soc. Rev.*, 3, 513 (1974).

value is based on the assumption that there is no tunnel correction in a related reaction with $k^H/k^D = 10$. The A^H/A^D factors, where established by more than two points, are also given.

This table, because it contains some large values for the rough lower limit of Q_H/Q_D, is, to some, more convincing than low A^H/A^D

values, since these are both more indirectly derived and are more subject to experimental error. However virtually every entry also shows an abnormal temperature dependence of the isotope effect, as shown by the low values of A^H/A^D in the table. In addition to entry 6, extremely large isotope effects have been reported in some other low temperature H atom transfer reactions by Williams and co-workers [23].

The correlation of large isotope effects with small A_H/A_D factors is impressive; only a few of the available data are presented but the correlation is typical. In particular Caldin and Mateo [11] have collected data on the solvent variation of rates and isotope effects on the reaction shown as entry 7 in the Table, finding numerous examples of low A^H/A^D (as low as 0.004 in one solvent, but as large as 1 in others) and isotope effects at 25°C as large as 50. The significance of the solvent effects will be discussed later. The correlation of tunnelling with low A^H/A^D is further supported by the observation that the proton transfer from phenyl sec-butyl ketone to acetate ion, in which a great deal of tunnelling is made impossible by the great endothermicity, has $A^H/A^D = 0.99$ [24]. We may take it as confirmed that Bell's conclusion is correct that low A^H/A^D factors are correlated with substantial tunnel corrections.

The final method based on an isotope effect makes use of the following qualitative argument. If Q_H is substantial, and Q_D is only slightly greater than unity, then Q_T will be only slightly closer to unity. We can write $k^H/k^T = (Q^H/Q^T)(k_s^H/k_s^T) = (Q^H/Q^D)(Q^D/Q^T)(k_s^H/k_s^D)^{1.442}$, where the exponential term comes from the Swain–Schaad [25] equation which is a reasonably rigorously derived relation between rates for hydrogen, deuterium, and tritium in the absence of tunnelling. Thus if Q^H/Q^D is quite different from unity, but Q^D/Q^T is virtually unity, then the tritium isotope effect will be smaller than that calculated from k^H/k^D by the Swain–Schaad equation, but if $Q_H/Q_T = (Q_H/Q_D)^{1.442}$ then there will be no deviation. Although the method has been proposed several times [26], it has not been found highly successful in examples 1, 2, and 3 of Table 1, studied by Lewis and Robinson [16]. The reason why this test is an insensitive indicator of tunnelling has been explained; it will only be detectable in the likely circumstance that k_s^H/k_s^D is comparable to or less than Q^H/Q^D. A similar conclusion has been reached by J. R. Jones [27], although he also finds one example of significant failure of the Swain–Schaad equation.

In summary, the isotope effect and its temperature variation is the most sensitive indication of tunnelling, and the results are internally consistent and not easily explained any other way.

10.5 The nature of the barrier and the tunnel correction

The proton and hydrogen atom transfers apparently have linear transition states, probably because this keeps the electrons in the orbitals of the attacking base or radical as far as possible from the reacting CH bond electrons, thus minimizing the interaction of the electrons, which must then be represented as electrons in the $n = 2$ shell of hydrogen. Hydride transfer reactions do not have this constraint, for the attacking electrophile has a vacant orbital, and thus these transition states need not be and probably are not linear [28]. The present discussion will be limited to proton-transfers and hydrogen-atom transfers with the assumption of linear geometry.

First we shall simplify the potential energy surface for these reactions by considering only one dimension, so that for the reaction A + HB → AH + B the two distances r_{AH} and r_{BH} define completely the location of all atoms (except for translation of the entire system). We make the conventional plot of potential energy contours in a plot of r_{AH} against r_{BH}, which then shows qualitatively a saddle point called the transition state. Figures 3a and 3b are sketches showing respectively a highly reagent-like transition state and a very symmetric transition state. Next we shall move the origin to the saddle point, which has the coordinates r_{BH}^0 and r_{AH}^0; we shall define $V = 0$ at the origin, and then approximate the surface in the neighbourhood of the origin by the quadratic expression shown in equation 8.

$$2V = k_1(r_{BH} - r_{BH}^0)^2 + 2k_{12}(r_{BH} - r_{BH}^0)(r_{AH} - r_{AH}^0)$$
$$+ k_2(r_{AH} - r_{AH}^0)^2 \tag{8}$$

Two contour maps of this equation resembling the surfaces of Figures 3a and 3b are shown in Figures 4a and 4b. All sections of these surfaces are either parabolas or hyperbolas. Although we cannot reproduce all the properties of the surface by this quadratic surface, we do get the saddle point [if $k_{12} > (k_1 k_2)^{\frac{1}{2}}$] and we can still recognize from the slope of the line representing the reaction coordinate the difference between the symmetric transition state (slope $= -1$) and the reagent-like transition state (slope $\ll -1$). Although the approximation looks very poor at points far from the origin, the approximation is the same in all respects as the harmonic approximation to the energies of stable molecules, and is also closely connected with Bell's use of the truncated parabola as a barrier model. The description of the curvature of the reaction coordinate in the neighbourhood of the transition state in terms of the imaginary frequency iv, as in

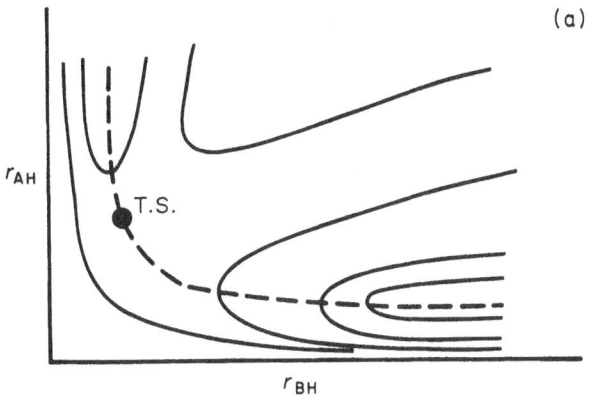

Figure 3a. Sketch of energy contours for a highly exothermic reaction with unsymmetrical transition state.

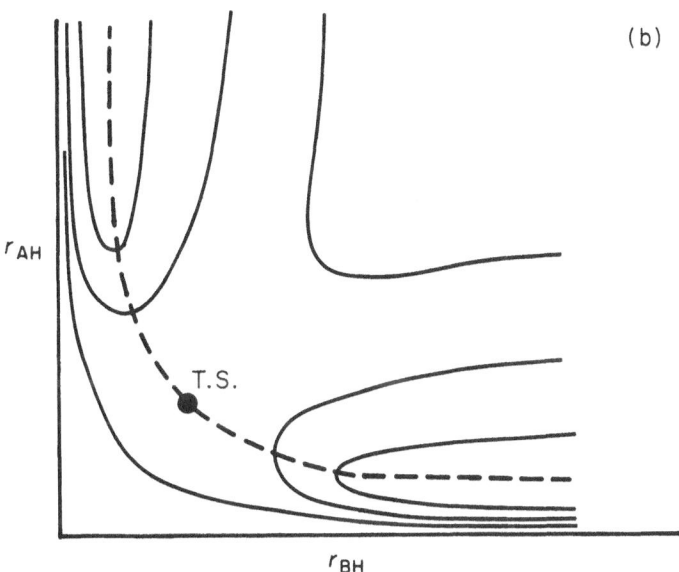

Figure 3b. Contours for a symmetric thermoneutral reaction.

Bell's or Wigner's calculation, also implies the quadratic approximation.

The feature of this quadratic representation which is of interest in our present discussion is the dependence of the surface on the values of k_1, k_2 and k_{12}. First, if $k_1 = k_2$, the slope of the line representing the reaction coordinate is exactly -1, irrespective of the value of k_{12}

329

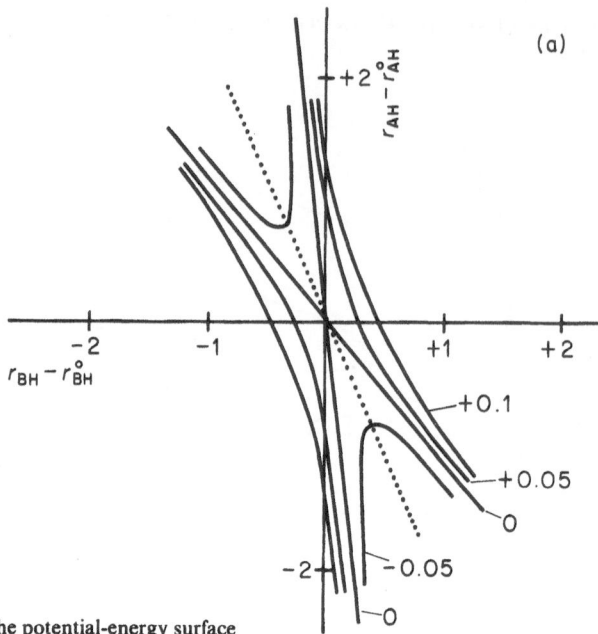

Figure 4a. The potential-energy surface

$$2V = (r_{BH} - r_{BH}^0)^2 + (r_{BH} - r_{BH}^0)(r_{AH} - r_{AH}^0) - 0.1(r_{AH} - r_{AH}^0)^2$$

in which $k_{12} = 1.58(k_1 k_2)^{\frac{1}{2}}$. The slope of the 'reaction-coordinate' dotted line is 2.24.

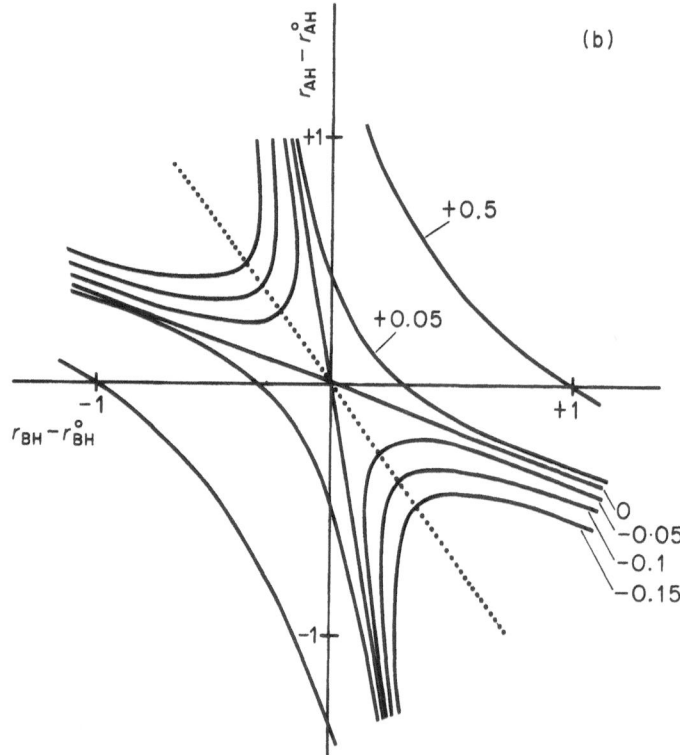

Figure 4b. The corresponding surface with a somewhat larger $k_{12} = 2.1(k_1 k_2)^{\frac{1}{2}}$ and more nearly similar k_1 and k_2: $2V = (r_{BH} - r_{BH}^0)^2 + 3(r_{BH} - r_{BH}^0)(r_{AH} - r_{AH}^0) + 0.5(r_{AH} - r_{AH}^0)^2$. The slope of the dotted 'reaction-coordinate' line is -1.18. This is clearly very similar to the plot with $k_1 = k_2$, with slope -1.0.

(so long as it is large enough to give a saddle point at all). For other values the slope depends on all three, as follows (for $k_1 > k_2$):

$$\text{slope} = \cot\left(\frac{1}{2}\tan^{-1}\frac{2k_{12}}{k_2 - k_1}\right)$$

For $k_2 > k_1$ this gives the slope of the line normal to the reaction coordinate. Thus, the slope is a large negative number for $k_1 \gg k_2$, but the larger k_{12}, for any given k_1/k_2, the closer the slope gets to -1.

The relevance of this to the problem of the tunnel correction is firstly that the curvature of the reaction coordinate (which we shall represent by v_3, following Bell; this is the same as v in equation 1) increases with k_{12}, and secondly that the isotope effect on v_3 is maximized with the slope of -1. If A and B are very heavy compared to H (or D), then with this slope of -1 the value of $v_3^H/v_3^D = 2^{\frac{1}{2}}$; but if the slope is a very large negative number, then $v_3^H \simeq v_3^D$, and no matter how large the tunnel correction it will be almost the same for H and D, and therefore will not lead to an isotope effect. (We may note that for realistic barriers for hydrogen-atom or proton transfers, the situation where Q_H is large but $Q_H \approx Q_D$ cannot be realized; the large tunnel corrections are associated instead with the reaction coordinates of slope close to -1). Thus large tunnel corrections on the isotope effect are associated with fairly symmetric transition states (with k_1 not very different from k_2) and large values of k_{12}.

Bell has constructed models for the interactions and concluded [1] that for proton transfers $k_{12} \geqslant 4(k_1 k_2)^{\frac{1}{2}}$, and $k_1/k_2 \leqslant 10$. With these restrictions the tunnel correction is always significant and is a maximum for thermoneutral proton transfers. This conclusion is essentially based on the conclusion by Marcus [29] that the activation energy for proton transfers in the favourable direction is largest with $\Delta G^0 = 0$; any other reasonable expression for barrier height would give the same qualitative conclusion. If Bell's conclusions about the probable range of k_{12} and k_1/k_2 are correct, then the isotope effect should be a maximum for the thermoneutral reaction, and tunnelling will account for most of the variation of isotope effect with ΔG^0 (or ΔH^0).

A consequence of this is (if the bending vibrations are ignored) that the isotope effects will be greater than the 'maximum' value as calculated on p. 326; indeed, most proton transfers have rather large isotope effects, with the lowest more than half the 'maximum' value, and furthermore, low A_H/A_D values (indicating substantial tunnel corrections) are not at all rare. However, very small isotope effects have been observed for proton transfers from carbonium ions [30].

The absence of large isotope effects in proton transfers between oxygen and nitrogen may arise from a different source.

Similarly, Willi [31] has constructed potential energy surfaces for some proton transfers which (exclusive of tunnelling) lead to the same 'maximum' isotope effect for all transfers, regardless of 'symmetry'.

The source of a significant value of k_{12} is a repulsion between A and B, and the one that leads to Bell's conclusions about the magnitude of k_{12} is an electrostatic repulsion. Yet this conclusion about the potential-energy surface is not general. Thus More O'Ferrall and Kouba [13] have made model calculations on isotope effects in proton transfers using potential energy surfaces constructed by the BEBO method. They find that the principal cause of variation in isotope effects is the increase of transition-state zero-point energy as the force constants k_1 and k_2 diverge, as originally proposed by Westheimer [32]. Like Bell, they find most tunnel correction when $k_1 = k_2$, but the tunnel corrections are in general smaller. The nature of the difference in force constants between this approach and that of Bell has been delineated sharply by Albery [33], who however has been concerned only with the k_s^H/k_s^D portion.

In this presentation we shall not attempt to compare the validities of the various theoretical treatments and assumptions; instead we shall search for evidence of an experimental nature bearing on this subject.

The first clue is that the tunnel correction is sensitive to steric effects. Thus the isotope effect in the reaction of 2-nitropropane with pyridine or with 3,4-dimethylpyridine is hardly more than half that with 2,6-dimethylpyridine or 2,4,6-trimethylpyridine [15, 34, 35], and the fact that there is substantial steric hindrance is demonstrated by the deviation of these last two bases from the Brønsted plot; they react too slowly by a factor of 5 [36, 37]. Since these first examples, there have been other examples of exaggerated isotope effects in sterically hindered proton transfers,* and the evidence for tunnelling has been reinforced.

The original explanation of this effect [34] can be restated in terms of the considerations of the quadratic representation, namely that steric hindrance is a repulsive interaction present in the transition state but absent in the initial and final states. It can therefore be

---

* The original work of Reference [15] was questioned as an experimental artifact, but in studies by quite experimentally different methods large tritium isotope effects have been found in the 2,4,6-trimethylpyridine-2-nitropropane reaction [16], and a large deuterium isotope effect is found in the 2,4,6-trimethylpyridine-methyl γ-nitrovalerate reaction (H. Wilson, J. D. Caldwell, and E. S. Lewis, *J. Org. Chem.*, **38**, 564 (1973). Other examples include [12] and E. Grovenstein, Jr., and F. C. Schmalstieg, *J. Amer. Chem. Soc.*, **89**, 5084 (1967).

expected to increase the value of k_{12}, the interaction force constant. As we have seen before, an increase in k_{12} increases the imaginary frequency v_3 and will also increase the ratio v_3^H/v_3^D, up to a maximum of $2^{\frac{1}{2}}$, hence increasing Q_H/Q_D and the isotope effect.

However, an alternative explanation can also be offered: In the highly polar proton-transfer reaction, there is extensive solvation at all times, and this changes along the reaction coordinate. In solution the solvent protons then can contribute significantly to the reduced mass of the imaginary vibration, and under these circumstances the effect of changing the transferring proton from H to D changes the reduced mass of the imaginary vibration by a factor much less than $2^{\frac{1}{2}}$ and the tunnel correction to the isotope effect Q_H/Q_D is not markedly different from unity. When the attacking base is sterically hindered, the solvation of the transition state is less effective, the solvent motion is reduced, v_3^H/v_3^D increases to nearly $2^{\frac{1}{2}}$, and Q_H/Q_D is correspondingly increased.

This explanation has been considered for sometime, but apparently the first published version is presented in a theoretical attack on solvation in proton transfers by Kurz and Kurz [38]. The problem has been extensively attacked by Caldin and Mateo [11] by studying solvent effects on reaction 9.

$$O_2N\left\langle\!\!\!\bigcirc\!\!\!\right\rangle CH_2 \cdot NO_2 + B \longrightarrow O_2N\left\langle\!\!\!\bigcirc\!\!\!\right\rangle \overset{|}{\underset{H}{C}}{-}NO_2 + BH^+ \qquad (9)$$

They observe that with several different bases (tetramethylguanidine, trbutylamine, and triethylamine) the deuterium isotope effect is much larger in the weakly-solvating solvent toluene than in the more polar acetonitrile, and lower A^H/A^D factors are also observed in toluene. This work has been greatly extended in other solvents and the results analysed by a complete Bell truncated-parabola treatment to evaluate the parameters a (the half-width of the parabola at the truncation) and E^H (the height of the barrier above the ground state (of the hydrogen compound)). They conclude that for the polar solvents acetonitrile and methylene chloride it is more reasonable to account for the decrease in tunnelling (as measured by A^H/A^D) by an increase in the effective mass of the hydrogen (from 1 to about 1.3) in the polar solvents, than to increase the value of a. Although these solvent changes also alter the symmetry of the transition state, this effect is not large (the equilibrium constant for reaction 9 changes from a minimum of 120 to a maximum of 6500 over the range of 10 solvents), so that the evidence for the conclusion that the solvents have a fundamental influence on the extent of tunnelling is convincing, and the motion of solvent molecules coupled with that of the proton is an attractive explanation.

If we accept the notion that tunnelling is almost universal, but is reduced in importance by solvent motion in many cases, the suggestion that nearly all the variation in k_H/k_D in proton transfers of different symmetry is attributable to tunnelling (as suggested by Bell, reference [1], p. 298) becomes more attractive.

The study of isotope effects in some hydrogen-atom transfer reactions introduces a new factor, however. In the reaction 2, the isotope effects vary strongly, they correlate with bond strengths and

$$R\cdot + HSR' \rightarrow RH + \cdot SR' \qquad (2)$$

are apparently a maximum in cases with $\Delta H \approx 0$ [39, 40]. Unlike the proton-transfer analogues, the range of isotope effects is very large; thus for $R = C_6H_5$ and $R' = C_6H_5$, $k_H/k_T = 1.4$ [40, 41] at 60°C, but for $R = (C_6H_5)_3C\cdot$, $R' = C_6H_5$, $k_H/k_T = 14.9$ at 25°C [42], with other values in between, covering a range of ΔH^0 from about $-37\ kcal\ mol^{-1}$ to about $+4\ kcal\ mol^{-1}$. The large value above probably has a perceptible tunnel correction, as shown both by the large isotope effect (compared to the 'maximum' value of 11) and $A^H/A^T \approx 0.17$, but it is clear that this tunnel correction is not likely to be enough to account for this large range. It appears that this variation in isotope effect is so large that we must use a Westheimer-type 'symmetry' argument, and that the transition state for the cases with low isotope effects must have considerable isotope-sensitive transition-state zero-point energy.

The atom-transfer results likewise bear upon the source of the steric enhancement of the isotope effect noted earlier. If the only effect of steric hindrance is to exclude solvent and thus prevent solvent motion along the transition state, then the much less strongly solvated radicals should abstract hydrogen with an important tunnel correction and the steric enhancement should disappear. The low values of isotope effect for reaction 2 with large $-\Delta H^0$ suggest that there is no large tunnel correction for the unsymmetric cases (and hence that the interaction force constant k_{12} is not very much greater than $(k_1 k_2^{\frac{1}{2}})$). The solvent in these reactions (thiophenol), although probably not inert, should not solvate the transition state strongly; the absence of extensive tunnelling is probably not attributable to solvent motion. When steric hindrance is deliberately introduced by the use of 2,4,6-trimethylthiophenol, the isotope effect k_H/k_T at

25°C now rises (from 14.9 with thiophenol) to 25.2 [20]. The steric enhancement of the isotope effect, to a magnitude giving strong evidence of tunnelling, is shown. Another possible case of steric contribution to the isotope effect in a hydrogen-atom transfer is reaction 10, where for various R_1 values the tritium isotope effects are

$$R_1\overset{.}{C}HCH_2\overset{\overset{O}{\|}}{P}(OMe)_2 + *H\overset{\overset{O}{\|}}{P}(OMe)_2$$

$$\rightarrow R_1C*H_2-CH_2\overset{\overset{O}{\|}}{P}(OMe)_2 + \overset{.}{O}P(OMe)_2 \quad (10)$$

rather large and random, although they do not exceed the 'maximum' value calculated for total loss of the P—H stretching vibration [43]; from models the transition states appear to be hindered. The conclusion is that the steric enhancement of the isotope effect, which is attributable to an increase in the tunnel correction, is as substantial in the un-charged hydrogen-atom transfers as it is in the strongly polar proton transfers. Confirming evidence for sterically enhanced isotope effects in hydrogen transfers in the gas phase is not yet available.

10.6 Conclusions

1. Tunnelling has a real influence on the rates of hydrogen transfer processes, within the restriction that the tunnel correction has no more meaning than an understandable deviation by nature from an artificial semi-classical model.

2. The criteria that have been used – curved Arrhenius plots, large isotope effects, and the excessive temperature-dependence of isotope effects – are all internally consistent and can best be explained by important tunnel corrections.

3. In proton or hydrogen-atom transfers, the tunnel correction is largest when the transition state is about halfway between reagent and product.

4. Electrostatic forces in proton transfers are important to tun-nelling. They not only contribute to the dimensions of the barrier, even in a three-particle one-dimensional approximation, by increasing the interaction force constant k_{12}; they are also responsible for the motion of the solvent along the reaction coordinate, which increases the effective mass of this imaginary vibration and reduces the isotope dependence v_3^H/v_3^D of the frequency of the imaginary vibrations.

5. Sterically-hindered transition states for hydrogen transfers have especially large tunnel corrections. This arises both from solvent exclusion and from the increase in the energy of the transition state and in the interaction force constant, and hence in the barrier curvature.

Acknowledgment

The experimental work from Rice University reported here was nearly all supported by a grant from the Robert A. Welch Foundation.

REFERENCES

[1] R. P. Bell, *The Proton in Chemistry*, 2nd Ed., Chapman and Hall, London and Cornell University Press, Ithaca, N.Y. (1973).

[2] J. H. Brewer, K. M. Crowe, F. N. Gygax, R. F. Johnson, D. G. Fleming, and A. Schenk, *Lawrence Berkeley Laboratory Preprint 1332* (1973) (submitted to *Physical Review A*). I thank Dr. Brewer for this preprint.

[3] S. Glasstone, K. S. Laidler, and H. Eyring, *The Theory of Rate Processes*, McGraw-Hill, New York, N.Y. (1941); a modern critique of the status of this theory is by R. Marcus, Chapter 2 in *Technique of Chemistry*, Vol. VI, Pt 1, ed. E. S. Lewis, Wiley-Interscience, New York, N.Y., 1974.

[4] E. Wigner, *Z. Physik. Chem.*, B, **19**, 203 (1932).

[5] (a) R. P. Bell, *Proc. Roy. Soc.*, A, **139**, 466 (1933); (b) *Trans. Faraday Soc.*, **55**, 1 (1959).

[6] (a) H. S. Johnson and D. Rapp, *J. Amer. Chem. Soc.*, **83**, 1 (1961); (b) J. D. Russell and J. C. Light, *J. Chem. Phys.*, **54**, 4881 (1971).

[7] R. J. LeRoy, E. D. Sprague, and F. Williams, *J. Phys. Chem.*, **76**, 546 (1972).

[8] R. P. Bell, J. A. Fendley, and J. R. Hulett, *Proc. Roy. Soc.*, A, **235**, 453 (1956).

[9] E. F. Caldin, *Chem. Rev.*, **69**, 135 (1969).

[10] A recent survey and critique of semiclassical calculations of isotope effects is W. H. Saunders, in *Technique of Chemistry*, Vol. VI Pt. 1, E. S. Lewis, Ed., Wiley-Interscience, New York, N.Y. (1974.) This and other recent treatments stem from J. Bigeleisen, *J. Chem. Phys.*, **17**, 675 (1949).

[11] E. F. Caldin and S. Mateo, *J. Chem. Soc., Faraday Trans. I*, in press.

[12] J. R. Keeffe and N. H. Munderloh, *J. Chem. Soc. Chem. Comm.*, 17 (1974).

[13] R. A. More O'Ferrall and J. Kouba, *J. Chem. Soc.*, B, 985 (1967).

[14] R. Stewart and R. van der Linden, *Disc. Faraday Soc.*, **29**, 211 (1960).

[15] E. S. Lewis and L. H. Funderburk, *J. Amer. Chem. Soc.*, **89**, 2322 (1967).

[16] E. S. Lewis and J. K. Robinson, *J. Amer. Chem. Soc.*, **90**, 4337 (1968).

[17] E. S. Lewis, J. M. Perry, and R. H. Grinstein, *J. Amer. Chem. Soc.*, **92**, 899 (1970).

[18] A. Bromberg, K. A. Muszkat, and E. Fischer, *Chem. Comm.*, 1352 (1968).

[19] A. Bromberg, K. A. Muszkat, and A. Warshel, *J. Chem. Phys.*, **52**, 1262 (1970).

[20] M. M. Butler, Ph.D. Thesis, Rice University, (1974).

[21] J.-T. Wang and F. Williams, *J. Amer. Chem. Soc.*, **94**, 2930 (1972).

[22] E. F. Caldin and S. Mateo, *J. Chem. Soc. Chem. Comm.*, 854 (1973).

[23] E. S. Sprague and F. Williams, *J. Amer. Chem. Soc.*, **93**, 787 (1971); A. Campson and F. Williams, *ibid.*, **94**, 7633 (1972).

[24] E. S. Lewis, J. D. Allen, and E. T. Wallick, *J. Org. Chem.*, **34**, 255 (1969).

[25] C. G. Swain, E. C. Stivers, J. F. Reuwer, and L. J. Schaad, *J. Amer. Chem. Soc.*, **80**, 5885 (1958).

[26] J. Bigeleisen, *Tritium in the Physical and Biological Sciences*, Vol. 1, International Atomic Energy Agency, Vienna, (1962), p. 161; K. B. Wiberg, *Physical Organic Chemistry*, John Wiley and Sons, Inc., New York, N.Y., (1964) p. 360.

[27] J. R. Jones, *Trans. Faraday Soc.*, **65**, 2138 (1969).

[28] M. F. Hawthorne and E. S. Lewis, *J. Amer. Chem. Soc.*, **80**, 4296 (1958); E. S. Lewis and M. C. R. Symons, *Quart. Rev.*, **12**, 230 (1958); G. A. Olah, Y. Halpern, J. Shen, and Y. K. Mo, *J. Amer. Chem. Soc.*, **93**, 1251 (1971) this is one of many papers by Olah concerned with triangular transition states for hydride transfer.

[29] R. A. Marcus, *J. Chem. Phys.*, **72**, 891 (1968).

[30] V. J. Shiner, Jr., *J. Amer. Chem. Soc.*, **75**, 2925 (1953); C. E. Boozer and E. S. Lewis, *ibid.*, **76**, 794 (1954).

[31] A. V. Willi, *Helv. Chim. Acta*, **34**, 1220 (1971).

[32] F. H. Westheimer, *Chem. Rev.*, **61**, 265 (1961).

[33] W. J. Albery, *Trans. Faraday Soc.*, **63**, 200 (1967).

[34] L. H. Funderburk and E. S. Lewis, *J. Amer. Chem. Soc.*, **86**, 2531 (1964).

[35] R. P. Bell and D. M. Goodall, *Proc. Roy. Soc.* (London), **A294**, 273 (1966).

[36] R. G. Pearson and F. V. Williams, *J. Amer. Chem. Soc.*, **75**, 3073 (1953).

[37] E. S. Lewis and J. D. Allen, *J. Amer. Chem. Soc.*, **86**, 2022 (1964).

[38] J. L. Kurz and L. C. Kurz, *J. Amer. Chem. Soc.*, **94**, 445 (1972).

[39] E. S. Lewis and M. M. Butler, *Chem. Comm.*, 941 (1971). Some of these results (not essential to the present argument) are revised in reference [20].

[40] W. A. Pryor and K. G. Kneipp, *J. Amer. Chem. Soc.*, **93**, 5584 (1971).

[41] E. S. Lewis and K. Ogino, unpublished work.

[42] E. S. Lewis and M. M. Butler, *J. Org. Chem.*, **36**, 2582 (1971).

[43] E. C. Nieh, *Ph.D. Thesis*, Rice University (1972).

11

Brian Capon

NEIGHBOURING GROUP PARTICIPATION

11.1 Introduction

Neighbouring-group participation in proton-transfer reactions is generally referred to as intramolecular general-acid or intramolecular general-base catalysis [1]. Two types are possible. The first involves an intramolecular proton transfer, either abstraction of a proton from a reacting functional group by an internal base or donation of a proton to one by an internal acid. The intramolecularly general-base catalysed enolization of the anion of 2-carboxy-acetophenone (see p. 370) and the intramolecularly general-acid catalysed hydrolysis of 2-methoxy-methoxybenzoic acid (see p. 342) are examples of reactions which are thought to involve processes of this type. In the second type the proton transfer is intermolecular with attack of a nucleophile on one functional group of a molecule being facilitated by proton abstraction by a second basic functional group. The intramolecularly general-base catalysed hydrolysis of the anion of phenyl salicylate (see p. 348) is a reaction which is thought to involve a process of this type.

As with intermolecularly general-acid and general-base catalysed reactions, it is frequently difficult to distinguish between kinetically equivalent mechanisms and eliminate all but one. Mechanisms which involve intramolecular nucleophilic catalysis are often kinetically equivalent to those which involve intramolecular general-acid or base catalysis. As these do not involve a proton transfer they will not be discussed in this chapter unless the evidence is fairly evenly balanced between them and the latter.

There are three recent reviews whose subject matter partly overlaps that of the present chapter, two entitled *Intramolecular Catalysis* [2, 3], and one entitled '*The Intramolecular Hydrogen Bond and the Reactivity of Organic Compounds*' [4].

11.2 Intramolecular general-acid catalysis in the hydrolysis of acetals and glycosides

The hydrolyses of acetals and glycosides derived from salicyclic acid occur much more rapidly than would be expected on the basis of the electronic effects of the *ortho*-carboxyl group. This was first demonstrated for the hydrolysis of o-carboxyphenyl β-D-glucoside, **1** [5, 6, 7]. The pH-rate profile for this reaction is of the form

$$\text{Rate} = k_{\text{obs}}[\text{total substrate}]$$

$$= k_1[\text{un-ionized form}] + k_2[\text{un-ionized form}] \times 10^{-\text{pH}} \quad (1)$$

with

$$k_{\text{obs}} = (k_1 + k_2 \times 10^{-\text{pH}})/(1 + K_a/10^{-\text{pH}})$$

where K_a, the dissociation constant for the carboxyl group, is $2.14 \times 10^{-4} \, \text{mol} \, \text{l}^{-1}$, $k_1 = 1.41 \times 10^{-3} \, \text{s}^{-1}$ and $k_2 = 5.49 \times 10^{-3} \, \text{l} \, \text{mol}^{-1} \, \text{s}^{-1}$ at 90.35°C. The second term on the right-hand side of equation 1 arises from the specific hydrogen-ion catalysed hydrolysis of the glycoside with the carboxyl group un-ionized, and the value of k_2 is of the magnitude expected for such a process. The first term arises from either a spontaneous hydrolysis of the un-ionized form or a specific hydrogen-ion catalysed hydrolysis of the ionized form. The second-order constant for the latter process would be $k_2^* = k_1/K_a = 6.59 \, \text{l} \cdot \text{mol}^{-1} \, \text{s}^{-1}$ which is over 1000 times greater than k_2 and over 600 times greater than the second-order rate constant for the hydrolysis of phenyl β-D-glucoside. As a result of the k_1-term in the rate law, the value of k_{obs} from the hydrolysis of o-carboxyphenyl β-D-glucoside **1** is 1.3×10^4 times that estimated for k_{obs} for hydrolysis of the *para*-isomer at pH 4.55. Similar high rates and similar pH-rate profiles are found with the hydrolysis of the α-anomer **2** [6] and with that of several other acetals and glycosides which have a carboxyl group *ortho* to a phenolic acetal group (see Scheme 1).

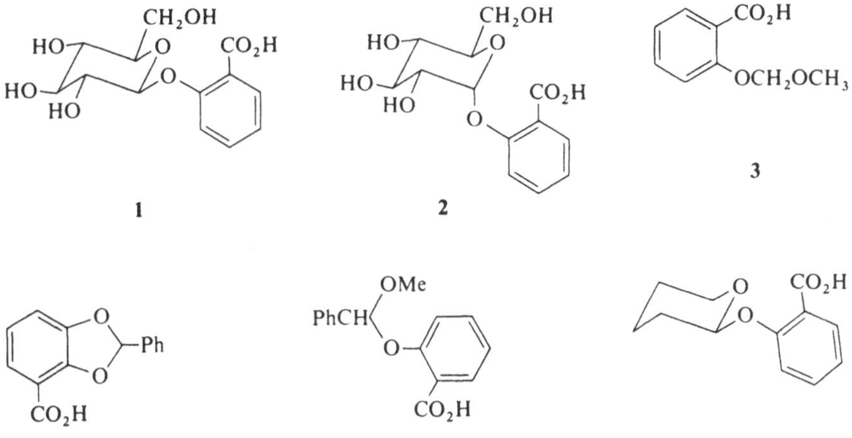

Scheme 1

The hydrolysis of 2-methoxymethoxybenzoic acid **3** was investigated in some detail [6]. Since this compound is an unsymmetrical acetal, two modes of bond fission are possible in the rate-determining step. It was shown by p.m.r. spectroscopy that methanolysis in CD_3OD occurs with fission of the aryloxy-carbon bond and hence it seems probable that hydrolysis does also. The possibility that the k_1-term arose from a specific hydrogen ion catalysed reaction of the ionized form with the carboxylate group acting as an intramolecular nucleophilic catalyst as shown in equations 2 and 3 was excluded by

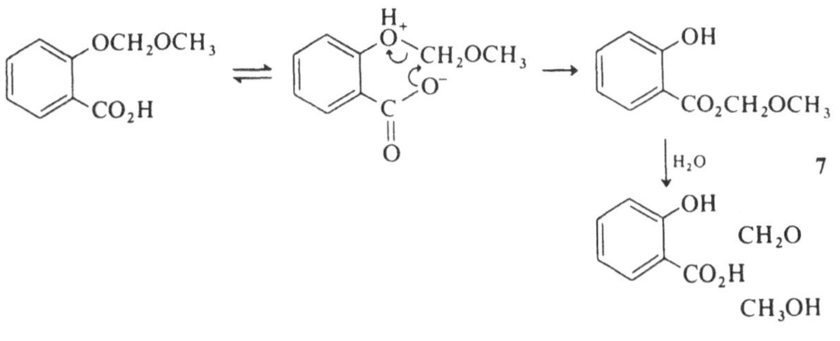

$$(2)$$

$$(3)$$

showing that **7** and **8** were not intermediates [6]. Mechanisms which involve intramolecular nucleophilic catalysis are also excluded by the observation that the hydrolysis of *0,0'*-benzylidene-2,3-dihydroxy-benzoic acid **4** is kinetically very similar to that of other *o*-carboxy-phenyl acetals. With this compound intramolecular attack by the carboxy-group on the acetal-carbon is not possible [8].

This leaves two reasonable explanations for the high value of k_1. Either the reactive species is the ionized form **9** and the high rate arises from an electrostatic effect of the carboxylate group in a specific-hydrogen-ion catalysed reaction (equation 4) or it is the un-ionized

$$(4)$$

form and the reaction involves intramolecular general-acid catalysis (equation 5). The former was favoured by Dunn and Bruice for the

$$(5)$$

342

reaction of 2-methoxymethoxybenzoic acid for the reason that the $\rho*$ value for the k_1 term for the hydrolysis of acetals **10** is the same as that for the k_2 term which definitely arises from a specific-acid

catalysed reaction [9]. However, since both mechanisms involve formation of the same carbonium-oxonium ion **11**, use of the $\rho*$ values based on the variation of R may not be a very sensitive method for distinguishing between them. In contrast Fife and Anderson have shown that the 'electrostatic-stabilization' mechanism is not possible for the hydrolysis of acetals **5** and **6** since the rate constants for the breakdown of the zwitterionic conjugate acids would have to be unreasonably large [10]. They therefore favoured intramolecular general-acid catalysis (for example equation 6) for the hydrolysis of

(6)

these compounds and suggested that there was a change in mechanism on going from the hydrolysis of **3** to that of **5** and **6**. This was considered to be reasonable since the hydrolysis of the acetals that are analogous to **5** and **6** but do not have a carboxyl group show inter-molecular general-acid catalysis whereas the corresponding analogue of **3** does not. However this argument is not very compelling since the absence of detectable intermolecular general-acid catalysis must arise because the free-energy of the transition state for the general-acid catalysed pathway is greater than that for specific-acid catalysed one. On changing from inter- to intra-molecular catalysis the free-energy of the former will be reduced but the pathway for specific acid catalysis will be unaltered. Therefore the free energy of the transition state corresponding to intramolecular catalysis could fall below that for the transition state corresponding to specific acid catalysis and hence intramolecular general-acid catalysis would be detectable when intermolecular general-acid catalysis was not.

The mechanism of equation 4 has been excluded by the work of Craze and Kirby who determined the effect of substituents on the rate of hydrolysis of a series of acetals **12**. The results were analysed by use of Jaffé's four parameter equation and ρ-values were determined for the effect of substituents on the phenolic and carboxyl group in the intramolecularly catalysed reaction [11]. The value of ρ_{phenol} was 0.89 ± 0.03 and that of $\rho_{carboxy}$ was 0.02 ± 0.08. The small value of ρ-carboxy indicates that there is relatively little proton transfer in the transition state and hence is inconsistent with the mechanism of equation 4 in which there is complete proton transfer. The value of $\rho_{carboxy}$, which corresponds to the Brønsted α-value, is smaller than normally found in the intermolecular general-acid catalysed hydrolysis of aryl acetals. This suggests that the intramolecularly general-acid catalysed hydrolysis of these salicyl acetals is not exactly analogous to the intermolecularly general-acid catalysed hydrolysis of aryl acetals. It is possible that the carboxyl group is twisted out of the plane of the benzene ring [11] but the evidence is not compelling on this point. Only three derivatives of salicylic acid which are not intramolecularly hydrogen bonded have had their structures determined by X-ray crystallography. In two of these, potassium and rubidium hydrogen *bis*-acetylsalicylate, the angle of twist is 34° [12, 13], but in aspirin itself, the angle is only 2° 02' [14]. The fact that this variation is found with similar compounds suggests that the angle may be determined by crystal-packing forces and that the plot of potential-energy against angle of twist for a molecule in solution may have a shallow minimum. The solvent isotope effect for these intramolecularly catalysed reactions varies very little over the compounds for which it has so far been measured and the variation of solvent isotope effect for intermolecularly catalysed reactions appears to be much larger (Table 1). The only reaction for which a direct comparison is possible is the hydrolysis of O,O'-benzylidene-2,3-dihydroxybenzoic acid **4** where the isotope effect ($k_H/k_D = 1.28$) is quite close to that for the acetic acid-catalysed hydrolysis of O,O-benzylidene catechol ($k_H/k_D = 1.4$ to 1.5) (Table 1). Unfortunately the absence of any other reactions for which a direct comparison can be made makes it uncertain whether this closeness is general or merely coincidental.

A direct comparison of the effectiveness of intra- and intermolecular catalysis is possible with three of the compounds in Scheme 1. Thus, the effective concentrations [3] of the internal carboxyl groups are about 100, 5000, and 10 000M for the hydrolyses of the compounds **4** [8], **6** [10], and **5** [10] respectively. The large values obtained for the last two are interesting as they show that high

Table 1

The solvent isotope effect (k_H/k_D) for some intra- and inter-molecularly catalysed hydrolyses

Intramolecular reactions	Intermolecular reactions

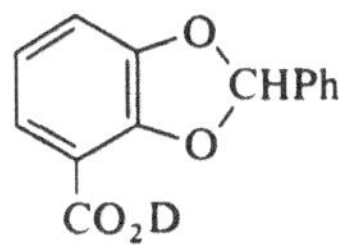

1.28 at 55° [a]

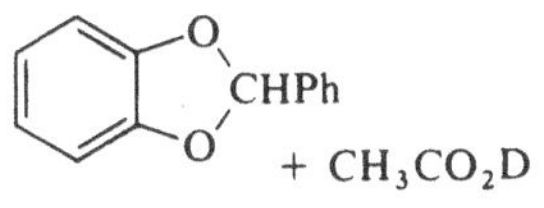

1.4 to 1.5 at 55° [b]

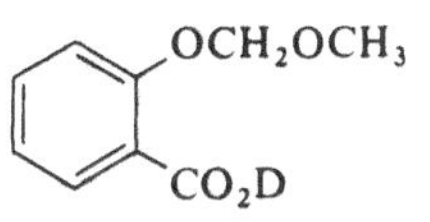

1.43 at 45° [a]
1.61 at 39° [c]

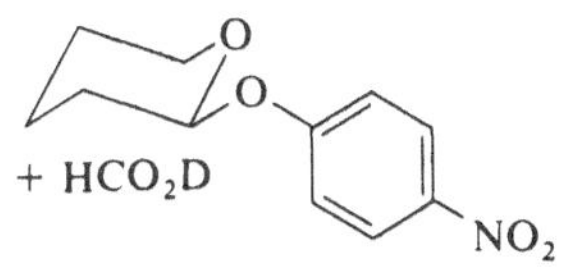

3.4 at 50° [d]

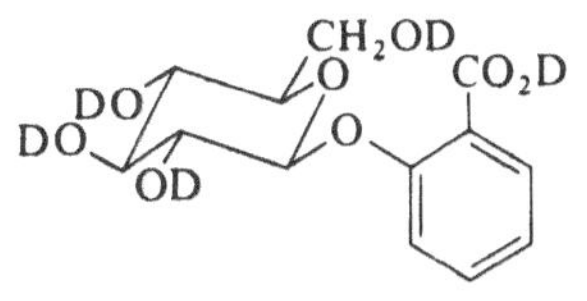

1.57 at 60.2° [e]

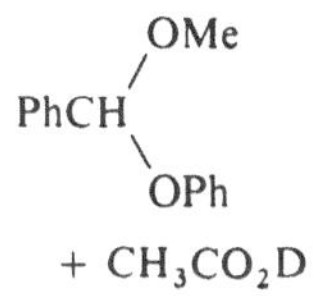

2.14 at 20° [f]

---

[a] B. Capon, M. I. Page, and G. H. Sankey, *J. Chem. Soc. Perkin II*, 529 (1972).
[b] B. Capon and M. I. Page, *J. Chem. Soc. Perkin II*, 522 (1972).
[c] G. A. Craze and A. J. Kirby, *J. Chem. Soc. Perkin II*, 61 (1974).
[d] T. H. Fife and L. H. Brod, *J. Am. Chem. Soc.*, **92**, 1681 (1970).
[e] B. Capon, M. C. Smith, E. Anderson, R. H. Dahm, and G. H. Sankey, *J. Chem. Soc. (B)*, 1038 (1969).
[f] E. Anderson and B. Capon, *J. Chem. Soc. (B)*, 1033 (1969).

effective molarities are possible for intramolecularly general-acid-catalysed reactions.

8-Methoxymethoxy-1-naphthoic acid is the only acetal, not a salicylic-acid derivative, for which there is any evidence for intra-molecular general-acid catalysis in its hydrolysis at present, and the rate enhancement is not very large. Thus at pH 4.10 the rate is only fifteen times greater than that estimated for the hydrolysis of 1-methoxymethoxy-naphthalene at this pH [6].

One factor which may be important in the hydrolysis of acetals derived from salicylic acid is that on ionization they give an intra-molecularly hydrogen-bonded anion and that this hydrogen bonding is partly developed in the transition state. The leaving group in the non-hydronium-ion catalysed hydrolyses of the phenyl acetals has a pK_a value of 9.98 and these reactions are immeasurably slow. With a 2-carboxyphenyl acetal, however, the leaving group as a pK_a of *ca.* 3 when proton transfer is complete, and hydrolysis is rapid. It is, of course, clear from the value of $\rho_{carboxy}$ [11] that proton transfer is not complete in the transition state but nevertheless even partial proton transfer should lead to a lowering of the pK_a.*

Attempts have been made to detect bifunctional catalysis in the reactions of acetals with two carboxyl groups but the results indicate that if this occurs the rate enhancement over that found with just intramolecular general-acid catalysis is relatively small [15, 16]. Thus the pH-rate profile for the conversion of the acid **13** into the corresponding phthalide is bell-shaped but at the maximum the rate constant is only three times greater than that for the hydrolysis of the acid **14** [15]. A larger effect is found in the reactions of the disalicyl-acetals **15** in dioxan-water to yield the corresponding dioxanone **16**

13

14

and salicylic acid. Bell-shaped pH-rate profiles were found and at the maximum the reaction of **15a** is 2.7×10^9 times faster than hydrolysis of the corresponding dimethyl ester. However most of this rate enhancement arises from intramolecular general-acid catalysis since the rate constant for reaction of the mono-anion is only 65 times greater than that for reaction of the un-ionized form. This factor is slightly larger for the *p*-nitro-compound **15b**, *viz.* 260, and it was also shown that the rate constant for the reaction of the mono-anion is 233 times greater than that for hydrolysis of the monoanionic *forms*

---

* Presumably there must be some proton transfer in the transition state, for if there were none it is difficult to see how a rate enhancement would arise.

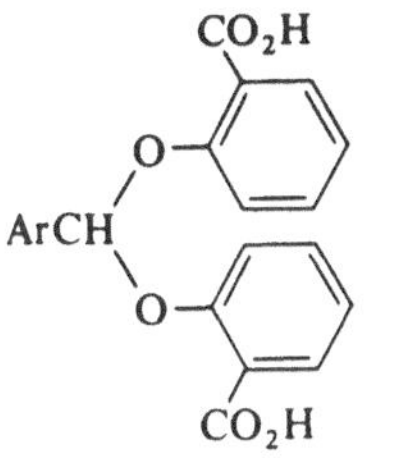

15 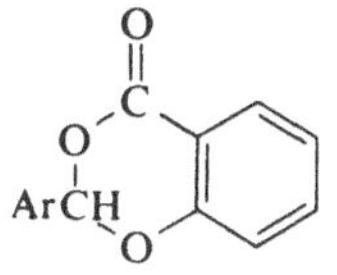   **16** 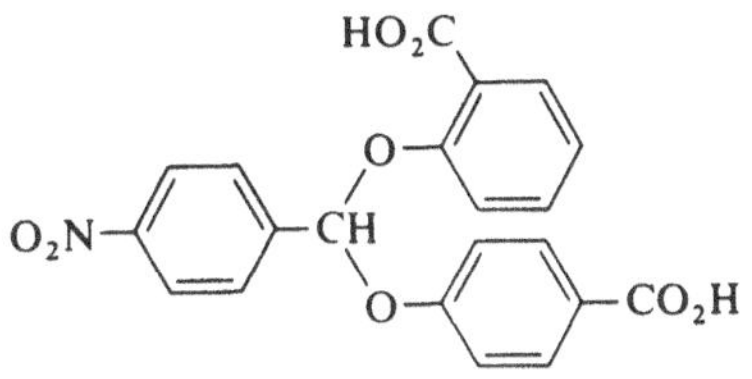   **17**

a Ar = C$_6$H$_5$
b Ar = O$_2$N · C$_6$H$_4$

of the acetal **17**. It is difficult to make an exact allowance for the difference in the substituent effects of the *ortho-* and *para*-carboxylate groups and to determine the exact proportions of the monoanionic forms but nevertheless there does appear to be a rate enhancement of 10 to 100 which can be attributed to nucleophilic assistance or electrostatic stabilization by the ionized carboxylate group acting concertedly with intramolecular general-acid catalysis by the un-ionized carboxyl group [16].

There is also evidence for the intramolecular bifunctional catalysis in the hydrolysis and methanolysis of 2-carboxyphenyl 2-acetamido-2-deoxy-β-D-glucoside **18** [17, 18]. The pH-rate profile for the hydrolysis is similar to that for hydrolysis of the corresponding glucoside, **1** (see p. 340) and at pH 3.67 and 65°C it reacts about twenty times faster. At 94°C the methanolysis occurs forty times faster and, significantly, occurs with more than 92.5 per cent retention of configuration, compared to 84 per cent inversion of configuration on methanolysis of the corresponding glucoside [18]. This suggests that the major

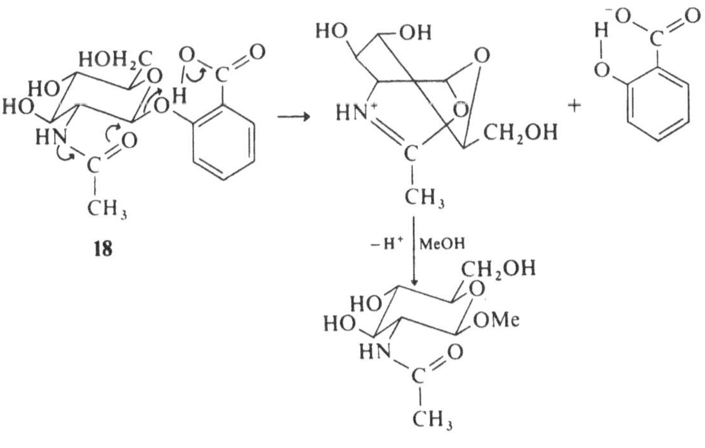

(7)

18

347

pathway for hydrolysis and methanolysis involves bifunctional catalysis (equation 7) but that the nucleophilic component contributes much less than the general-acid component.

11.3 Intramolecular general-acid-base catalysis by phenolic groups in reactions of derivatives of carboxylic acids

Most of the examples of this class of reaction are found with catechol esters or with esters and amides of salicylic acids. Examples of intra-molecular nucleophilic catalysis are also known [19] but these fall outside the scope of this chapter.

11.3.1 Intramolecular catalysis in reactions of mono-esters of catechol

The pH-rate profile for the hydrolysis of catechol monobenzoate in the pH range 7.28 to 9.75 is sigmoid with the rate proportional to the concentration of ester with ionized hydroxyl group [20]. The rate of hydrolysis in this pH range is several hundred times larger than when the hydroxyl group is methylated. This rate enhancement cannot arise from nucleophilic catalysis since nucleophilic attack by the ionized phenolate group on the ester group would only lead to a degenerate rearrangement. Mechanisms involving intramolecular general-acid catalysis or general-base catalysis as symbolized by **19** and **20** therefore seem most likely. That involving general base

19   20

catalysis was preferred since if **19** were followed intramolecular catalysis for reaction with other nucleophiles without a detachable proton might be expected and such catalysis was not found with imidazole. The solvent-isotope effect for the hydrolysis, $k(H_2O)/k(D_2O) = 1.8$ at 25°C, is consistent with intramolecular general-base catalysis.

Similar behaviour is found with catechol monoacetate [21] and catechol monocinnamoate [22]. With catechol monosuccinate intra-molecular general base catalysis by the phenolate group takes over from intramolecular nucleophilic catalysis by the carboxylate group as the main mode of hydrolysis at pH values at which the phenolic group is ionized [23].

The rate of methanolysis of catechol monobenzoate is also enhanced and reaction probably proceeds by a mechanism analogous to that for the hydrolysis [24].

Intramolecular general-base catalysis by the ionized phenolic group occurs in the hydrolysis of ethyl 2-hydroxy-5-nitrophenyl carbonate. The pH-rate profile shows two plateau regions corresponding to hydrolysis of the ionized and un-ionized forms respectively, with the rate constant in the former about fifty times greater than in the latter. Again the rate of reaction with imidazole is not enhanced and the most likely mechanism for hydrolysis is one involving intramolecular general-base catalysis as symbolized by **21** [25]. The solvent-isotope effect $k(H_2O)/k(D_2O) = 2.3$ at 30°C is consistent with this mechanism. Similar behaviour is found in the hydrolysis of ethyl 2-hydroxy-phenyl carbonate which has been investigated less extensively [26].

21   **22**

Although catechol esters are not especially reactive towards imidazole they are reactive towards primary and secondary amines and this has led to their use in peptide synthesis. A mechanism involving intramolecular general-base catalysis as symbolized by **22** seems most likely [27].

11.3.2 *Intramolecular catalysis in reactions of esters of salicylic acid*

The kinetics of the hydrolysis of salicylate esters, which have the phenolic group in the acid-moiety, is similar to that for catechol esters. Thus the pH rate profile for the hydrolysis of phenyl salicylate between pH 6.58 and 11.50 is sigmoid with the rate proportional to the concentration of the form with the phenolic group ionized and at pH 7.26 phenyl salicylate is hydrolysed about 60 times faster than phenyl *o*-methoxybenzoate. Similar behaviour is found with *p*-nitrophenyl 5-nitrosalicylate [28]. Mechanisms involving intramolecular general-acid catalysis (for example **23**) and intramolecular general-base catalysis (for example **24**) are again possible but the latter seems

23

24

more likely since if the former were operative the rate of reaction with other nucleophiles should be enhanced, and this was found not to be so for the reactions of p-nitrophenyl-5-nitrosalicylate with sulphite, azide, and imidazole [28] and for phenyl salicylate with imidazole [20]. The solvent isotope effect, $k(H_2O)/k(D_2O) = 1.8$ at 59.2°C, and the ρ-value $(+0.80)$ for the hydrolysis of a series of substituted phenyl salicylates are consistent with a mechanism involving intramolecular general-base catalysis. The introduction of a second ortho-hydroxyl group, as in phenyl 2,6-dihydroxy-benzoate, does not lead to bifunctional catalysis [29].

The reaction of p-nitrophenyl 5-nitrosalicylate with ethanethiol also has a sigmoid pH-rate profile but at present it is not possible to decide whether the reaction involves intramolecular general-acid or general-base catalysis [30].

The reactions of salicylate esters in aprotic solvents are also intramolecularly catalysed. Thus the reactions of phenyl salicylate and phenyl o-methoxybenzoate with n-butylamine in acetonitrile are both second-order in amine but phenyl salicylate reacts 132 times faster [31]. These reactions are also catalysed by triethylenediamine and the rate constant for the reaction which is first-order in n-butylamine and first-order in triethylenediamine is over one hundred times greater for phenyl salicylate than for phenyl o-methoxybenzoate. The reaction of phenyl o-methoxybenzoate, but not that of phenyl salicylate, is catalysed by n-butylamine hydrochloride, which suggests that the phenolic hydroxyl group of the latter acts as an internal acid catalysis. Since the rate-limiting step in the aminolysis of esters in acetonitrile solutions is the breakdown of tetrahedral intermediate [32], a reasonable mechanism for the reaction of phenyl salicylate is one in which breakdown of the tetrahedral intermediate is catalysed intermolecularly by a second molecule of amine and intramolecularly by the phenolic hydroxyl group as shown in 25. The reaction of

25

methyl salicylate with n-butylamine in dioxan is also second-order in amine and is much faster than that of methyl p-hydroxybenzoate [33]. Presumably a similar mechanism is followed. In contrast the reaction of methyl salicylate with aniline in nitrobenzene is first-order in amine and occurs at the same rate as that of phenyl o-methoxy-benzoate and slightly more slowly than that of methyl p-hydroxy-benzoate [34].

11.3.3 Intramolecular catalysis in reactions of other esters with phenolic hydroxyl groups

Higuchi and his co-workers have shown that the pH-rate profile for the hydrolysis of the mono-acetate of hexachlorophene is sigmoid with the rate proportional to the concentration of the ionized form and that below pH 8 the reaction is 500 times faster than hydrolysis of the di-acetate. This indicates that the reaction is intramolecularly catalysed and again general-acid 26 or general-base 27 catalysis is possible. Less convincing is the claim that the hydrolysis of the monosuccinate is bifunctionally catalysed as symbolized by 28. The

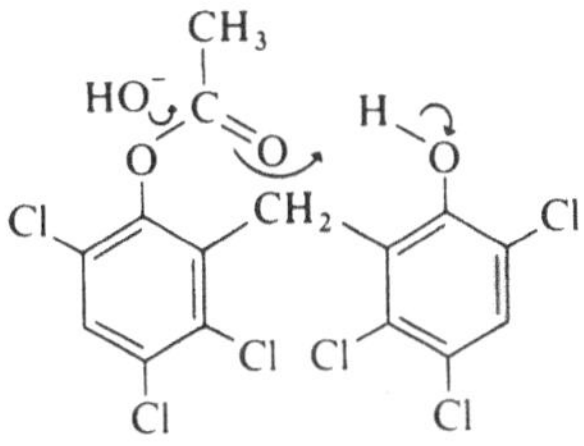

26

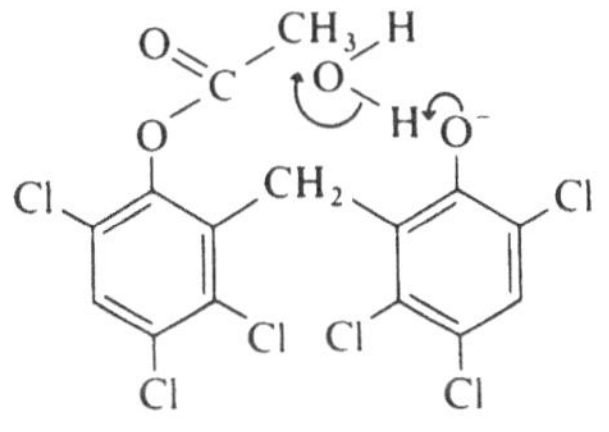

27

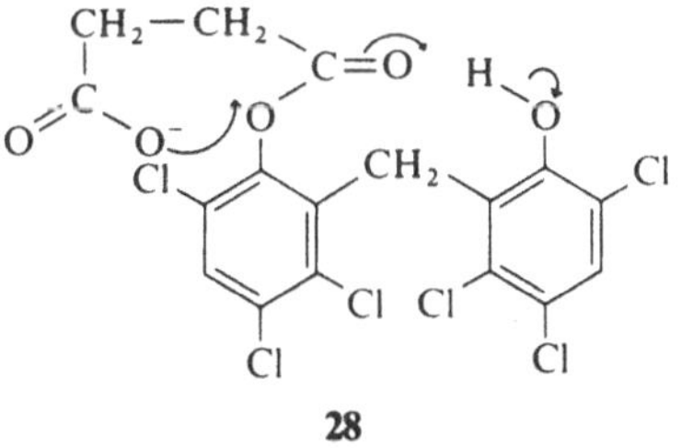

28

difference in rate between the forms with the phenolic group ionized and un-ionized is less than ten and this could arise from just an electronic effect [35].

11.3.4 *Intramolecular catalysis in reactions of amides and imides of salicylic acid*

In the pH-range 6 to 8 the hydrolysis of salicylamide is faster than that of benzamide and the pH-profile is sigmoid. The plateau rate constant is 17.5 times that for 5-nitrosalicylamide [36] but this result does not allow a distinction to be drawn between possible mechanisms which involve intramolecular general-acid and intramolecular general-base catalysis [20].

Imide **30** is an intermediate in the hydrolysis of ester **29** and its hydrolysis occurs over 1000 times faster than that of benzimide. The

pH-rate profile is sigmoid with an apparent pK_a of 6.08 which is close to the measured pK_a (6.38) [37]. Similar behaviour is found with the N-methyl compound [38]. Again, kinetically equivalent mechanisms involving intramolecular general-acid or general-base catalysis are possible but it is not possible at present to distinguish between them.

The *ortho*-hydroxy group of N-methylsalicylamide also catalyses exchange of the nitrogen-bound proton. The rate of this reaction can be measured by p.m.r. spectroscopy by studying the coalescence of the signal of the methyl group which under conditions of slow exchange is split by coupling to the nitrogen-bound proton. The nitrogen-bound proton of *o*-methoxy-N-methylbenzamide undergoes exchange 3 to 4 times more slowly than that of benzamide which itself reacts about 4 times more slowly than N-methylsalicylamide. All the reactions are first-order in hydroxide ion and the greater rate of reaction of N-methylsalicylamide may arise from processes symbolized by **31** or **32** [39].

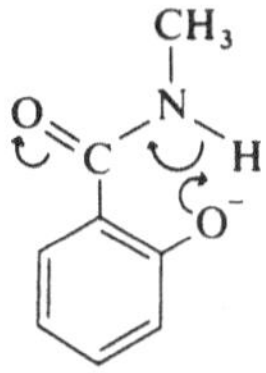

31 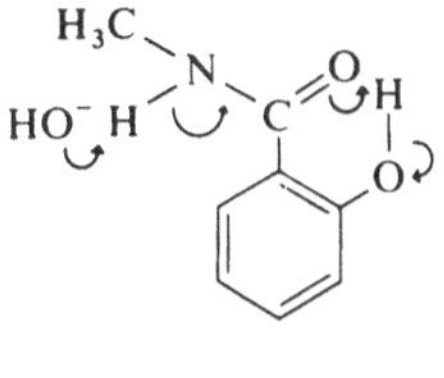   32

11.4 Intramolecular general-acid catalysis by alcoholic hydroxyl groups in reactions of esters

When an alcoholic hydroxyl group is in a suitable situation to hydrogen-bond to an ester group it may facilitate its hydrolysis or methanolysis. One of the most favourable stereochemical situations for an interaction of this type is when a hydroxyl and an acyloxy group have a 1,3-diaxial arrangement. The rapid solvolysis of an ester group in this situation was first demonstrated by Henbest and Lovell [40] who studied the cholesteryl and coprostanyl derivatives shown in Table 2 and determined the percentage reactions after

Table 2

Kinetics of solvolysis of some steroidal esters

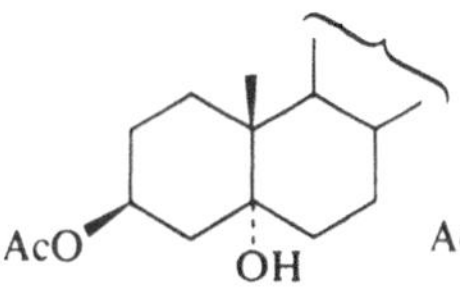

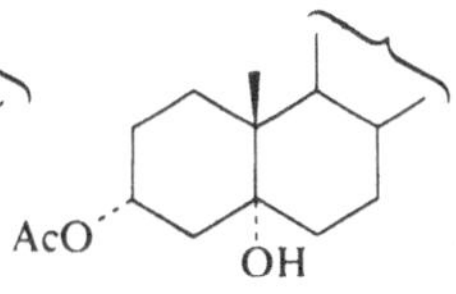

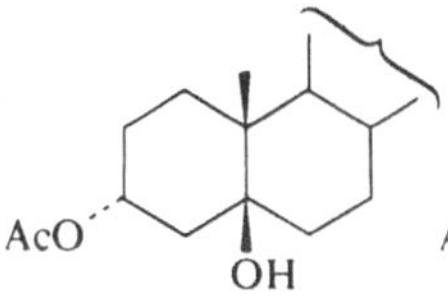

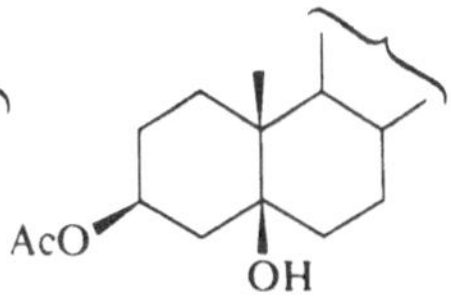

	33	34	35	36
a	18	70	13	78
b	16	100	12	180
c	—	—	—	300

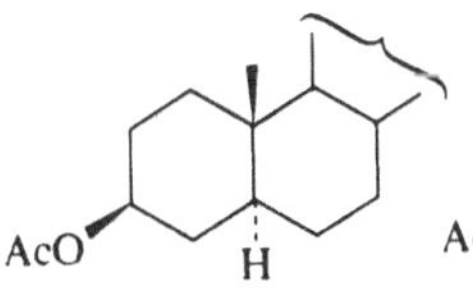

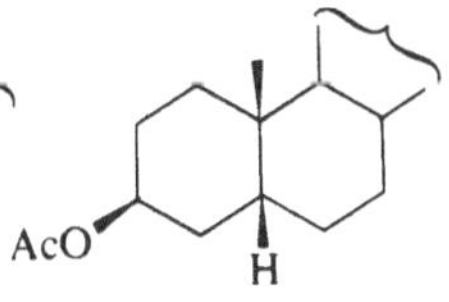

a	<2	—
b	<1	—
c	—	1

[a] Percentage hydrolysis after treating 60 mg with 7 mg of $KHCO_3$ in a mixture of benzene (1 ml), methanol (4 ml) and water (0.5 ml) for 65 hours at 20°C (ref. 40).

[b] Approximate relative rate constants based on *a*.

[c] Relative rate constants for hydrolysis in 3:1 triethylamine-triethylammonium buffer 0.1 *M* at 40°C; from reference 41.

353

treatment under a standard set of conditions as shown. This enables the approximate relative rate constants to be determined. Compound **34** and **36**, which have the acetoxy and hydroxyl groups in the 1,3-diaxial arrangement, react 10^2 to 10^3 times faster than compounds without the hydroxyl group, but all of this rate enhancement cannot be ascribed to intramolecular catalysis since the isomeric compounds **33** and **35**, which do not have a diaxial arrangement, also react faster. Since equatorial acetoxy-compounds normally react about 6 times faster than equatorial ones, correction of the relative rates of esters **33** and **35** by this factor suggests that the rate enhancement ascribable to intramolecular catalysis of compounds **34** and **36** is 40–60 fold.

Various mechanisms are possible for the intramolecular catalysis [43] and those generally favoured involve intramolecular general-acid catalysis with the hydroxyl group donating a proton to either the 'ether' or 'carbonyl' oxygen of the acetoxy group.

Similar rate enhancements have been reported for the solvolysis of esters of a number of other natural products [41, 44–48] and a rather smaller one for the hydroxysis of *exo*-2-acetoxy-*syn*-7-hydroxy-norbornane which is hydrolysed about 16 times faster than *exo*-2-acetoxynorbornane [48].

There is also some evidence for intramolecular catalysis in the hydrolysis of mono-esters of 1,2-dihydroxycyclopentane [48, 49, 50] and 3,4-dihydroxytetrahydrofuran [49] but the rate enhancements are small. Thus *trans*-1-acetoxy-2-hydroxycyclopentane is hydrolysed 5.5 times more rapidly than *trans*-1-acetoxy-2-methoxy-cyclopentane and *cis*-1-acetoxy-2-hydroxycyclopentane is hydrolysed slightly more rapidly still (1.7 times). It has been suggested that these small rate enhancements do not arise from intramolecular catalysis but from a 'microscopic medium' effect [49].

The hydrolysis of esters with the hydroxyl group in the acyl portion has been less widely studied [51, 52]. Here a facile intramolecular *exo*-cyclic displacement (see p. 12 of ref. [2]) is sometimes a possible alternative pathway to mechanisms which involve intramolecular general-acid or general-base catalysis.

The alkaline hydrolysis of ethyl *cis*-2-hydroxycyclopentanecarboxylate is 1.8 to 11 times faster than that of ethyl cyclopentanecarboxylate but this rate difference probably does not arise from intramolecular catalysis since the *trans*-isomer for which such catalysis is unlikely reacts even faster. The solvent dependence of the rates of both the *cis*- and *trans*-2-hydroxy esters in aqueous dioxan is much smaller than that for ethyl cyclopentane-2-carboxylate and this was tentatively attributed to solvent sorting [53].

11.5 Intramolecular catalysis by amino groups in reactions of esters and amides

A number of examples of intramolecular catalysis by amino groups in the hydrolysis and aminolysis of esters and amides have been discovered. These will now be reviewed and the difficulties in deciding whether the catalysis is nucleophilic or general-base catalysis will be emphasized.

11.5.1 Reactions of 2-(4-imidazolylphenyl) esters

Intramolecular catalysis in the hydrolysis of 2(4-imidazolyl)phenyl acetate was discovered by Schmir and Bruice [54]. The pH-rate profile is sigmoid with the rate proportional to the concentration of the unprotonated form and the rate enhancement is substantial, but not larger than a factor of 1000 [55]. Schmir and Bruice explained their results by a mechanism which involved intramolecular nucleophilic catalysis (equation 9) [54]. A ρ-value of 1.67 for the hydrolysis of a

$$(9)$$

series of 2-imidazolyl-phenyl benzoates was held to support a mechanism analogous to equation 9 [56]. However Bruice and Felton [57] later showed that this mechanism is unlikely since the solvent isotope effect for the hydrolysis of 2-(4-imidazolyl)phenyl acetate, $k(H_2O)/k(D_2O) = 3.23$, is much larger than would be expected for a rate-limiting nucleophilic attack by the imidazole group. Therefore on the basis of this result and the value of $T\Delta S^{\ddagger}$ (-10.8 kcal mol^{-1}) a mechanism involving intramolecular general-base catalysis 37 was preferred. Nevertheless it was realized later [55] that these results could also be accommodated by a mechanism which involves intramolecular nucleophilic catalysis if formation of the N-acetyl compound were a rapid and reversible process as shown in equation 10. Now $k_0 = K_1 \times k_2$, and k_2 should show a sizable isotope effect since the solvent isotope effect for the hydrolysis of N-acetylimidazole is probably quite large, and the overall entropy of activation might be strongly negative since the rate-determining (k_2) step is now bimolecular. Hence all the experimental results appear to be consistent

355

with either mechanism [37 or equation 10]. This situation is similar to that found with the hydrolysis of aspirin (see p. 362) and has been discussed by Kirby and Fersht [2]. In the hydrolysis of aspirin anion the concentration of the mixed anhydride is too low for the pathway which involves it as an intermediate to be followed. However the concentration of the N-acetylimidazole, 38, in equilibrium with

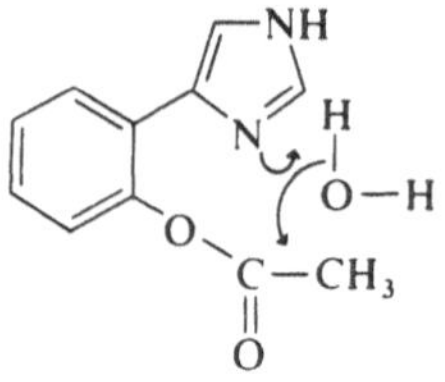

37

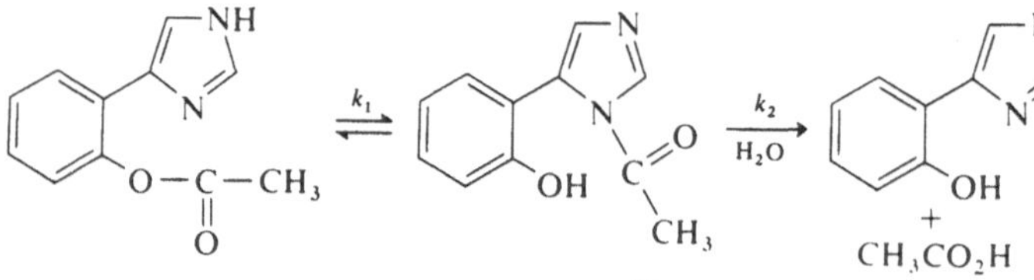

(10)

38

2-(4-imidazolyl)phenyl acetate is probably much higher and the equilibrium constant for its formation may be as great as 0.01 to 0.02 [2]. Since the rates of hydrolysis of N-acetylimidazole and 2-(4-imidazolyl)phenyl acetate are similar, this means that the pathway of equation 10 would be reasonable if the hydrolysis of 38 were enhanced by a factor of 50 to 100 by either intramolecular catalysis or the substituent effect of the 2-hydroxyphenyl group. There is no evidence to suggest whether this is possible, and the simplest explanation for the kinetics results is the mechanism symbolized by 37, but the mechanism of equation 10 has not been completely excluded.

Recently Rogers and Bruice have extended these investigations to the hydrolysis of 2-(2-imidazolyl)phenyl acetate [58].

The rate of reaction of 2-(4-imidazolyl)phenyl acetate with hydrazine is enhanced by a factor of about 20 and mechanisms analogous to 37 and equation 10 are possible [55].

11.5.2 Reactions of 8-quinolyl and 2-pyridyl esters

The pH-rate profile for the hydrolysis of 8-acetoxyquinoline is sigmoid between pH 2 and 8 with the rate proportional to the concentration of the unprotonated form [59, 60, 61, 55, 62]. At pH 5 the rate is about 500 times greater than that for 6-acetoxyquinoline [61,

55] and substantially greater than that for 7-acetoxyquinoline [60]. The electronic effects of the ring-nitrogen in the 6- and 8-positions should be similar and, in accordance with this, the 6-acetoxyquinoline reacts about 3 times faster with hydroxide ion than 8-acetoxyquinoline. Therefore the faster spontaneous hydrolysis of the latter must arise from intramolecular catalysis. The solvent isotope effect, $k(H_2O)/k(D_2O) = 2.35$ at 55°C, and $T\Delta S^{\ddagger}$ $(= -8.70 \text{ kcal mol}^{-1})$ are similar to these found for the hydrolysis of 2-(4-imidazolyl)phenyl acetate (see above) and so analogous mechanisms [39 and equation 11] are possible. Again the situation is similar to that found with aspirin and as pointed out by Kirby and Fersht if the reaction proceeds via an N-acetyl intermediate the rate law requires that this be species 40, which would be present at much lower concentration than 41. If it is

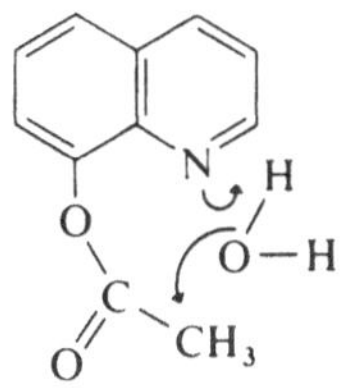

39

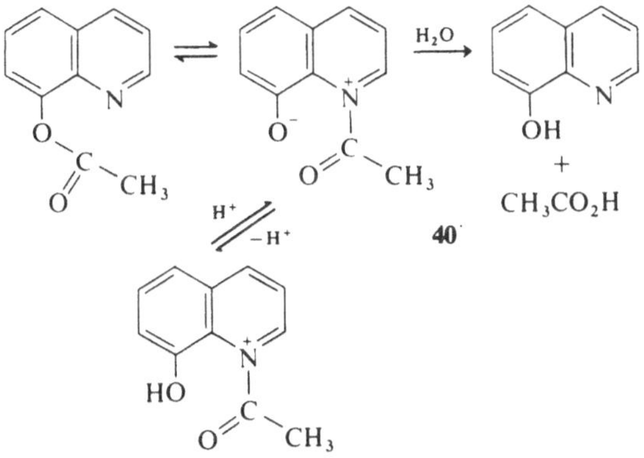

(11)

40

41

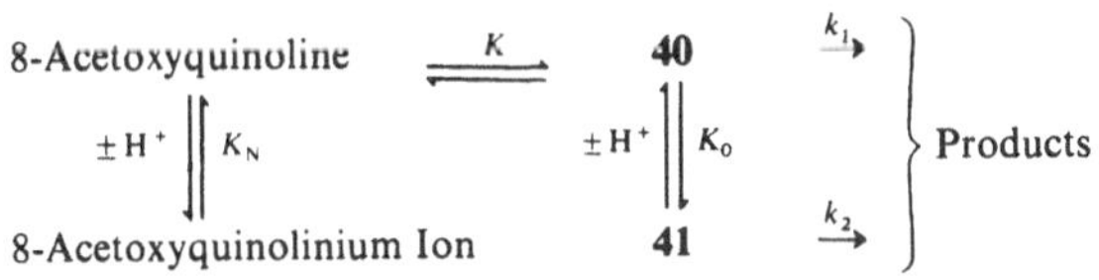

$$k_S = k_1 K$$

$$k_{SH} = k_2 K K_N / K_0$$

$$\therefore \quad k_S / k_{SH} = (k_1/k_2)/(K_N/K_0)$$

$$K_N/K_0 = 10^4 - 10^5 \text{ and if } k_S/k_{SH} > 10$$

$$k_1/k_2 > 10^5 - 10^6$$

Scheme 2

357

assumed that the rate constant for the hydrolysis of the unprotonated form of 8-acetoxyquinoline is at least 10 times greater than that for the protonated form) ($k_S/k_{SH} > 10$) then the kinetic analysis shown in Scheme 2 indicates that the ratio of the rate constants k_1/k_2 must be at least 10^5 to 10^6 which seems unreasonably large. Therefore a mechanism involving intramolecular general-base catalysis is most reasonable.

The rates of the second-order reactions of 8-acetoxyquinoline with primary and secondary amines are enhanced [63–65] but by much smaller factors (1.5 to 10) and these reactions may also involve intramolecular catalysis. 8-Quinolyl esters, like catechol esters, have been used in peptide synthesis [63–65].

Intramolecular catalysis also occurs in the aminolysis of 2-pyridyl acetate by n-butylamine in chlorobenzene [66]. The rate constant for the reaction which is first-order in amine is substantially enhanced and that for the reaction which is second-order in amine is enhanced slightly (see Table 3). Now intramolecular nucleophilic catalysis

Table 3

Rate constants for reactions with n-butylamine in chlorobenzene at 25°C

	$k_2 \, 1 \cdot \text{mol}^{-1} \, \text{s}^{-1}$	$k_3 (1^2 \, \text{mol}^{-2} \, \text{s}^{-2})$
4-Pyridyl acetate	0.0	4.0
2-Pyridyl acetate	2.5	24.0

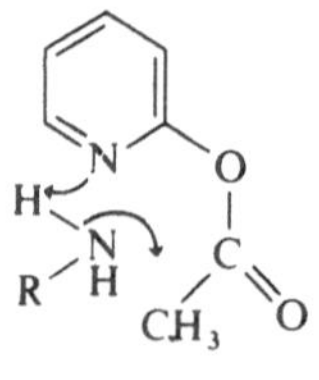

would involve reaction via a four-membered ring and therefore is less likely. A mechanism involving intramolecular general-base catalysis **42** seems most reasonable. 2-Pyridyl thio-esters have been used in peptide synthesis and give a product in which the amount of racemization is low [67].

42

11.5.3 *The reactions of diamines with amides and esters*

For the intramolecular catalysis in the aminolysis of N-acetyl-imidazole by diamines [68, 69] the catalytic group is in the same molecule as the nucleophile, and the problem of distinguishing between nucleophilic and general base catalysis does not arise. The rate law for the aminolysis of N-acetylimidazole by monoamines has both terms of first and of second order in amine, but in the aminolysis by ethylene diamine the term which is first order in amine is relatively more important. The rate constant for this reaction is 186 times greater than that for a monoamine of comparable basicity and a similar rate enhancement is found for the reaction with 1,3-diaminopropane. Rather smaller rate enhancements are found with 1,4-diaminobutane and 1,5-diaminopentane (see Table 4). Intra-molecular catalysis is also found in the aminolysis of methyl formate

Table 4

Rate enhancements and effective molarities in the reactions of diamines with acetyl imidazole and methyl formate

Amine	pK_a	Rate enhancement over RNH_2	Effective molarity
Acetylimidazole			
1,2-Diaminoethane	10.18	186	0.55
1,3-Diaminopropane	10.93	118	0.94
1,4-Diaminobutane	11.17	18	0.20
1,5-Diaminopentane	11.20	19	0.25
Methyl formate			
1,2-Diaminoethane	10.28	25	0.5
1,3-Diaminopropane	11.02	100	0.6

with 1,2-diaminoethane and 1,3-diaminopropane but not in the reaction of diamines with phenyl acetate [70, 71]. This behaviour presumably arises because the aminolysis of phenyl acetate is less dependent on general base catalysis (inter- or intra-molecular) than that of N-acetylimidazole and methyl formate which have poorer leaving groups. Although the rate enhancements arising from intra-molecular catalysis in the reactions of 1,2-diaminoethane and 1,3-diaminopropane are quite large, if the comparison is made with the uncatalysed aminolysis by a monoamine, the effective molarities obtained by comparison with an analogous intermolecularly catalysed aminolysis are small. As discussed on p. 379, this may arise from an unfavourable strain energy in the transition state for the intra-molecular reaction or because there is only a small entropy loss on forming the transition state in the intermolecular reaction.

The mechanism for intramolecular catalysis favoured by Page and Jencks involves proton removal from the reversibly formed tetra-hedral intermediate **43** to yield **44** which breaks down rapidly to products.

There also appears to be a slight rate enhancement in the reaction of the monoprotonated form of 1,3-diaminopropane with N-acetyl-imidazole. This may arise from an intramolecularly acid catalysed breakdown of the tetrahedral intermediate **45**.

The reaction of p-nitrophenyl acetate with 1,3-diaminopropane in chlorobenzene is intramolecularly catalysed [72]. The reaction of n-butylamine with p-nitrophenyl acetate in chlorobenzene is second order (k_3) in amine and no reaction which is first order in amine was detected. In contrast, the reaction with 1,3-diaminopropane has both terms of first and of second-order in amine, and the rate constant for the reaction which is first-order in amine is at least 300 times greater than that for the undetected reaction of n-butylamine, while the rate constant for the reaction which is second-order in amine is about 14 times greater than the analogous reaction of n-butylamine [72]. Since breakdown of the tetrahedral intermediate is rate-limiting in the aminolysis of aryl esters in aprotic solvents [32], the intramolecular catalysis presumably functions in this step, possibly as symbolized by **46**.

11.6 Intramolecular general-base catalysis by carboxylate groups in reactions of carboxylic esters

The most thoroughly studied reactions for which this type of behaviour has been proposed are the hydrolysis of aspirin and other O-acylsalicylic acids [73–78]. The pH-rate profiles for these reactions show that the anions react at enhanced rates, and a careful comparison of the rates of hydrolysis of the un-ionized forms with those of the corresponding methyl esters has shown that these are also enhanced, but usually to a smaller extent [79]. A possible mechanism for the reactions of the anions involves intramolecular nucleophilic catalysis with the anhydrides **47** as intermediates, and there is good evidence that such

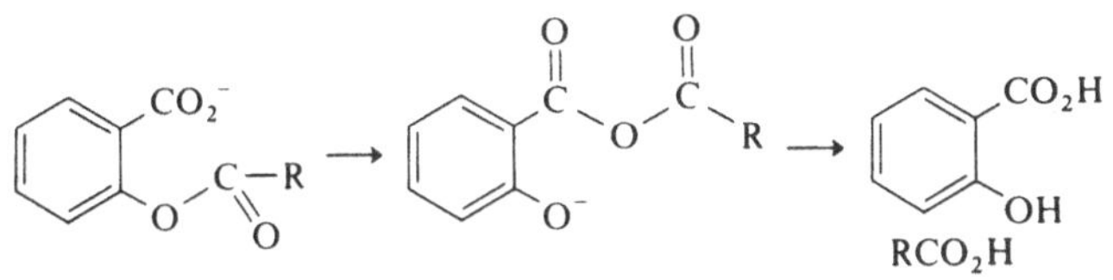

47

anhydrides are formed rapidly and reversibly from O-acyl salicylic acids. Thus Kemp and Thibault have shown that the labelled salicyl salicylate **48** undergoes equilibration of the label 10 000 times faster than it is hydrolysed at pH 8.2. This must involve formation of anhydride **49**, analogous to **47**, which may revert to salicyl salicylate

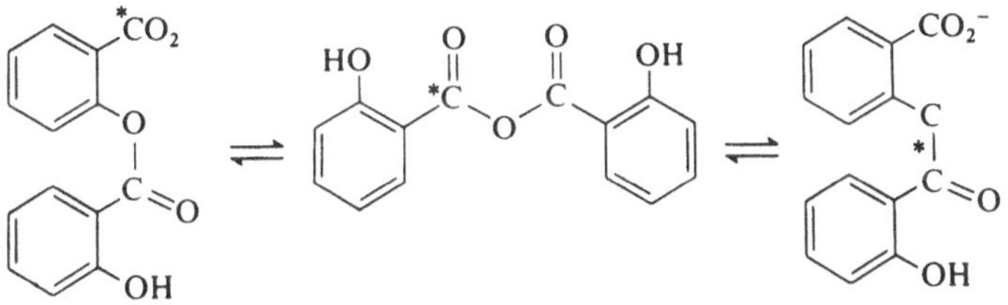

48   **49**

with the label in either the ester or in the carboxyl group [80]. Also, the mono-anion of 3-acetoxyphthalic acid **50** is hydrolysed 6000 times faster than the acetylsalicylate ion in a reaction in which anhydride **51** is an intermediate. Here the introduction of an internal carboxyl group provides a new efficient pathway for the decomposition of the mixed acetic salicylic anhydride [81].

So in the hydrolysis of O-acylsalicylates we have the familiar situation [82] of a substrate which is in rapid equilibrium with a second compound, and we have to decide whether the latter lies on the reaction pathway or whether the equilibrium is parasitic.

It appears that the only anion for which the pathway via the mixed acetic salicylic anhydride is important is that of 3,5-dinitroaspirin [83]. Here the rate enhancement is not large. The rate constant $(2.68 \times 10^{-2} \text{ min}^{-1})$ at 39°C is only 7.3 times greater than that for the hydrolysis of the methyl ester $(6.65 \times 10^{-5} \times 55.5 = 3.69 \times 10^{-3} \text{ min}^{-1})$ [83]; the corresponding ratio for the hydrolysis of aspirin itself is 175 [77, 79]. Hydrolysis of the anion of 3,5-dinitroaspirin in oxygen-18-enriched water leads to 39 per cent incorporation of the label into the product, 3,5-dinitrosalicylic acid, and solvolysis in 50 per cent v/v aqueous methanol yields 60 per cent of methyl 3,5-dinitrosalicylate [83]. These results can be explained only if the reaction proceeds, at least partially, via the mixed anhydride **52**. Three pieces of evidence

suggest that hydrolysis of this anhydride is the rate-limiting step. These are the solvent isotope effect, $k_{\text{H}}/k_{\text{D}} = 2.05$, $\Delta S^{\ddagger} = -22.6$ cal K^{-1}, and the observation that the rate is faster in aqueous methanol than in water.* There is some evidence that attack by

---

* It should be noted that similar values are found for the hydrolysis of aspirin and so cannot be used to exclude this mechanism for the latter reaction.

methanol and by water on the anhydride are intramolecularly catalysed, but quantitative estimation of the rate enhancement is difficult. Thus the reactions of the anion of 3,5-dinitroaspirin with anionic nucleophiles (for example acetate) are 3 to 7 times slower than with the methyl ester, and it could be argued [83] that this is because the reaction of the intermediate anhydride with water is intramolecularly catalysed whereas the reactions with the anionic nucleophiles are not. However this difference could also arise from an electrostatic effect in the reaction between two anionic species.* Also reaction of the mixed anhydride 53 in 50 per cent aqueous methanol yields only 15 per cent of methyl 2-methoxy-3,5-dinitrobenzoate compared to 60 per cent of methyl 3,5-dinitrosalicylate formed from 3,5-dinitroaspirin. It could be argued [83] that this arises from intramolecular catalysis 54, which is more efficient than intramolecular catalysis, 55. However, as the proportion of methyl acetate formed in the two reactions is not known, there is some ambiguity. The greater proportion of attack of methanol at the aroyl carbon could therefore arise either from a relatively greater reactivity of anhydride 52 at aroyl carbon, compared to acetyl carbon, or from its greater overall reactivity towards methanol compared to water than is the case for the uncharged anhydride 53.

54   55

Attempts to evaluate the rate enhancement arising from the presumed intramolecular catalysis lead to conflicting results. One way is to compare the rate constant for reaction of acetic 2-methoxy-3,5-dinitrobenzoic anhydride with acetate ($25.4 \, l \, mol^{-1} \, min^{-1}$ at 39°C) with that for reaction of 3,5-dinitroaspirin ($7.9 \times 10^{-3} \, l \, mol^{-1} \, min^{-1}$) with acetate. On the assumption that the latter reaction proceeds via a mixed anhydride and that this reacts at the same rate as acetic 2-methoxy-3,5-dinitrobenzoic anhydride with acetate, the concentration of the intermediate is calculated to be 0.03 per cent [83]. On this basis the rate of hydrolysis of the mixed anhydride 52 is 17

---

*It should be noted that the proposed attack of acylate ion at the salicoyl carbon, unlike attack of water, would not lead to hydrolysis and that attack at the salicyl carbonyl group would be more adversely affected by an electrostatic effect than attack at the acetyl group.

times faster than that of anhydride **53**. However, as shown below, this is probably an overestimate.

It has also been shown that hydrolysis of undissociated 3,5-dinitroaspirin proceeds via the mixed anhydride [79]. This reaction occurs about 30 times faster than that of the ionized form. That this reaction also proceeds via a mixed acetyl salicoyl anhydride was indicated by the observation that reaction in 50 per cent aqueous methanol yields about 12 per cent of methyl 3,5-dinitrosalicylate, which is identical with the percentage of methyl 2-methoxy-3,5-dinitrobenzoate formed from acetic 2-methoxy-3,5-dinitrobenzoic anhydride. The hydrolysis of 3,5-dinitroaspirin may therefore be formulated as shown in Scheme 3. As discussed by Fersht and Kirby [83], the rate constant for the

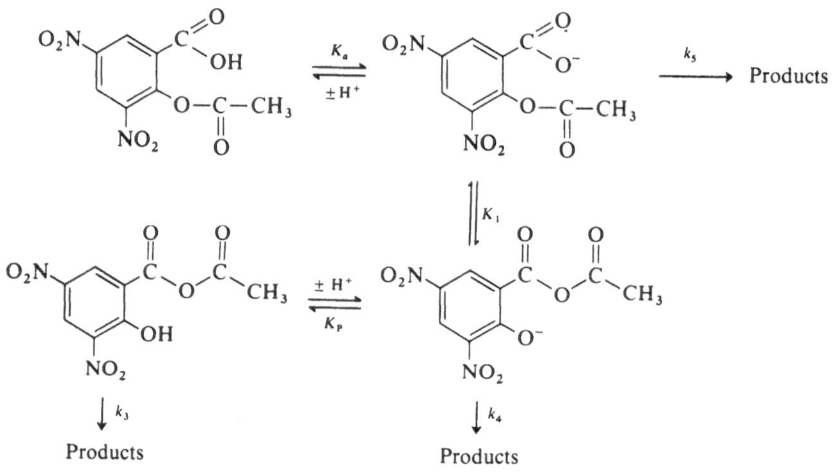

Scheme 3

hydrolysis of the ionized form [k(ionized)] may contain a contribution from k_5 so that its value sets an upper limit on the value of k_4.

$$k(\text{ionized}) \geqslant k_4 \times K_1$$

and

$$k(\text{un-ionized}) = k_3 K_1 K_a / K_P$$

$$\therefore \quad \frac{k(\text{ionized})}{k(\text{un-ionized})} \geqslant \frac{k_4}{k_3} \times \frac{K_a}{K_P}$$

The value of K_a was estimated by Fersht and Kirby to be 10^{-2} M and that of K_P to be between 10^{-3} and 10^{-4} M [83]. The latter seems reasonable since the dissociation constant of 2,4-dinitrophenol is 8.8×10^{-5} M at 28°C [84] and that of picric acid is 2.12 ($pK_a = 0.327$) [85]. Therefore $K_a/K_P \geqslant 10^{-2}$ and since $k(\text{ionized})/k(\text{un-ionized}) = 2.68 \times 10^{-2}/0.746 = 0.0359$.

$$0.0359 \geqslant \frac{k_4}{k_3}(\geqslant 10^{-2})$$

$$\therefore \quad k_4/k_3 \leqslant 3.59$$

It is therefore clear that any intramolecular catalysis of the hydrolysis of the ionized form of the mixed acetic salicylic anhydride can produce only a small rate enhancement. This is reasonable since the ratio in the rates of hydrolysis of the ionized to un-ionized forms of p-nitrophenyl 5-nitrosalicylate is about 30 [28], and a reduced catalytic effect would be expected with a less basic dinitrophenolate ion. It has recently been claimed, on the basis of the variation of $k(\text{ionized})$ with atom fraction of deuterium for hydrolysis in H_2O-D_2O mixtures, that the acetic salicylic anhydride reacts with 39 per cent attack at salicyl carbon and 61 per cent attack at acetyl carbon [86]. This conclusion does, however, depend on several unproven assumptions.

To decide the mechanism of hydrolysis of the ionized and un-ionized forms of aspirin and monosubstituted aspirins is more difficult because these lead to no incorporation of label into the product salicylic acid when reaction is carried out in ^{18}O-enriched water, and there is no formation of the methyl salicylate when reaction is carried out in aqueous methanol [77]. Therefore either the reaction does not proceed via a mixed anhydride or this anhydride reacts exclusively by attack at the acetyl group.

It is uncertain whether it is justifiable to extrapolate the conclusions drawn for the hydrolysis of the undissociated form of 3,5-dinitroaspirin to that for the other undissociated forms. Thus, if the effects of the 3- and 5-nitro-groups are taken to be additive, the calculated rate constant for hydrolysis of 3,5-dinitroaspirin is 0.128 min^{-1} at 39°C which is only about a sixth of the actual rate constant 0.746 min^{-1}. The corresponding figures for the ionized form as 9.79×10^{-3} and 2.68×10^{-2} with the ratio 1/2.7. In the latter case there is a change in mechanism on going from the mono-substituted to 3,5-dinitroaspirin, and hence it cannot be automatically assumed that there is no change with the un-ionized forms. This makes it difficult to come to a definite decision on the mechanism of hydrolysis of undissociated aspirin and its monosubstituted derivatives since most of the other evidence is ambiguous. Thus the solvent isotope effect, $k_H/k_D = 1.3$, and the small value of $\rho_{carboxy} \approx 0$ are very similar to those found with other reactions involving intramolecular general acid catalysis by an *ortho* carboxyl group [acetal (see p. 344) and phosphate hydrolysis (see p. 376)] and hence cannot be used to exclude a mechanism involving such a process. A possible argument in favour of a mechanism involving intramolecular general-acid catalysis is based on the report that un-ionized aspirin reacts with weakly basic nucleophiles but not with the strongly basic ones, methylamine and piperidine [87]. This could be explained if reaction proceeded as shown in **54**, since general acid catalysis might be necessary for attack of the weak nucleophiles but not for the strong ones. However, it might be difficult to detect a reaction between un-ionized aspirin and strongly basic nucleophiles since a low pH is required to obtain a reasonable concentration of the former and a high pH to obtain a reasonable concentration of the latter. On the assumption that a 10 per cent increase in k_{obs} over that for the spontaneous hydrolysis would have been observed, the upper limits for k_{cat} for methylamine and piperidine are $8.3 \, l \cdot mol^{-1} \, min^{-1}$ and $11.8 \, l \cdot mol^{-1} \, min^{-1}$ respectively. These are smaller than the observed k_{cat} for hydroxylamine which is $17 \, l \cdot mol^{-1} \, min^{-1}$. So whether these results can be used to exclude a mechanism involving a rapidly and reversibly formed anydride as an intermediate depends on whether it is reasonable for methylamine and piperidine to react 0.5 to 0.7 times as fast as hydroxylamine with it. It is known that the reaction of hydroxylamine is almost as fast or faster than that of some strongly basic amines with acetyl derivatives which have good leaving groups. For instance, hydroxylamine reacts faster than ethylamine and almost as fast as piperidine with 1-acetoxy-4-methoxypyridine [88]. So, if this were so for the mixed acetic salicylic anhydride and if there were an error in the estimated rate constants for reaction with methylamine and piperidine (perhaps a 20 per cent increase in rate would be needed for catalysis by methylamine or piperidine to be detected) the results would still be compatible with the intervention of this compound as an intermediate.

If therefore seems that, whereas the experimental evidence for the hydrolysis of the un-ionized form of 3,5-dinitroaspirin can only be explained by a mechanism involving the mixed anhydride as intermediate, that for the un-ionized forms of mono-substituted aspirins and aspirin itself admits both this mechanism and one involving intramolecular general-acid catalysis.

On the other hand, there does definitely seem to be a change in mechanism on going from the hydrolysis of the mono-ionized form of 3,5-dinitroaspirin to that of the ionized form of mono-substituted aspirins and aspirin itself. The hydrolysis of aspirin may be written as shown in Scheme 4. Since it is uncertain how much of the hydrolysis

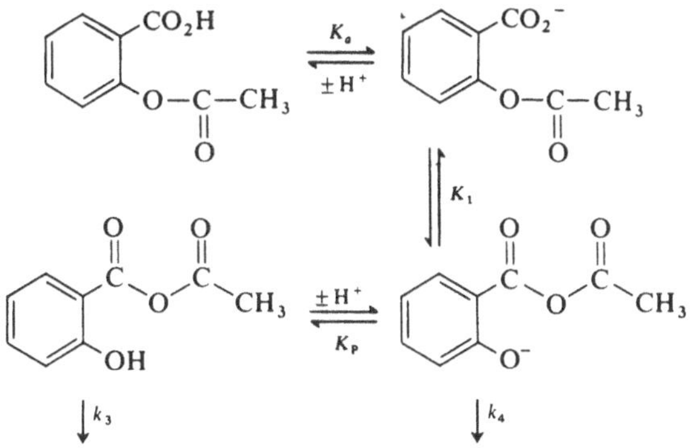

Scheme 4

of the un-ionized form proceeds via the anhydride, the measured rate constant sets an upper limit for the rate constant for reaction via this pathway.

$$k(\text{un-ionized}) \geqslant k_3 K_a K_1 / K_p$$

with $k(\text{un-ionized}) = 0.59 \times 10^{-4} \text{ min}^{-1}$ at 39°C. Let us denote the rate constant for the reaction of the ionized form which proceeds via the anhydride by $k(\text{ionized})^*$. Then

$$k(\text{ionized})^* = k_4 K_1$$

and

$$k(\text{un-ionized})/k(\text{ionized})^* \geqslant (k_3/k_4) \times (K_a/K_p)$$

and if one now assumes that all the reaction proceeds via this pathway $k(\text{ionized})^* = k(\text{ionized}) = 6.65 \times 10^{-4} \text{ min}^{-1}$ [77] and

$$\frac{0.59 \times 10^{-4}}{6.65 \times 10^{-4}} \geqslant \frac{k_3}{k_4} \times \frac{2.0 \times 10^{-4}}{1.6 \times 10^{-9}}$$

taking $pK_a = 3.69$ [78] and $pK_p = 8.80$ which is the pK_a of phenyl salicylate [20].

$$\therefore \quad k_4 > k_3 \times 1.4 \times 10^6$$

Hence the assumption that all the reaction of the ionized form of aspirin proceeds via the anhydride requires the ionized form of the anhydride to be hydrolysed more than 10^6 times faster than the un-ionized form. This seems to be an excessively large rate enhancement. It is conceivable that the hydrolysis of the ionized form could be intramolecularly catalysed and it is possible (see p. 378) that such catalysis would be more effective with an eight-membered cyclic transition state than with a six-membered one, but even so a rate enhancement of greater than 10^6 would be unprecedented for this kind of reaction (*cf.* ref. 86). Therefore a mechanism which involves intramolecular general base catalysis by the carboxylate groups seems to be the most reasonable. This mechanism is also consistent with the effect of substituents in the hydrolysis of substituted aspirins [77]. It has been claimed that the observation that the linearity of the plot of k(ionized) against atom fraction of deuterium for the hydrolysis in H_2O-D_2O mixtures is also evidence for this mechanism [89a], but it appears that more complex mechanisms could lead to linear plots as well [89b].

In contrast to the behaviour of aspirin, 2-carboxyphenyl mesitoate undergoes methanolysis in methanol which contains tris(hydroxymethyl)aminomethane to yield 100 per cent methyl salicylate. This reaction therefore proceeds via the mixed mesityl salicyl anhydride [90a].

Recently it has been shown by Kirby and Lloyd that the hydrolysis of the anion of 2-carboxyphenyl-4-hydroxybutyrate is 485 times faster than that of phenyl 4-hydroxybutyrate and hence it probably proceeds with concerted intramolecular general-base catalysis by the carboxylate group and intramolecular nucleophilic-catalysis by the hydroxyl group [90b].

11.7 Intramolecular catalysis in the mutarotation of aldoses

The mutarotation of α-D-glucose-6-phosphate **55a** was the first example of the mutarotation of an aldose for which convincing evidence for intramolecular catalysis was reported [91, 92]. This reaction occurs very rapidly and its rate was measured by studying the oxidation of the formed β-D-6-phosphate with glucose-6-phosphate-dehydrogenase. The physical property that was measured was thus the decrease in absorbance of the coenzyme, triphosphopyridine nucleotide at 340 nm. At 37°C and pH 7.0 the rate constant for conversion into the β-isomer was 240 times greater than that for the conversion of α-D-glucose into β-D-glucose. However, the effective molarity of the internal phosphate group was quite small, *ca.* 2M.

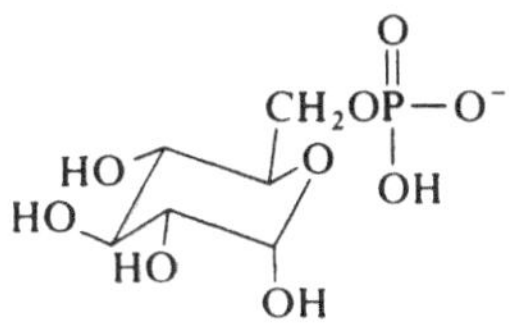

55a 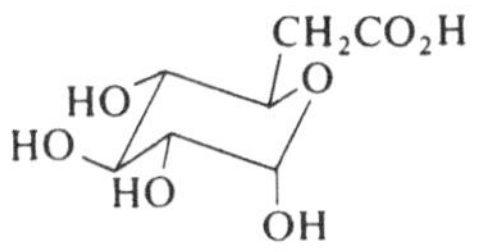   56 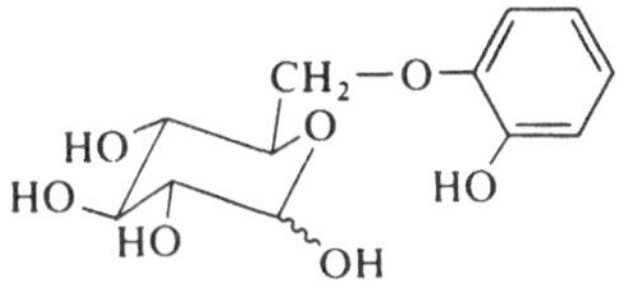   57

A more detailed investigation was carried out of the mutarotation of 6-deoxy-D-*gluco*hepturonic acid **56**. The pH-rate profile for this reaction is sigmoid and the rate constant for the mutarotation of the ionized form is about nine times greater than that for the un-ionized form and 8.3 times greater than that estimated from a linear-free-energy plot of the rate constants for the spontaneous mutarotation of some 6-substituted glucoses. Therefore the rate enhancement which can be ascribed to intramolecular catalysis is fairly small. A much larger effect is found in the mutarotation of **57**. The second-order rate constant for the mutarotation of the un-ionized form catalysed by hydroxide ion is 2700 times greater than that for the mutarotation of 6-O-phenyl-D-glucose. If this rapid hydroxide-ion-catalysed mutarotation is regarded as being the kinetically equivalent spontaneous mutarotation of the ionized form, the effective molarity of the internal phenolate group is 14M. However the most likely reaction mechanism for all three of these intramolecularly catalysed mutarotations appears to be one in which the catalytic group provides general-acid catalysis in a reaction of the form in which the hydroxyl group at C-1 is ionized (*e.g.* **58**) [93].

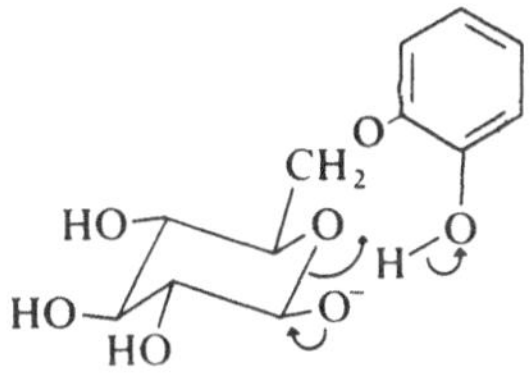

58

11.8 Intramolecular catalysis in enolization reactions and in ionization reactions of nitroalkanes

The enolization of ketones involves abstraction of a proton from the α-carbon and a number of examples in which this is facilitated by an internal acidic or basic group is known. Frequently the rate of enolization is determined by measuring the rate of loss of a small concentration of tri-iodide ion in the presence of a large concentration of

369

ketone under conditions where enolization is the rate-limiting step in the iodination. This method has a number of disadvantages; thus product analysis is difficult and it is susceptible to the presence of low concentrations of highly reactive impurities. Nevertheless it is the most widely used method and it has been used in most of the examples reported below.

11.8.1 Catalysis by carboxyl and carboxylate groups

Three examples of intramolecular catalysis in the enolization of aromatic ketones with o-carboxylate groups have been reported. One of the problems associated with studying these reactions is that the un-ionized forms exist mainly as cyclic tautomers which are not ketonic and therefore do not undergo enolization. Bell, Cox and Henshall studied the enolization (measured as iodination) of o-acetyl-benzoic acid and made allowance for the ring-chain tautomerism by determining the 'true' dissociation constant of the chain-form [94]. They found that the anion reacted about 300 times faster than the anion of p-carboxyacetophenone but the undissociated form did not appear to be especially reactive. The effective concentration of the internal carboxylate group was determined to be 5M by comparing the rate constant for the intramolecular reaction with that estimated for an analogous intermolecular reaction catalysed by a carboxylate ion of the same basic strength. The deuterium isotope effect for the intramolecular reaction was $k_H/k_D = 5.4$ which is substantially greater than that found for acetate and pyridine catalysis for which $k_H/k_D = 3.9$ and 4.1 respectively. Presumably the intramolecular reaction has a transition state as **59** and it was suggested 'that the higher value for the intramolecular process may depend upon a more favourable distance between the carbon and oxygen atoms between which the proton is transferred'.

59   **60**

Harper and Bender studied the iodination of o-isobutyrylbenzoic acid [95] but it is difficult [96] to interpret their pH-rate profile as they did not make allowance for any ring-chain tautomerism. Nevertheless it is clear that the anion reacts at a high rate which can be attributed to intramolecular catalysis and the effective molarity of the internal carboxylate group was calculated to be 56M, using the benzoate-catalysed deuterium-exchange of isobutyrophenone as the model intermolecular reaction. The larger factor found here compared to that found with o-acetylbenzoic acid may in part arise from the differences in the model intermolecular reaction, or more likely to a gem-dimethyl effect [97]. Possibly there is a release of steric strain on going to a cyclic transition state with the dimethyl compound which is not found with o-acetylbenzoic acid [98].

The ketone 60 is probably also enolized with intramolecular catalysis, as it undergoes a pH-independent racemization, the rate of which is not increased very much by added base [99]. This would involve a seven-membered cyclic transition state, but unfortunately it is not possible to estimate the rate enhancement quantitatively.

There is also evidence that intramolecular catalysis occurs in the enolization of aliphatic keto-acids. Thus Bell and Fluendy studied the iodination of the keto-acids, $CH_3 \cdot CO(CH_2)_nCO_2H$ and obtained the rate constants given in Table 5 [100]. The high values for the compounds with $n = 2$ and 3 suggest that they react with intramolecular catalysis and this was later [101] supported by the observation that the values of $k[A^-]$ are respectively 150 and 560 times greater than the rate constants for the iodination of the corresponding ethyl esters and that the values of $k[HA]$ are 7 and 100 times greater.

Table 5

Rate constants for iodination of a series of keto-acids $CH_3 \cdot (CH_2)_nCO_2H$ at 25°C

n	2	3	4	5	11
$10^8 k[A^-]^a$	29.8	179	7.2	3.4	7.2
$10^8 k[HA]^b$	1.4	28	6.6	—	—

[a] Rate constant for the spontaneous iodination of $CH_3 \cdot CO(CH_2)_nCO_2^-$: units: s^{-1}.
[b] Rate constant for the spontaneous iodination of $CH_3 \cdot CO(CH_2)_nCO_2H$; units: s^{-1}.

These results are best explained in terms of intramolecular catalysis as symbolized by 61 and 62.

61

62

Intramolecular catalysis also occurs in deprotonation (measured as iodination in 54 per cent aqueous t-butyl alcohol of the 4-nitro-valerate ion **63** which is 200 times faster than that of the un-ionized form. The primary deuterium isotope effect for this process is $k_H/k_D = 5.5 \pm 0.5$ [102].

The spontaneous enolization of acetoacetic acid (measured as bromination) is 60 times faster than that of ethyl acetoacetate and 10^5 to 10^6 times faster than that of levulinic acid but the acetoacetate anion is unreactive. Also the glycollate- and acetate-catalysed enolization of acetoacetic acid is 30 to 40 times faster than the ana-logously catalysed enolization of ethyl acetoacetate [103]. A mechanism as symbolized by **64**, where B is a water molecule or an acetate or a glycollate ion seems reasonable [101]. Similar rate enhancements were found in the enolization (studied as bromination) of **65** and **66** [104].

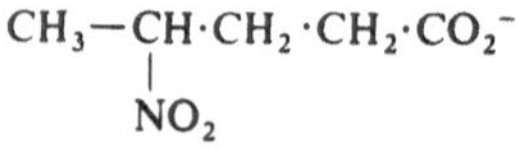

CH$_3$—CH·CH$_2$·CH$_2$·CO$_2^-$
|
NO$_2$

63

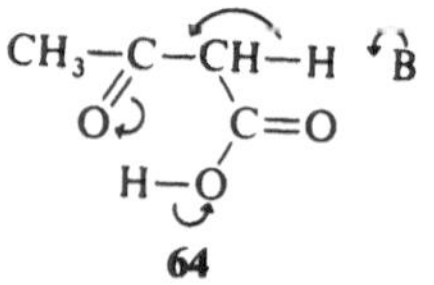

64

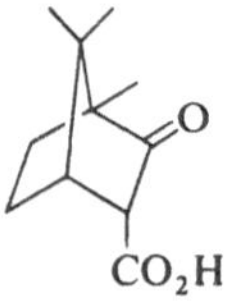

65

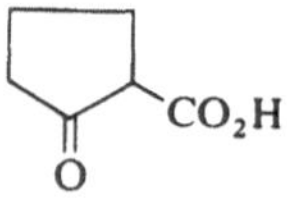

66

A particularly detailed investigation of the enolization of **67** (measured as iodination) has been reported by Bell and Page [104]. The pH-rate profile for this reaction has a sigmoid portion where the rate is proportional to the concentration of the un-ionized form. No analogous spontaneous reaction was detected in the enolization of the corresponding methyl ester. If this reaction were a specific hydrogen-ion catalysed reaction of the ionized form the rate constant for this process would be 180 times greater than that for the hydrogen-ion catalysed enolization of the un-ionized form and 230 times greater

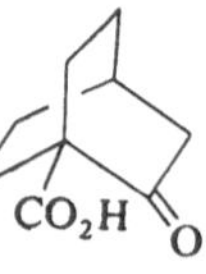

67

372

than that for the methyl ester. The acetate-catalysed enolization of the un-ionized form of **67** is also accelerated, since the catalytic constant is 22 times greater than that for the acetate-catalysed enolization of the methyl ester. Possible transition states for the spontaneous enolizations are **68** and **69**, with the rate enhancement arising respectively from intramolecular general-acid or electrostatic stabilization and that for the acetate-catalysed reaction could be similar with an acetate ion replacing the water molecule [104].

68   69

11.8.2 *Catalysis by amino groups*

Some work has to be done on the enolization of ketones with tertiary amino groups [105, 106]. The reactions of these compounds with iodine are complex processes which consume several moles of iodine per mole of ketone. Nevertheless, the pH-rate profile for enolization of 4-dimethylaminobutan-2-one appears to be sigmoid with the rare proportional to the concentration of the deprotonated form. The plateau rate constant and that for the enolization of 4-diethylamino-butanone and 5-ethylaminopentan-2-one are seven to eight powers of ten greater than the rate constants for the spontaneous enolization of simple aliphatic ketones. The results suggest that these compounds react with intramolecular catalysis. A mechanism as symbolized by **70** seems unlikely since the plateau rate constant for the enolization of **71** is similar to that for the enolization of **72** [105]. Instead, the kinetically equivalent process, involving reaction of the protonated form with hydroxide ion seems more likely. If this is correct, reaction via the six-membered cyclic transition state **73** is faster than via the

70   71   72   73

seven-membered one **74**. Despite the large rate enhancement calculated when comparison is made with uncatalysed reaction, the effective molarities obtained from comparison with an analogous inter-molecular reaction are quite small, in the range 5×10^{-2} to $1M$.

74

11.8.3 *Catalysis by amino-acids and by diamines*

In aqueous solution isobutyraldehyde-2-*d*, $Me_2CD \cdot CHO$, undergoes deuterium exchange in the presence of primary aliphatic amines by a third-order process which involves formation of the protonated Schiff base, followed by base catalysed de-deuteriation [107]. Attempts to detect an intramolecularly catalysed de-deuteriation, as symbolized by **75** in the presence of glycine, β-alanine, δ-aminovaleric acid, and ε-aminocaproic acid were unsuccessful [108]. Possibly this is because the Schiff base has the *trans*-configuration **76**. No intramolecular

75   **76**   **77**

catalysis was detectable in the de-deuteriation of $Me_2CD \cdot CHO$ when diamines, $Me_2N(CH_2)_2NH_2$, and their monoprotonated forms were used as catalysts, but $Me_2N(CH_2)_3NH_3^+$ is about 60 times more effective for the de-deuteriation of acetone-d_6 than expected from its basic strength and *cis-* and *trans*-2-(dimethylaminomethyl)cyclo-pentylamine are respectively about 35 to 150 and 17 to 75 times more effective. Here there is no *cis-trans* isomerism and intramolecular catalysis as symbolized by **77** is possible [109]. It appears that, with these compounds, formation of the imine is partly rate-limiting and that this step is itself intramolecularly catalysed (see p. 375). Small rate enhancements are found in the de-deuteriation of isobutyraldehyde-2-d catalysed by 1-dimethyl-amino-8-amino-2-octyne [110] and poly-ethylene imines [111, 112]. These may arise from intramolecular catalysis. With these amines intramolecular de-deuteriation is sterically possible even with the unfavourable *trans*-configuration of the Schiff base.

11.9 Intramolecular catalysis in the formation of imines

The best-authenticated examples of this are found in the work of Hine and his co-workers on the reaction of acetone with some dimethylamino-alkylamines. As mentioned above, there is evidence that intramolecular carbon-deprotonation sometimes occurs in the resulting iminium ions and formation of these ions is also sometimes intramolecularly catalysed. The largest rate enhancement is found with

$Me_2 \overset{+}{N}H \cdot CH_2 \cdot CH_2 \cdot NH_2$. The second-order constant for the formation of imine and for iminium ion from this species and acetone, as measured by trapping with hydroxylamine, is about 1000 times larger than expected from its pK_a value and the Brønsted plot for reactions of a series of mono-amines. Smaller rate enhancements are found with $Me_2 \overset{+}{N}H(CH_2)_3NH_2$ and with the analogous conjugate acid of *trans*-2-(dimethylaminomethyl)cyclopentylamine. These effects seem to be too large to be attributed to an activity-coefficient effect arising from the charge-difference and are most reasonably explained as arising from intramolecular catalysis. Since the rate limiting step is dehydration of the carbinolamine, intramolecular general-acid catalysis, as symbolized by **78**, would explain the results [113].

78

There have been claims [114] that the hydrolysis of imines derived from salicylaldehyde are intramolecularly catalysed but it seems that these are probably incorrect [115, 116].

11.10 Intramolecular general-acid catalysis in the hydrolysis of phosphate and sulphate esters

The most thoroughly investigated examples of this behaviour are found in the hydrolysis of salicyl phosphate and salicyl sulphate. The pH-rate profile for the hydrolysis of salicyl phosphate is bell-shaped with the reaction whose rate is proportional to the concentration of the di-ionized species being particularly rapid [117]. Thus the rate constant for this reaction is $0.370 \, h^{-1}$ at 37.2°C [117] compared to $0.147 \, h^{-1}$ for hydrolysis of the di-ionized form of p-carboxyphenyl phosphate at 80°C [118]. The most stable di-ionized form of salicyl phosphate is **79** but this is clearly not the reactive form, since salicyl phosphate is hydrolysed 10^5 to 10^6 times faster than the dianion of methyl salicyl phosphate [119]. This observation and the fact that the *ortho*-carboxyl group is necessary for the high reactivity suggest that the reactive species is **80**. The effect of substituents on the rate of

hydrolysis has been determined by Bromilow and Kirby and the results analysed using Jaffe's four-parameter equation to yield values for ρ_{phenol} and $\rho_{carboxy}$ of 1.74 and -0.99 respectively [119]. If, as discussed above, the stable form of the dianion is **79** and the reactive form **80**, the value of k(di-ionized) is equal to $K_1 \times k_2$ so that the values of ρ_{phenol} and $\rho_{carboxy}$ are the sums of the ρ values for the two steps, indicated by superscripts 1 and 2.

$$\rho_{phenol} = \rho^1_{phenol} + \rho^2_{phenol}$$

$$\rho_{carboxy} = \rho^1_{carboxy} + \rho^2_{carboxy}$$

Presumably ρ^1_{phenol} is small so that $\rho_{phenol} \approx \rho^2_{phenol}$, but $\rho^1_{carboxy}$ should be approximately -1 since K_1 is the equilibrium constant for a process which resembles the reverse of the ionization of the carboxyl group. Therefore $\rho^2_{carboxy} = ca \, 0$. Thus, as in the ionization of the analogous acetals (see p. 344), there appears to be little proton transfer in the transition state.

The requirement for the carboxyl group to lie in the plane of the aromatic ring in order to provide intramolecular catalysis appears to be more stringent here than in the hydrolysis of the acetals. Thus the dianion of salicyl phosphate is hydrolysed about thirty three times faster than that of 1-carboxy-2-naphthyl phosphate (at 37.2°C) [120], while 2-methoxymethoxybenzoic acid is hydrolysed only about six times faster than 2-methoxymethoxy-1-naphthoic acid (at 65°C) [121]. A larger steric effect still is found in the hydrolysis of 6-methylsalicyl phosphate whose dianion is hydrolysed at less than 1/200th the rate for salicyl phosphate [119].

The solvent isotope effect for the hydrolysis of the dianion is $k(H_2O)/k(D_2O) = 0.96$ [122], but this value is difficult to interpret since, as discussed above, the rate constants are complex quantities.

The most reasonable mechanism therefore seems to be that symbolized by **81** with a substantial amount of C—O bond breaking and relatively little proton transfer (see footnote on p. 346). This process yields metaphosphate ion which would be rapidly hydrated.

Intramolecular catalysis may occur in the hydrolysis of the dianion of 8-carboxy-1-naphthyl phosphate, but the rate enhancement is not as large as that found in the hydrolysis of salicyl phosphate. The rate constant is only about 7.5 times greater than that for hydrolysis of the dianion of m-carboxyphenyl phosphate [118]. This result is similar to that found with the analogous acetals (see p. 345) and presumably the smaller rate enhancement arises from the greater difficulty of proton transfer via a seven-membered ring in this stereochemical situation.

Intramolecular catalysis has also been detected in the hydrolysis of salicyl sulphate [123]. The pH-rate profile for this reaction has a sigmoid portion, with the reaction whose rate is proportional to the concentration of the form with an un-ionized carboxyl group being particularly rapid. At pH 3 the hydrolysis of salicyl sulphate is over 100 times faster than that of p-carboxyphenyl sulphate. The most reasonable explanation for these observations is that reaction proceeds with intramolecular catalysis, as symbolized by **82**, to yield

81

82

sulphur trioxide. This is consistent with the solvent isotope effect $k(H_2O)/k(D_2O) = 1.2$ and with the observation that the same ratio methyl sulphate/sulphate is obtained on solvolysis in methanol-water mixtures as is obtained in the acid-catalysed solvolysis of *p*-carboxy-phenyl sulphate [124]. Similar intramolecular catalysis occurs in the hydrolysis of 2[4(5)-imidazolyl]phenyl sulphate [125].

11.11 The effect of ring size on the rates of intramolecular proton transfers

The reactions described in the previous sections involve intramolecular proton transfers via cyclic transition states and the dependence of reaction rate on ring size differs from what is found in reactions which involve classical neighbouring-group participation, *i.e.* intramolecular nucleophilic displacements. This behaviour has been discussed recently by Hine and his co-workers [109] and by Schowen and Gandour [86, 126]. In some stereochemical situations the optimum ring size is eight. Presumably this is largely the result of a balance of enthalpy and entropy effects, as in intramolecular nucleophilic displacements [97]. The discussion of Hine and his co-workers on the intramolecular proton transfer in reactions of the Schiff bases derived from acetone and diamines is particularly clear [109]. In this initial state four of the atoms which are going to be part of the cyclic transition state are coplanar (see **83**) and three of these remain coplanar in

$$Me_2N-(CH_2)_{n-6}$$

83

the product. This and the requirement that the bond to the deuteron which is transferred lie in a perpendicular plane, places certain steric restrictions on the transition state, such that a linear $N-D-C$ bond and complete staggering of the methylene groups can be achieved only if the ring is nine-membered. In fact, maximum velocity is obtained when it is eight-membered. So presumably the additional loss of internal rotational entropy on having a nine-membered ring is less favourable than having a non-linear $N-D-C$ bond or some eclipsing in the cyclic transition state with eight members. However, in reactions where the developing bond is not part of the ring, such as **84** or **85**, a six-membered ring appears to be most highly favoured, although it is possible that these reactions involve a larger ring size in which a water molecule acts as a solvent bridge.

Gandour and Schowen have discussed ring-size in the presumed general-base-catalysed hydrolysis of the anion of the mixed acetic 3,5-dinitrosalicylic anhydride which is an intermediate in the hydrolysis of 3,5-dinitroaspirin [86, 126]. On the basis of model building, they suggested that intramolecular catalysis occurs via an eight-membered ring 86, but not via a six-membered one 87. Although their demonstration that an eight-membered cyclic transition state permits a linear proton transfer, whereas a six-membered one does not is undoubtedly correct, it is difficult to accept all their conclusions, since the evidence for intramolecular catalysis in the hydrolysis of the anhydride is not very strong (see p. 365), and it is not certain that hydrolysis of 3,5-dinitroaspirin proceeds completely via the anhydride [83]. Further, there is quite good evidence that the hydrolysis of esters of salicylic acid proceeds via a transition state analogous to (87) (see p. 350), and indeed it is not at all clear how stringent the requirement for a linear proton transfer is. Hine and his co-workers have suggested that 'The force constant for bending such a hydrogen bond should not be very large' [109].

In some situations intramolecular catalysis via a five-membered cyclic transition state appears to be possible. Thus Page and Jencks in their investigation of the reactions of N-acetylimidazole with diamines obtained maximum rate enhancements with 1,2-diamino-ethane and 1,3-triaminopropane, corresponding to five- and six-membered cyclic transition states if no solvent molecule is involved in the proton transfer [69] (see p. 359). The effective molarities for these reactions are small (ca 1M). This may arise because the transition states of the intramolecularly catalysed reactions have an unfavourable strain energy or because this transition state is 'loose' and is formed in the intermolecular reaction with only a small entropy loss, the loss of translational entropy being offset by entropy associated with internal rotation and other low-frequency motions in the transition state [69, 127]. In the latter situation the rate enhancement arising from intramolecularity will only be small. Presumably not every general-acid or general-base catalysed reaction has a 'loose' transition state since some intramolecularly catalysed reactions show high effective molarities (*cf. e.g.* p. 344).

REFERENCES

[1] M. L. Bender, *J. Amer. Chem. Soc.*, **79**, 1258 (1957).

[2] A. J. Kirby and A. R. Fersht, *Prog. Bioorg. Chem.*, **1**, 1 (1971).

[3] B. Capon, *Essays in Chemistry*, 3, 127 (1972).

[4] I. D. Sadekov, V. I. Minkin, and A. E. Lutskii, *Uspekhi Khim.*, **39**, 380 (1970); *Russian Chemical Reviews*, **39**, 179 (1970)

[5] B. Capon, *Tetrahedron Letters*, 911 (1963)

[6] B. Capon, M. C. Smith, E. Anderson, R. H. Dahm, and G. H. Sankey, *J. Chem. Soc. (B)*, 1038 (1969).

[7] D. Piszkiewicz and T. C. Bruice, *J. Amer. Chem. Soc.*, **90**, 2156 (1967).

[8] B. Capon, M. I. Page, and G. H. Sankey, *J. Chem. Soc. Perkin II*, 529 (1972).

[9] B. M. Dunn and T. C. Bruice, *J. Amer. Chem. Soc.*, **93**, 5725 (1971).

[10] T. H. Fife and E. Anderson, *J. Amer. Chem. Soc.*, **93**, 6610 (1971).

[11] G.-A. Craze and A. J. Kirby, *J. Chem. Soc. Perkin II*, 61 (1974).

[12] S. Grimvall and R. F. Wengelin, *J. Chem. Soc (A)*, 968 (1967).

[13] L. Manojlović and J. C. Speakman, *J. Chem. Soc. (A)*, 971 (1967).

[14] P. J. Wheatley, *J. Chem. Soc.*, 6036 (1964).

[15] B. Capon and M. I. Page, *J. Chem. Soc. Perkin II*, 2057 (1972).

[16] E. Anderson and T. H. Fife, *J. Amer. Chem. Soc.*, **95**, 6437 (1973).

[17] D. Piszkiewicz and T. C. Bruice, *J. Amer. Chem. Soc.*, **90**, 2156 (1968).

[18] R. L. Foster, Ph.D. Thesis, University of Leicester, 1969.

[19] B. Capon, S. T. McDowell, and W. V. Raftery, *J. Chem. Soc. Chem. Comm.*, 389 (1971); *J. Chem. Soc. Perkin II*, 1118 (1973); A. Tsuji, T. Yamana, and Y. Mizukami, *Chem. Pharm. Bull.*, **20**, 2528 (1972); J. E. C. Hutchins and T. H. Fife, *J. Amer. Chem. Soc.*, **95**, 2282 (1973).

[20] B. Capon and B. C. Ghosh, *Tetrahedron Letters*, 1707 (1964); *J. Chem. Soc. (B)*, 472 (1966).

[21] B. Hansen, *Acta Chem. Scand.*, **17**, 1375 (1963).

[22] Y. Shalitin and S. A. Bernhard, *J. Amer. Chem. Soc.*, **86**, 2291 (1964).

[23] L. E. Eberson and L.-Å. Svensson, *J. Amer. Chem. Soc.* **93**, 3827 (1971); *Acta Chem. Scand.*, **26**, 2631 (1972).

[24] R. Biggins and E. Haslam, *J. Chem. Soc.*, 6883 (1965).

[25] T. H. Fife and J. E. C. Hutchins, *J. Amer. Chem. Soc.*, **94**, 2837 (1972).

[26] J. G. Tillett and D. E. Wiggins, *Tetrahedron Letters*, 911 (1971).

[27] J. H. Jones and G. T. Young, *Chem. Comm.*, 35 (1967); *J. Chem. Soc.* (C) 436 (1968); Y. Trudelle, *J. Chem. Soc. Chem. Comm.*, 639 (1971).

[28] M. L. Bender, F. J. Kézdy, and B. Zerner, *J. Amer. Chem. Soc.*, **85**, 3017 (1963).

[29] T. Maugh and T. C. Bruice, *J. Amer. Chem. Soc.*, **93**, 3237 (1971).

[30] A. Williams, E. C. Lucas, and K. T. Douglas, *J. Chem. Soc. Perkin II*, 1493 (1972).

[31] F. M. Menger and J. H. Smith, *J. Amer. Chem. Soc.*, **91**, 5346 (1969).

[32] F. M. Menger and J. H. Smith, *J. Amer. Chem. Soc.*, **94**, 3824 (1972).

[33] R. L. Snell, W.-K. Kwok, and Y. Kim, *J. Amer. Chem. Soc.*, **89**, 6728 (1967).

[34] N. T. Vartak, N. L. Phalnikar, and B. V. Bhide, *J. Indian Chem. Soc.*, **24**, 131A (1947).

[35] T. Higuchi, H. Takechi, I. H. Pitman, and H. L. Fung, *J. Amer. Chem. Soc.*, **93**, 539 (1971).

[36] T. C. Bruice and D. W. Tanner, *J. Org. Chem.*, **30**, 1668 (1965).

[37] R. M. Topping and D. E. Tutt, *J. Chem. Soc.* (B), 1346 (1967).

[38] R. M. Topping and D. E. Tutt, *J. Chem. Soc.* (B), 104 (1969).

[39] F. M. Menger and G. Saito, *J. Amer. Chem. Soc.*, **95**, 6838 (1973).

[40] H. B. Henbest and B. J. Lovell, *Chem. and Ind.*, 278 (1956); *J. Chem. Soc.*, 1965 (1957).

[41] S. M. Kupchan, S. P. Eriksen, and M. Friedman, *J. Amer. Chem. Soc.*, **88**, 343 (1966).

[42] S. M. Kupchan, S. P. Eriksen, and M. Friedman, *J. Amer. Chem. Soc.*, **84**, 4159 (1962); see E. L. Eliel, N. L. Allinger, S. J. Angyal and G. A. Morrison, *Conformational Analysis*, Interscience, New York, 1965, pp. 73–74

[43] See B. Capon, *Quart. Rev.*, **18**, 59 (1964).

[44] R. West, J. J. Korst, and W. S. Johnson, *J. Org. Chem.*, **25**, 1976 (1960).

[45] S. M. Kupchan and W. S. Johnson, *J. Amer. Chem. Soc.*, **78**, 3864 (1956); S. M. Kupchan and W. S. Johnson, and S. Rajagopalan, *Tetrahedron*, **7**, 47 (1959).

[46] S. M. Kupchan, S. P. Eriksen, and Y. T. Shen, *J. Amer. Chem. Soc.*, **85**, 350 (1963).

[47] S. M. Kupchan, S. P. Eriksen, and Y.-T. S. Liang, *J. Amer. Chem. Soc.*, **88**, 347 (1966).

[48] S. M. Kupchan, J. H. Block, and A. C. Isenberg, *J. Amer. Chem. Soc.*, **89**, 1189 (1967).

[49] T. C. Bruice and T. H. Fife, *J. Amer. Chem. Soc.*, **84**, 1973 (1962).

[50] H. G. Zachau, and W. Karau, *Chem. Ber.*, **93**, 1830 (1960).

[51] M. J. Allen, *J. Chem. Soc.*, 4904 (1960); 4252 (1966).

[52] T. Yamanaka, A. Ichihara, K. Tanabe, and T. Matsumoto, *Tetrahedron*, **21**, 1031 (1965).

[53] B. Capon and M. I. Page, *J. Chem. Soc.* (B), 741 (1971).

[54] G. L. Schmir and T. C. Bruice, *J. Amer. Chem. Soc.*, **80**, 1173 (1958).

[55] S. M. Felton and T. C. Bruice, *J. Amer. Chem. Soc.*, **91**, 6721 (1969).

[56] U. K. Pandit and T. C. Bruice, *J. Amer. Chem. Soc.*, **82**, 3386 (1960).

[57] T. C. Bruice and S. M. Felton, *Chem. Comm.*, 907 (1968).

[58] G. A. Rogers and T. C. Bruice, *J. Amer. Chem. Soc.*, **96**, 2463, 2473 (1974).

[59] D. Elliot, L. C. Howick, B. G. Hudson, and W. K. Noyce, *Talanta*, **9**, 723 (1962).

[60] C. R. Wasmuth and H. Freiser, *Talanta*, **9**, 1059 (1962); R. H. Barca and H. Freiser, *J. Amer. Chem. Soc.*, **88**, 3744 (1966).

[61] T. C. Bruice and S. M. Felton, *J. Amer. Chem. Soc.*, **91**, 2799 (1969).

[62] E. J. Billo, R. P. Graham, and P. G. Calway, *Talanta*, **17**, 180 (1970).

[63] H.-D. Jakubke and A. Voigt, *Chem. Ber.*, **99**, 2419 (1966); H.-D. Jakubke, A. Voigt and S. Burkhardt, *Chem. Ber.*, **100**, 2367 (1967).

[64] J. H. Jones and G. T. Young, *Chem. Comm.*, 35 (1967).

[65] F. Weygand, A. Prox, and W. König, *Chem. Ber.*, **99**, 1451 (1966).

[66] F. Smith and J. W. Watson, *J. Chem. Soc. Chem. Comm.*, 786 (1969).

[67] K. Lloyd and G. T. Young, *Chem. Comm.*, 1400 (1968); *J. Chem. Soc.* (C), 2890 (1971).

[68] W. P. Jencks and K. Salvesen, *J. Chem. Soc. Chem. Comm.* (1970), 548.

[69] M. I. Page and W. P. Jencks, *J. Amer. Chem. Soc.*, **94**, 8818 (1972).

[70] R. W. Huffman, A. Donzel, and T. C. Bruice, *J. Org. Chem.*, **32**, 1973 (1967).

[71] R. F. Pratt and J. M. Lawlor, *Chem. Comm.*, 522 (1968).

[72] H. Anderson, C.-W. Su, and J. W. Watson, *J. Amer. Chem. Soc.*, **91**, 482 (1969).

[73] L. J. Edwards, *Trans. Faraday Soc.*, **46**, 723 (1950); **48**, 696 (1952).

[74] M. L. Bender, F. Chloupek, and M. C. Neveu, *J. Amer. Chem. Soc.*, **80**, 5384 (1958).

[75] E. R. Garrett, *J. Amer. Chem. Soc.*, **79**, 3401 (1957).

[76] E. R. Garrett, *J. Org. Chem.*, **26**, 3660 (1961).

[77] A. R. Fersht and A. J. Kirby, *J. Amer. Chem. Soc.*, **89**, 4853 (1967).

[78] A. R. Fersht and A. J. Kirby, *J. Amer. Chem. Soc.*, **89**, 4857 (1967).

[79] A. R. Fersht and A. J. Kirby, *J. Amer. Chem. Soc.*, **90**, 5826 (1968).

[80] D. S. Kemp and T. D. Thibault, *J. Amer. Chem. Soc.*, **90**, 7154 (1968).

[81] A. R. Fersht and A. J. Kirby, *J. Amer. Chem. Soc.*, **90**, 5833 (1968).

[82] D. S. Kemp, *J. Amer. Chem. Soc.*, **90**, 7153 (1968).

[83] A. R. Fersht and A. J. Kirby, *J. Amer. Chem. Soc.*, **90**, 5818 (1968).

[84] See G. Kortüm, W. Vogel, and K. Andrusson, *Dissociation Constants of Organic Acids in Aqueous Solution*, Butterworths, London (1961), 453.

[85] D. J. G. Ives and P. G. N. Moseley, *J. Chem. Soc.* (B), 757 (1966).

[86] R. D. Gandour and R. L. Schowen, *J. Amer. Chem. Soc.*, **96**, 2231 (1974).

[87] T. St. Pierre and W. P. Jencks, *J. Amer. Chem. Soc.*, **90**, 3817 (1968).

[88] W. P. Jencks and M. Gilchrist, *J. Amer. Chem. Soc.*, **90**, 2622 (1968).

[89a] S. S. Minor and R. L. Schowen, *J. Amer. Chem. Soc.*, **95**, 2279 (1973).

[89b] A. J. Kresge, *J. Amer. Chem. Soc.*, **95**, 3065 (1973).

[90a] H. D. Burrows and R. M. Topping, *J. Chem. Soc. Chem. Comm.*, 1389 (1970).

[90b] A. J. Kirby and G. J. Lloyd, *J. Chem. Soc. Perkin II*, 637 (1974).

[91] J. M. Bailey, P. H. Fishman, and P. G. Pentchev, *J. Biol. Chem.*, **243**, 4827 (1968).

[92] J. M. Bailey, P. H. Fishman, and P. G. Pentchev, *Biochemistry*, **9**, 1189 (1970).

[93] B. Capon and R. B. Walker, *J. Chem. Soc. Chem. Comm.*, 1323 (1971); *J. Chem. Soc. Perkin II*, 1600 (1974).

[94] R. P. Bell, B. G. Cox, and J. B. Henshall, *J. Chem. Soc. Perkin II*, 1232 (1972).

[95] E. T. Harper and M. L. Bender, *J. Amer. Chem. Soc.*, **87**, 5625 (1965).

[96] J. K. Coward and T. C. Bruice, *J. Amer. Chem. Soc.*, **91**, 5339 (1969).

[97] Cf. B. Capon, *Quart. Rev.*, **18**, 109 (1964).

[98] Cf. L. Eberson and H. Welinder, *J. Amer. Chem. Soc.*, **93**, 5821 (1971).

[99] C. Rappe and M. Bergarder, *Acta Chem. Scand.*, **23**, 214 (1969).

[100] R. P. Bell and M. A. D. Fluendy, *Trans. Faraday Soc.*, **59**, 1623 (1963).

[101] R. P. Bell and P. de Maria, *Trans. Faraday Soc.*, **66**, 930 (1970).

[102] H. Wilson and E. S. Lewis, *J. Amer. Chem. Soc.*, **94**, 2283 (1972).

[103] K. J. Pedersen, *J. Phys. Chem.*, **38**, 601, 999 (1934).

[104] R. P. Bell and M. I. Page, *J. Chem. Soc. Perkin II*, 1681 (1973).

[105] J. K. Coward and T. C. Bruice, *J. Amer. Chem. Soc.*, **91**, 5339 (1969).

[106] R. P. Bell and B. A. Timini, *J. Chem. Soc. Perkin II*, 1518 (1973).

[107] J. Hine, B. C. Menon, J. Mulders, and J. P. Idoux, *J. Org. Chem.*, **32**, 3850 (1967).

[108] J. Hine, B. C. Menon, J. Mulders, and R. L. Flachskam, *J. Org. Chem.*, **34**, 4083 (1969).

[109] J. Hine, M. S. Cholod, and J. H. Jensen, *J. Amer. Chem. Soc.*, **93**, 2321 (1971); J. Hine, M. S. Cholod and R. A. King, *ibid.*, **96**, 835 (1974).

[110] J. Hine, J. L. Lynn, J. H. Jensen, and F. C. Schmalstieg, *J. Amer. Chem. Soc.*, **95**, 1577 (1973).

[111] J. Hine, E. F. Glod, R. E. Notari, F. E. Rogers, and F. C. Schmalstieg, *J. Amer. Chem. Soc.*, **95**, 2537 (1973).

[112] J. Hine and R. L. Flachskam, *J. Org. Chem.*, **39**, 863 (1974).

[113] J. Hine, M. S. Cholod, and W. K. Chess, *J. Amer. Chem. Soc.*, **95**, 4270 (1973).

[114] R. L. Reeves, *J. Org. Chem.*, **30**, 3129 (1965); D. S. Auld and T. C. Bruice, *J. Amer. Chem. Soc.*, **89**, 2090, 2098, 4250, 4251 (1967); J. Hoffmann, J. Klicnar, V. Štěrba, and M. Večeřa, *Coll. Czech. Chem. Comm.*, **35**, 1387 (1970).

[115] W. Bruyneel, J. J. Charette, and E. de Hoffmann, *J. Amer. Chem. Soc.* **88**, 3808 (1966); E. de Hoffmann, *Coll. Czech. Chem. Comm.*, **36**, 4115 (1971).

[116] B. Capon, M. J. Perkins, and C. W. Rees, *Organic Reaction Mechanisms* (1965), 243–244; (1966), 318–319; (1967), 315.

[117] J. D. Chanley, E. M. Gindler, and H. Sobotka, *J. Amer. Chem. Soc.*, **74**, 4347 (1952).

[118] J. D. Chanley and E. Feageson, *J. Amer. Chem. Soc.*, **77**, 4002 (1955).

[119] R. H. Bromilow and A. J. Kirby, *J. Chem. Soc. Perkin II*, 149 (1972).

[120] J. D. Chanley and E. M. Gindler, *J. Amer. Chem. Soc.*, **75**, 4035 (1953).

[121] B. Capon, E. Anderson, N. S. Anderson, R. H. Dahm, and M. C. Smith, *J. Chem. Soc. (B)*, 1963 (1971).

[122] M. L. Bender and J. M. Lawlor, *J. Amer. Chem. Soc.*, **85**, 3010 (1963).

[123] S. J. Benkovic, *J. Amer. Chem. Soc.*, **88**, 5511 (1966).

[124] S. J. Benkovic and P. A. Benkovic, *J. Amer. Chem. Soc.*, **90**, 2646 (1966).

[125] S. J. Benkovic and L. K. Dunikoski, *Biochemistry*, **9**, 1390 (1970).

[126] R. D. Gandour, *Tetrahedron Letters*, 295 (1974).

[127] M. I. Page, *Chem. Soc. Rev.*, **2**, 295 (1973).

12

Myron L. Bender and Ferenc J. Kézdy

PROTON-TRANSFER IN
ENZYMATIC CATALYSIS

12.1 Introduction

Enzymatic catalysis has fascinated biochemists and physical and organic chemists alike for several generations. By their specificity and their catalytic efficiency enzymes are even today the paragons of homogeneous catalysis, especially when catalysis in aqueous media at neutral pH values is concerned. Thus, the mechanism of enzymatic catalysis is the subject of more intense study than ever before.

At the early stages of the study of enzymatic catalysis many fanciful mechanisms were proposed, each of which attributed to the catalyst some special, hitherto unobserved, chemical property, such as the ability to act at a distance, or the ability to form unorthodox chemical bonds. The number of theories accounting for the catalytic properties of enzymes was narrowed down considerably when it was demonstrated that all enzymes are proteins. Furthermore, when other chemical compounds, such as cofactors or metal ions are also required for catalysis, enzymatic action is still intimately related to the intact tertiary structure of the protein.

The next major step toward the understanding of the mechanism of enzymatic action came from the accumulation of kinetic and physico-chemical evidence showing that all enzymes form a complex with the substrate or substrates and that it is this complex alone which is able to yield reaction products. In such a complex the substrate should be able to interact only with a very limited number of chemical functions of the protein, all of them localized in a relatively small region of the protein molecule, the 'active centre' or 'active site' of the enzyme. The requirement for complex formation, together with the extreme sensitivity of the enzymatic reaction rate to even slight structural alterations of the substrate or of the protein molecule immediately pointed to proximity and orientation effects as the major factors responsible for specificity and catalytic efficiency. The importance of these factors is reaffirmed over and over again with each new enzyme studied.

The search for other catalytic factors common to all enzyme-catalysed reactions continued. Medium effects and a special micro-environment of the active site were found to contribute somehow, but abundant data in organic model systems showed that the magnitude of the rate-accelerating effect of these factors must be rather modest. In the same way, electrostatic catalysis was not found to be all-important in enzymatic catalysis. In non-enzymatic reactions the best and most efficient catalytic processes involve covalent catalysis, *i.e.*, the distribution of the high activation energy of the uncatalyzed reaction into several low-energy steps through the formation or one or more unstable intermediates between the substrate and the catalyst. If enzymatic catalysis obeys the same principles, then the amino-acid side chains of the protein and perhaps even the peptide backbone should participate in the chemical events of catalysis. The unity of organic and enzymatic catalytic reactions was indeed established when it was found that enzymes can and do form covalent intermediates with their substrates and that the formation and de-composition of these intermediates occur in full agreement with the reactivities displayed by the same chemical functions involved in organic model reactions. For example, in the catalysis of decarboxyla-tion by acetoacetate decarboxylase, a primary amine of a lysine side chain forms a Schiff base with the ketone function of the subs-trate, or in the case of the whole class of proteases, which cleave peptide bonds, the alcohol group of a serine side chain acts as an acyl-acceptor, forming an ester intermediate in the hydrolysis of an amide.

Unfortunately, in proteins the number of reactive amino-acid side-chain functions is extremely limited and the peptide chain is notoriously inert toward most reagents. At best, the alcohols, thiols, phenols, amines and carboxylates of the protein could act as nucleo-philes toward appropriate electrophilic centres of the substrate. Thus covalent bond formation between enzyme and substrate is limited in scope, and, to make the rather poor nucleophiles more efficient, further activating processes are needed. This activating action was found to occur in the numerous acid and base functions carried by all proteins. The types of functions found in proteins are listed in Table 1, together with the pK values which they normally display when attached to intact proteins. The ionizable amino-acid side-chain groups can act either as proton donors or proton acceptors, and they are, as a rule, well exposed to the solvent, the pH of the medium determining their state of ionization.

In addition to the readily ionizable amino-acid side chains listed in Table 1, there are several amino-acid moieties in enzymes, which do not contain ionizable hydrogens. These include amino acids such as alanine, valine, and phenylalanine among others, that contain only carbon-hydrogen bonds in either aliphatic or aromatic side chains, and thus do not engage in proton transfer except in very unusual circumstances. Proton transfers in enzymatic reactions can also occur from co-enzymes (co-factors) which can be considered as tightly bound substrates. Typical examples of co-enzymes that are involved in proton transfer are pyridoxal phosphate and thiamine pyrophosphate.

The first indication of the importance of proton-transfer reactions in enzyme catalysis come from the observation that the rate of most enzyme-catalysed reactions displays a relatively simply, sigmoidal or bell-shaped pH dependence. Thus enzymatic reactions require a small number of acids in a definite state of ionization. Later mechanistic studies indeed confirmed that in many cases these acids and bases – usually identifiable from the pK values of the pH-rate profile – act as proton donors and proton acceptors in the rate-limiting step of the catalytic process. Since in biological systems enzymatic reactions occur almost invariably near neutrality, where oxonium and hydroxide ion concentrations are at a minimum, it is not surprizing to find that enzymes make extensive use of general acid and general base catalysis.

The near-neutral pH of the reaction medium imposes a severe limitation on the pK of the acidic function which can act as an efficient general acid/general base. This is a direct consequence of the inverse relationship which exists between the catalytic efficiency and the ionizability of the catalyst. In the following derivation this relationship will be demonstrated for the case of general base catalysis only; the same considerations can also be used for general acid catalysis.

The catalytic rate constants (k_b) for general base catalysis obey the equation

$$k_b = G_b(1/K_{gb})^\beta$$

or

$$\log k_b = A + \beta p K_{gb} \tag{1}$$

where $0 < \beta < 1$.

The pK_{gb} of the catalyst also determines it degree of ionization and hence the concentration of the catalytically active species (C_B), according to the equation

$$C_B = \frac{C_0}{1 + [H^+]/K_{gb}} \tag{2}$$

Thus the experimentally observable, pH-dependent catalytic rate constant $(k' = k_b \cdot C_B)$ at a given analytical concentration of the catalyst (C_0) can be expressed by combining equations 1 and 2:

$$\log k' = A + \log C_0 + \beta pK_{gb} - \log(1 + [H^+]/K_{gb}) \tag{3}$$

or

$$k' = gbC_0 \frac{K_{gb}^{-\beta}}{1 + [H^+]/K_{gb}} \tag{4}$$

For any given pH and β there is one value of $K_{gb}(=K_{max})$ for which k' is at a maximum. K_{max} can be found by differentiating equation 4 with respect to K_{gb} and solving the resulting equation for the value of K_{gb} which yields $(dk'/dK_{gb}) = 0$. In this way we find:

$$K_{max} = [H^+]\frac{1-\beta}{\beta}, \text{ for } \beta < 1 \tag{5}$$

i.e., the optimal general base catalysis occurs when the pK_{gb} value of the catalyst is as close as possible to the pH value of the reaction medium. These considerations are of value for $0.2 < \beta < 0.8$, the usually observed range for general base catalysis.

Figure 1 further illustrates the occurrence of maximal catalysis when $pK_{gb} \simeq pH$. The theoretical $\log k'$ vs. pK_{gb} curves shown were calculated from equation 3, with $pH = 7$ and $A = -3$. When β varies from 0.3 to 0.7, a shift of less than $0.5 pK$ unit occurs in the value of pK_{max} on either side of the vertical line representing $pK_{gb} = pH$.

Two limiting cases are interesting, namely $\beta = 0$ and $\beta = 1$. For $\beta = 0$, equation 4 predicts that the prototropic solvent will be the most efficient catalytic species, since it is the species present in the highest concentration in the basic form. For $\beta = 1$, all bases with $pK_{gb} > pH$ will be equally efficient. Again, the solvent is present in the highest concentration and the fastest reaction is specific base catalysis by the lyate ion [1].

In the light of the preceding considerations, the approximately neutral pH of biological media should restrict general acid-general base catalysis in enzymatic reactions to the amino-acid side chains

388

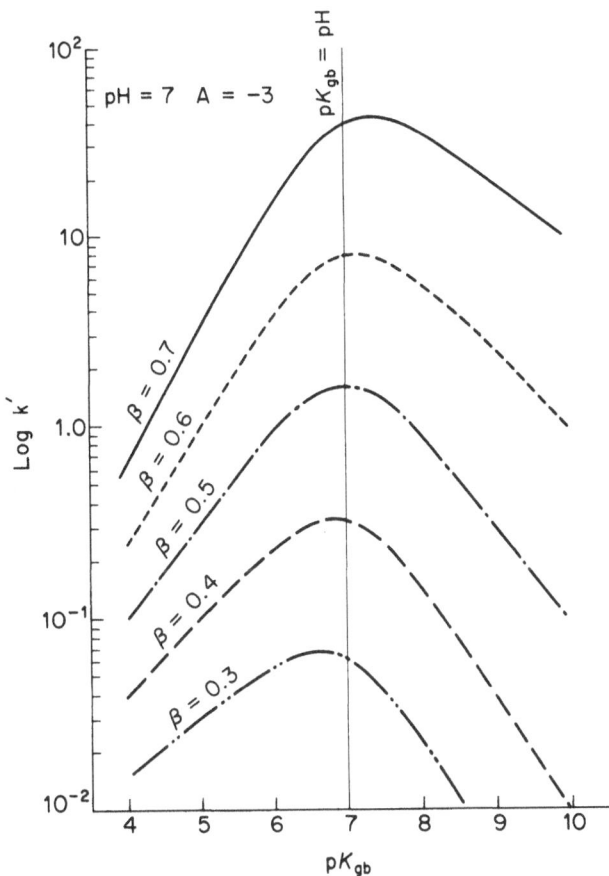

Figure 1. The dependence of the rate constant of a general-base catalysed reaction at pH = 7 on the pK_{gb} of the catalyst, according to equation 3.

possessing pK values between 4 and 10. This was indeed found to be the case; the groups most often occurring in enzymes as general acid-general base catalysts are the imidazole group of histidine, and the carboxylate ions of aspartic or glutamic acids. In some enzymes such as acetoacetate decarboxylase, the ε-amino group of lysine acts as a general acid-general base catalyst, but even in these cases the pK value of the catalytic amino group is lowered from the expected pK = 10.8 to a value close to 7, presumably because of some factor in the environment.

As a corollary, enzymes acting in acidic media, *e.g.* acidic proteases, would be expected to use general base catalysts of pK values in the acidic range. Carboxylates or imidazoles of slightly lower-than-usual pK were indeed found to act as catalysts in these enzymes.

The neutrality of the reaction medium and the pK_{gb} value of the catalysing base imposes some restrictions on the choice of the nucleophile too. At present, no quantitative theory exists relating the reactivity of the nucleophile to the efficiency of general base catalysis for the same nucleophile. It is well established, however, that increasing the reactivity of the nucleophile (n_i) decreases the sensitivity of the nucleophilic reaction to general base catalysis (β_i), according to the equation

$$n_i C_5 = \beta_0 - \beta_i \qquad (6)$$

where C_5 and β_0 are constants [2].

On the other hand, n_i is grossly proportional to the pK_N value of the protonated nucleophile:

$$n_i = \gamma p K_N \qquad (7)$$

although the proportionality factor (γ) decreases toward the region of high pK_N values [3]. On the basis of equations 3, 6 and 7 one is then able to predict, that toward a given electrophilic substrate, and for the same general base catalyst, an increase in the pK_N value of the nucleophile will first result in an increase in the rate of the reaction. At high pK_N values, however, the decrease of β_i, combined with a decreasing value of γ, renders the general base-catalyzed reaction less and less efficient. Thus, for a given substrate and a given general base, a pK_N value must exist which results in maximal catalytic rates.

Experimentally, it was found that for serine proteases, which act in the pH range of 7 to 8, the general base catalyst is imidazole with $pK = 7$. The serine alcohol has an estimated $pK_N = 12.5$, and with this value we calculate $\Delta pK = pK$ (acceptor) $- pK$ (donor) $= -5.5$ pH units. In the same way, for acidic sulfhydryl proteases $(pK_N = 8.5)$, the general base has $pK_{gb} = 3.5$, and thus again $\Delta pK = -5$ pH units. It would thus appear that $\Delta pK = -5$ is a suitable value for maximizing the rates of general base-catalysed enzymatic amide hydrolyses. That this choice of nucleophile and catalyst is not excessively sensitive to the nature of the leaving group of the substrate is shown by the fact that serine proteases are excellent catalysts of the hydrolysis of a variety of acyl derivatives with a wide range of leaving groups.

The preceding analysis concerning the acidity of nucleophiles and acid-base catalysts in enzymes shows that for the enzymatic catalysis of a given type of reaction not all acidic functions of the protein are appropriate. In fact, the choice in each pH range is limited to perhaps not more than two different amino-acid side chains and for this reason the mechanistic diversity found in a class of enzymes catalysing the same type of reaction is extremely limited. In dwelling on this aspect of structure-function relationships in enzymatic reactions our intention was certainly not to show the power of teleological argument. Rather, we hoped to illustrate the prominent place occupied by acid-base catalysis in enzymic processes and the less importance of most other mechanistic details with respect to proton transfer.

A survey of reactions catalysed by enzymes reveals that probably all types of proton transfers observed in organic reactions also occur in enzymic reactions. The transfer can occur to or from oxygen, nitrogen, sulfur or carbon and it may consist of a simple acid-base reaction – as in the case of epimerases – or it may be intimately coupled with other bond-making, bond-breaking or electron transfer processes. The donors or acceptors can be linked covalently to the peptide chain of the enzyme, or they can be situated on substrate or cofactor molecules. Finally, hydrogen can also be transferred enzymatically by mechanisms involving hydride ions or hydrogen atoms.

Rather than attempting an exhaustive review of all possible enzymatic proton transfers, which are manifold, we would like to discuss in the following a few well-documented enzymatic mechanisms involving different types of prototropic reactions. Of these, enzymatic amide- and ester-hydrolyses will be treated the most thoroughly, since at the present time they are the most extensively studied and their corresponding non-enzymatic model systems provide the most solid foundation for discussion. These examples will be preceded by a review of experimental techniques often utilized in mechanistic investigations of enzymatic reactions.

Excellent reviews exist on several important aspects of enzymatic proton transfers. For the enzymology of carbon-acid reactions the reader is referred to the chapter by Rose in *The Enzymes* [4]; an interesting discussion of proton transfers in biological *redox* reactions can be found in Hamilton's review [5].

12.2 *Experimental methods*

The methods generally applied to the detection, identification and the mechanistic study of enzymatic proton transfers are essentially the same as the ones developed for the study of organic reactions. Their use for the study of enzymatic processes is not without difficulties, however, and adaptations are often required to contend with the structural complexity of the protein catalyst. A complete three-dimensional structure is still not available for most enzymes and thus the identification of the protein side chains involved in catalysis has to rely heavily on active-site directed chemical modification methods. The use of physical methods is further complicated by the fact that the physical properties of the catalytically active amino-acid side chains are in general indistinguishable from those of some twenty or thirty identical side chains located in the same protein molecule. Thus the use of spectroscopic and magnetic resonance methods has met most success in cases where groups of unique physical properties were present in the cofactor or substrate, or when such groups were introduced at the active site by chemical or isotope labelling. Carbon[13] nuclear magnetic resonance spectroscopy appears to be a promising tool for the study of the state of protonation of enzymatic catalytic groups. It has been applied recently [6] to the determination of the ionization of an imidazole residue at the active site of α-lytic protease, by using histidine enriched in ^{13}C at position 2 of the imidazole ring. The pH-dependence of the chemical shift and of the coupling constant of such a system should in principle allow one to determine the state of protonation of the imidazole, provided that the results are compared for interpretation with those of an appropriate model system. Since the choice of the latter is not trivial, the method is still not absolute and the interpretation somewhat debatable.

The rate of the enzyme-catalysed reaction is an accurate measure of the interaction of the substrate with only the active-site moiety of the protein and for this reason a variety of kinetic methods occupy a place of choice among the chemical methods.

The pH-rate profile of the enzymatic reaction yields directly the pK of the acids and bases most likely to be involved in the catalytic process. The known pK values of ionizable groups of proteins (Table 1) then allow one to identify the particular catalytic groups. The imidazole group is often involved in enzymatic proton transfer and a host of pH-rate constant profiles show an inflection point at about pH 7, the pK of the histidine group at the active site of many enzymes. The pK value from the pH-rate profile does not necessarily identify the catalytic function and in most cases further methods, such as selective chemical modification, are necessary to eliminate the possibility of abnormal pK values. It has been proposed for example, that instead of a carboxylate, an imidazole of pK = 4 would act as a general base catalyst at the active site of the sulfhydryl protease papain [7]. Although the results of the X-ray crystallography of papain [8] slightly favour the involvement of histidine, strong arguments can be raised in

Table 1

Amino acid side chain on protein	Acid function	pK
N-terminal	α-NH$_3^+$	7.8
C-terminal	α-CO$_2$H	3.8
Aspartic acid	β,γ-CO$_2$H	4.4
Glutamic acid		4.6
Histidine	imidazolium	7.0
Cysteine	—SH	8.7
Tyrosine	—C$_6$H$_4$OH	9.6
Lysine	ε-NH$_3^+$	10.4
Serine	β-OH	13
Threonine		
Arginine	—NH(CNH$_2$)NH$_2^+$	12.5
Peptide	R·CO·NHR'	14.8

favour of the carboxylate group [9], and the question is still unsettled. Enzymatic reactions can also be sensitive to the ionization of groups not directly involved in the catalytic process. For example, the ionization of an α-amino group has been shown to affect under certain conditions the rate of α-chymotrypsin-catalysed reactions, by interfering with the binding of the substrate to the enzyme [10]. Finally, the requirement of an enzymatic group in the acidic or base form is still mechanistically ambiguous, because of the kinetic equivalence of nucleophilic and general base catalyses or that of general acid and specific acid-nucleophilic catalyses.

In the case of enzymes reacting at measurable rates with a wide variety of substrates, structure-reactivity correlations are useful to establish mechanistic similarities with model reactions involving proton transfers [11]. As with most other methods applied to enzyme mechanisms, use of this criterion alone can be misleading. For α-chymotrypsin, for example, a limited series of substrates can be found which shows reactivities not inconsistent with the active-site imidazole acting as a nucleophile [12], whereas overwhelming evidence from all other methods shows that the imidazole acts as a general base [13, 14].

The most important piece of evidence for proton transfer in enzymatic systems comes from the application of deuterium oxide kinetic solvent isotope effects. If the enzymatic rate-limiting step involves a proton transfer in the transition state, a solvent isotope effect (k_{H_2O}/k_{D_2O}) of 2 to 3 is observed, in good agreement with values observed in non-enzymatic proton transfer reactions [13]. The application of solvent isotope effects to enzymatic catalysis has been criticized [11] since there is a possibility that the change in solvent from water to deuterium oxide will lead to a change in conformation of the enzyme and hence a change in its catalytic activity. Until now, no such conformational change is documented for any enzyme. Moreover, in thc case of α-chymotrypsin at least, direct experimental evidence shows that the conformation of the active site is independent of the isotopic composition of the medium: the reaction of the alkylating agent N-tosyl phenylalanyl chloromethylketone with the N-3 nitrogen of the active site-histidine is an S_N2-type reaction, unlikely to involve any proton transfers in the rate-limiting step. This reaction indeed occurs without any measurable kinetic solvent isotope effect [15], whereas the pH-dependence of the reaction shows the expected 0.5 pH unit shift in the pK of the imidazole upon changing the solvent from water to D_2O. In the light of these data, observation of a kinetic isotope effect in the enzyme-catalysed reaction together with a normal isotope effect on the pK value of the catalytic group provide clear evidence for the enzymatic proton transfer.

In enzymatic reactions involving carbon-acid substrates primary deuterium isotope effects provide convincing evidence for rate-limiting proton transfers. Larger primary isotope effects were found, for example, in the reaction of aldolases, with light and deuteriated dihydroxyacetone phosphate as the substrate [16].

Finally, the use of isotopes in carbon-acid substrates is an invaluable tool for the determination of the stereochemistry of the enzymatic proton transfer. In contrast to organic reactions, stereospecific proton transfers are the rule, rather than the exception, in enzymatic reactions, owing to the inherently asymmetric nature of the protein surface. An example is the pair of isotopic exchange reactions between dihydroxyacetone phosphate and tritiated water catalysed by the enzymes aldolase and triose phosphate isomerase [17]. In the two cases a different α-hydrogen of the ketone is exchanged with water, leading to the two discrete monotritiated derivatives **1** (labelled by the isomerase) and **2** (labelled by the aldolase):

$$
\begin{array}{cc}
\begin{array}{c}
\text{H} \\
| \\
\text{T} \blacktriangleright \text{C} \blacktriangleleft \text{OH} \\
| \\
\text{C}=\text{O} \\
| \\
\text{CH}_2\text{OPO}_3\text{H}^-
\end{array}
&
\begin{array}{c}
\text{H} \\
| \\
\text{HO} \blacktriangleright \text{C} \blacktriangleleft \text{T} \\
| \\
\text{C}=\text{O} \\
| \\
\text{CH}_2\text{OPO}_3\text{H}^-
\end{array} \\
\mathbf{1} & \mathbf{2}
\end{array}
$$

394

These two exchange reactions indicate stereospecific formation of the corresponding enolate ion intermediates on the surfaces of the two enzymes.

Proton abstraction in pyridoxal-phosphate reactions is now also thought to be governed by stereospecific interaction with the enzymes. With the same substrates, a proton can be removed, a carboxyl group can be removed, or a carbon–carbon bond may be broken. It is postulated that one of these three possibilities occurs according to a strict stereochemistry and thus the reaction takes a different course depending on the stereochemistry of individual enzymes.

12.3 Examples of enzymatic proton transfers

12.3.1 α-Chymotrypsin: general base-nucleophilic catalysis

Chymotrypsin is the most-studied member of the serine protease family of enzymes. The enzyme-catalysed hydrolytic reaction has been shown to occur in at least three kinetically distinguishable steps. The first of these consists of a very fast, diffusion-controlled formation of a non-covalent enzyme-substrate complex, followed by an acylation step. In the latter the acyl group of the substrate is covalently attached to a serine alcohol of the active site with the concomitant release of the amine of an amide substrate, or the alcohol of an ester substrate. In a final deacylation step the acyl-enzyme intermediate is hydrolysed by water, thus regenerating the free enzyme and releasing the carboxylic acid:

$$E + S \rightleftharpoons E \cdot S \rightarrow E - A + P_1 \rightarrow E + A$$

From the experimental point of view the deacylation step is the easiest to study, since no binding step is associated with it. The mechanistic information most pertinent to a description of the deacylation step includes the following:

1. The acyl-enzyme is an ester which was shown by both chemical and physical methods to result from the attachment of the acyl moiety of the substrate to Ser-195 of the enzyme. 2. A base of $pK_a = 7$ is required for the reaction. 3. This base is the imidazole ring of His-57. 4. The deacylation is a nucleophilic reaction and a series of substituents in the acyl group yields a Hammett ρ-constant of $+1.6$ [18]. 5. The reaction is first order with respect to the nucleophile, as determined from the kinetics of the methanolysis reaction [19]. 6. The nucleophile reacts in the protonated form with the acyl-enzyme, as evidenced by the pH-dependence of the reaction of acetyl-, iso-

propoxyphosphonyl-, and diethylphosphoryl-chymotrypsins with the nucleophiles isonitrosoacetone, glycinehydroxamic acid and phenylacetohydroxamic acid [20–23]. All of these reactions exhibit bell-shaped curves depending on two groups, one with $pK_a = 7$ and the other displaying the pK_a value of the nucleophile. 7. The reactivity of amine and alcohol nucleophiles depends only slightly on their basicities [24]. 8. No detectable intermediate was observed in the deacylation step, although indirect evidence suggests the existence of at least one [25–27]. 9. A kinetic solvent isotope effect of 2 to 3 is associated with the deacylation step, as discussed above. 10. X-ray crystallographic data show that His-57 is hydrogen bonded to the carboxylate group of Asp-102 [28]. At the same time, deacylation pH-rate profiles show that the reaction is insensitive to the ionization of any carboxylic acid in the pH range 2 to 9 [29].

The components of the transition state of the deacylation reaction must then include the acyl-serine ester, an unprotonated imidazole group immobilized by a hydrogen bond, and a molecule of water as the acyl acceptor. The rate-determining proton transfer, which must shift a proton from the water molecule (or from another nucleophile) to the imidazole base, is presumably concerted with attack of the nucleophile on the acyl carbon, and perhaps the shift of the proton from imidazole to carboxylate in the hydrogen bond. The reaction scheme shown in Figure 2 takes into account all the experimental data enumerated above [30]. The mechanism shown involves the formation of a tetrahedral intermediate. In spite of considerable effort expended, the existence of this intermediate is still based only on indications from structure-activity relationships [25–27, 31], since isotopic oxygen exchange studies in the deacylations of cinnamoyl-carbonyl-^{18}O-chymotrypsin and p-nitrobenzoyl-carbonyl-^{18}O-chymotrypsin yielded negative results [32]. The mechanism depicted in Figure 2 utilizes the unique ability of imidazole to serve simultaneously as a proton donor and a proton acceptor. The concerted nucleophilic attack and a proton transfer enhance the kinetic efficacy of the catalyst. At the same time, all transition states are neutral, in agreement with experimental data showing that the reaction is rather insensitive to variations in ionic strength and dielectric constant. Finally, the mechanism possesses a high degree of symmetry: the serine nucleophile and the alcohol or amine leaving group of the acylation step are mechanistically equivalent to the water nucleophile and the serine leaving group of the deacylation step.

Sulfhydryl proteases, such as papain, ficin and bromelain, follow essentially the mechanism outlined in Figure 2. The acyl acceptor is

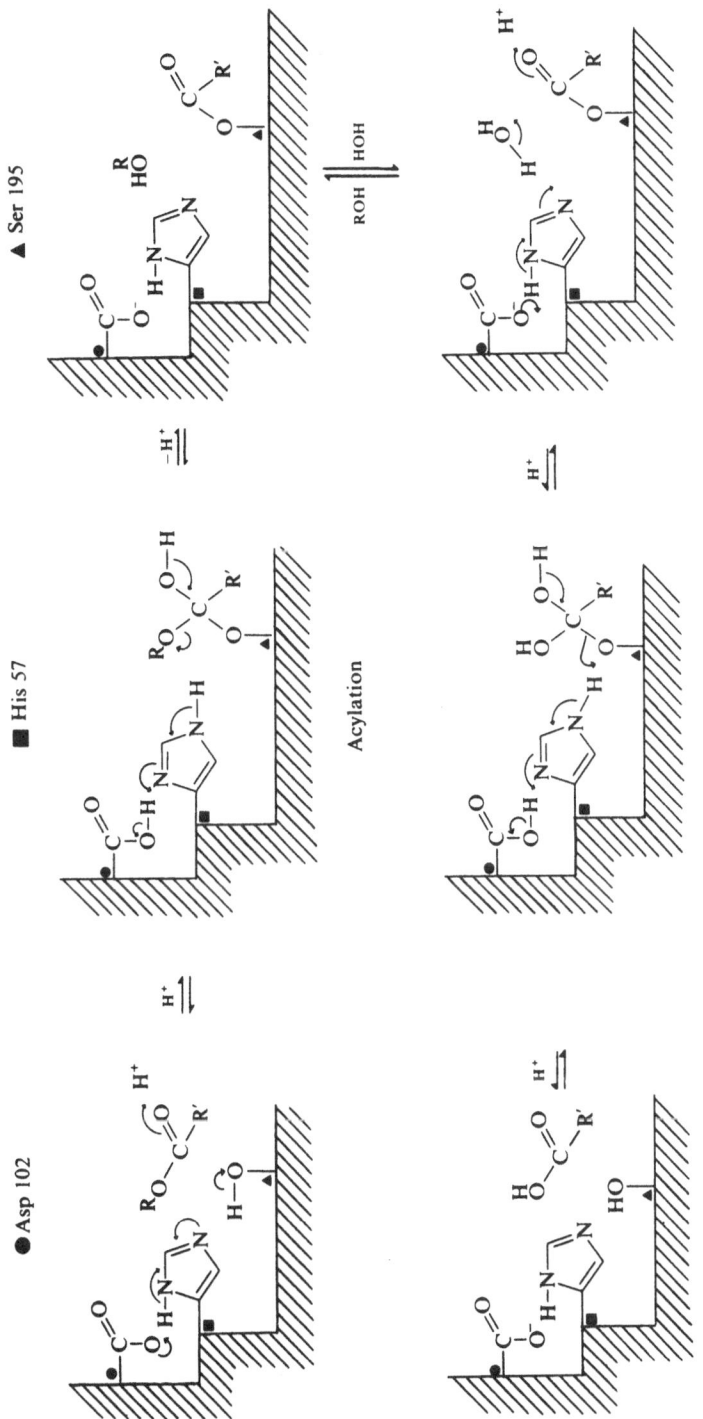

Figure 2. The mechanism of α-chymotrypsin-catalysed hydrolysis of an ester substrate. From reference 30(b).

the sulfhydryl group of a cysteine moiety at the active site, as shown by chemical inhibition studies [33, 34], by the pH-dependence of the catalytic reaction ($pK_a = 8.4$ appears in acylation but not in de-acylation [35]), and by spectroscopic identification of a thiol ester in the acyl-enzyme [36, 37]. Papain-catalysed reactions show again a substantial deuterium oxide kinetic solvent isotope effect, implying a rate-determining proton transfer [37]. The chemical nature of the general base catalyst was discussed above. Since thiol esters react more readily with amines than do oxygen esters, papain is a better catalyst for transpeptidation than is chymotrypsin.

$$AAD-NH_3{}^+$$

$$K_E \updownarrow$$

$$AAD-NH_2 + CH_3 \cdot CO \cdot CH_2 \cdot CO_2{}^- \quad \underset{k_{-1}}{\overset{k_1}{\rightleftharpoons}}$$

$$AAD-N{=}\underset{\diagup CH_3}{C}{-}CH_2CO_2{}^-$$

$$H^+ \Updownarrow K_{ES_1}$$

$$AAD-N^+H{=}\underset{\overset{|}{CH_3}}{C}CH_2CO_2{}^-$$

$$\downarrow k_2$$

$$AAD-N^+H_2\underset{\overset{|}{CH_3}}{C}{=}CH_2 \underset{H^+}{\overset{K_{ES_1}}{\rightleftharpoons}} AAD-NH\underset{\overset{|}{CH_3}}{C}{=}CH_2 + CO_2$$

$$H^+ \downarrow k_3$$

$$AAD-N{=}\underset{\overset{|}{CH_3}}{C}CH_3 \underset{H^+}{\overset{K_{ES_1}}{\rightleftharpoons}} AAD-N^+H{=}\underset{\overset{|}{CH_3}}{C}CH_3$$

$$k_4 \downarrow H_2O$$

$$AAD-NH_2 + CH_3COCH_3$$

12.3.2 *Acetoacetate decarboxylase, Schiff-base intermediate*

Enzymatic proton transfers to and from nitrogen atoms will be illustrated with the mechanism of acetoacetate decarboxylase, one of the many enzymes which form a Schiff base (imine) as a catalytic intermediate. Acetoacetate decarboxylase (AAD) from *Clostridium acetobutylicum* catalyses the formation of acetone from acetoacetic acid. Chemical and kinetic evidence demonstrates the presence of an essential ε-amino group of lysine at the active site. In the presence of the substrate, a Schiff base salt of acetone has been trapped by boro-hydride reduction. By use of radioactive 3-^{14}C-acetoacetate, the peptide containing the label was isolated after acid hydrolysis and shown to contain N$^\varepsilon$-isopropyl-lysine [38]. The enzyme also catalyses the de-deuteriation of acetone-d_6 in water [39] and the carbonyl-oxygen exchange between acetone or acetoacetate and water [40]. The pH-rate profile of the enzymatic reaction is bell-shaped, indicating the requirement for a base of p$K = 5.5$ and an acid of p$K = 6.5$ [41]. The base was identified as the ε-amino group of lysine acting as the nucleophile in the Schiff-base formation, whereas the acid of p$K = 6.5$ must act as a general acid catalyst in the protonation of the enamine intermediate [39, 41, 42]. The reaction mechanism scheme shown on the opposite page is consistent with all observations [39].

12.3.3 *Carboxypeptidase and oxaloacetate decarboxylase, metallo-enzymes*

A large number of enzymatic reactions require metal ions as one of the essential components of the catalytic process. A variety of metal ions can be involved, ranging from the alkali metals Na$^+$ and K$^+$ through the divalent ions Ca^{2+}, Mg^{2+}, Zn^{2+}, Fe^{2+}, Cu^{2+}, Co^{2+}, Ni^{2+} to the rarely occurring vanadium and molybdenum ions. Their role can be purely structural, but most often they are tightly bound to the active site where they participate in the catalytic reactions. Such participation can be limited to a stereospecific ligand formation with the substrate, for example, with the phosphate groups. In the case of redox enzymes the metal ion acts as an electron transfer agent by undergoing a reversible change between two states of oxidation.

Most often, however, as in metal-ion catalysis of organic reactions, the role of the metal ion directly participating in the enzymatic reaction is generally that of a Lewis acid; a substitute for the proton, but more efficient, especially at neutrality. These reactions always involve a simultaneous proton transfer between the substrate and the enzyme and it is in this sense that one can discuss metal-ion catalysed enzymatic proton transfers. As an example, the mechanism of carboxypeptidase-catalysed reactions will be considered.

Carboxypeptidase is a metalloenzyme containing one atom of zinc per protein molecule. It catalyses the hydrolysis of C-terminal peptide bonds in proteins and oligopeptides and that of carboxyl esters of α-hydroxyacids. The rate constant of the reaction shows a kinetic solvent isotope effect of 2 for the ester substrate O-(*trans*-cinnamoyl)-L-β-phenyllactate, but only 1.33 ± 0.15 for the peptide N-(N-benzoylglycyl)-L-phenylalaninate [43]. The three-dimensional structure of the enzyme, determined by X-ray crystallography [44], shows a globular protein containing a zinc ion coordinated with two histidines and a glutamate carboxylate ion. The active site also contains a carboxylate ion (Glu-270), a phenolic group (Tyr-248) and a guanidino group (Arg-145), the latter shown to form a salt bond with the free carboxylate ion of the substrate. The simplest mechanism accounting for all present structural and kinetic information is shown in the following scheme [43, 45]:

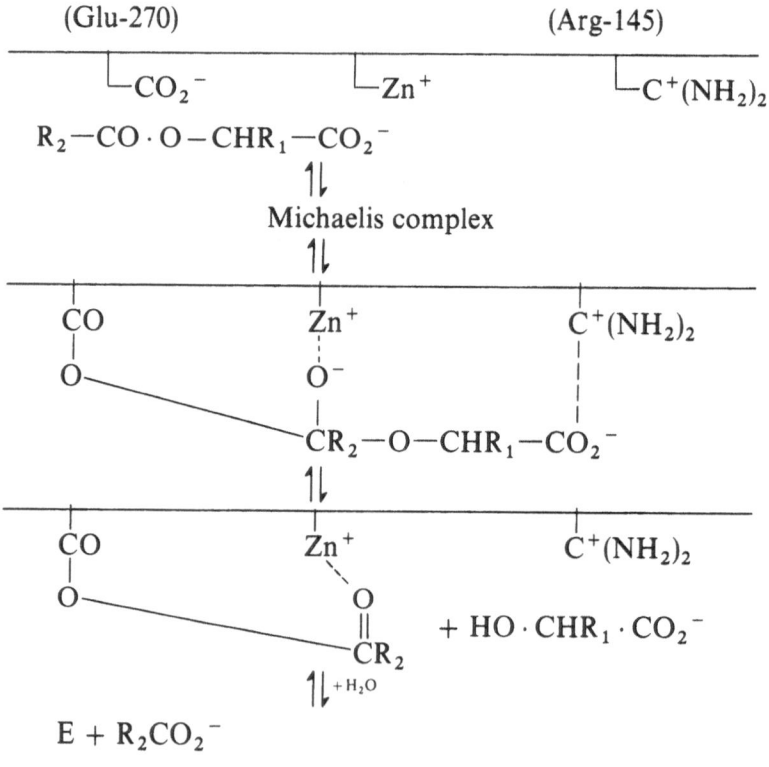

Thus the reaction proceeds through the transient formation of an acyl-anhydride intermediate, with simultaneous release of the terminal carboxylic acid. In a second step the anhydride then rapidly reacts with water. It is possible, that the phenol of Tyr-248 acts as a general base in the deacylation step, by abstracting a proton from water, and perhaps even as a general acid in the first step, by protonating the leaving group. No definite experimental proof is yet available on this point. It is clear, however, that the role of the metal ion is limited to polarizing the carbonyl function as a general acid and at the same time to contribute to the rigid orientation of the peptide or ester bond so as to facilitate the formation of the tetrahedral intermediate indicated in the scheme. Such a polarization is necessary in view of the rather weak nucleophilicity of the attacking carboxylate ion. The metal ion does not appear to play any direct role in the proton transfer or to interact with any other proton donor or acceptor participating in the reaction.

12.3.4 *Oxaloacetate decarboxylase from Micrococcus lysodeicticus* [46]

This enzyme catalyses the decarboxylation of the β-ketoacid oxaloacetate, with the same stoichiometry as acetoacetate decarboxylase. The former however, requires a Mn^{2+} ion for activity and is insensitive to the action of sodium borohydride. This duality of mechanism is not unlike the one observed for enzymatic aldol condensation, where enzymes of Class 1 react by forming Schiff-base intermediates, whereas enzymes of Class II show metal ion requirements [47]. Oxaloacetate decarboxylase from cod also catalyses the reduction by borohydride of the enzymatic reaction product pyruvate. This is evidenced by the accumulation of D-lactate in presence of enzyme, reducing agent, and manganous ions. It has been proposed that both reduction and decarboxylation occur by way of an enzyme-metal ion-substrate complex in which the metal ion acts as an electron sink, thereby stabilizing the enolate ion formed in the decarboxylation reaction [48]:

$$E \quad Mn^{++} \quad \begin{array}{c} {}^{-}O{-}CO \\ | \\ {}_{-}O{-}C{=}CH_2 \end{array}$$

The duality of hydrolytic enzyme mechanisms, chymotrypsin and carboxypeptidase-, and that of decarboxylation by metal ion and

Schiff base enzymes, reflects essentially the sensitivity of carbonyl functions toward acid or base catalysis. Whereas the protein side chains provide adequate *nucleophiles* and *bases* for catalytic activity, the superiority of metal ions as *acidic* catalysts in comparison with protons is amply demonstrated by metallo-enzymes. A further reason for the occurrence of numerous metalloenzymes might be sought in the multidentate nature of metal complexes. The precise stereochemical positioning of several reaction components in the same complex provides an easy optimization of proximity and orientation effects.

12.3.5 *Mandelic acid racemase, a simple acid-base reaction*

On the basis of presently available information, mandelic acid racemase might display the simplest enzymatic proton transfer mechanism. The enzyme, isolated from *Pseudomonas putida*, catalyses the epimerization reaction [49]:

$$H-\underset{\underset{C_6H_5}{|}}{\overset{\overset{CO_2^-}{|}}{C}}-OH \rightleftharpoons HO-\underset{\underset{C_6H_5}{|}}{\overset{\overset{CO_2^-}{|}}{C}}-H$$

without a cofactor requirement.

Deuterium exchange with the solvent does occur during the reaction, but at a rather slow rate. A symmetrical reaction intermediate must exist, since the rate of incorporation of tritium from the solvent into D-madelate as the substrate yields equimolar amounts of D- and L-product [50]. Thus the data are consistent with the formation of an α-carbanion intermediate with an enzymatic base group acting as the proton acceptor [50]. The proton transfer has to be rate-limiting, as indicated by the approximately 5-fold primary isotope effect for deuterium. In the enzyme-substrate complex, the epimerization occurs with a rate constant of the order of $10^3 \, s^{-1}$.

12.3.6 *Lysozyme, general acid catalysis*

Lysozyme hydrolyses the $\beta(1 \rightarrow 4)$ glycosidic linkages of mucopolysacharides which are constituents of the cell wall of certain microorganisms. In addition to the natural substrate, which is composed of N-acetylmuramic acid and N-acetylglucosamine, the enzyme also hydrolyses chitin (poly-N-acetyl-β-glucosamine) and its degredation products and a number of β-aryl di-N-chitobiosides. The complete three-dimensional structure of the enzyme shows a globular protein, roughly the shape of a butterfly, with one half containing a good deal of helix, and the other mostly β-pleated sheet. The active site is between these two halves and has been described as a cleft or groove adapted to the long-chain polymer natural substrate.

The catalytically active functional groups have been suggested to be the carboxyl groups of aspartic acid-52 and glutamic acid-35 [51]. The mechanism of lysozyme action, the enzyme whose structure is the best defined, is still somewhat uncertain. In total, three mechanisms have been proposed for lysozyme action: (1) the neighbouring N-acetylamino group of the substrate acts as a nucleophile and aspartic acid as a proton donor; (2) the carboxylate anion of glutamic acid-36 further assists the N-acetylamino group by acting as a proton acceptor; (3) the carboxylate anion of aspartic acid-52 acts as a nucleophile, and the carboxylic acid of glutamic acid-35 acts as a proton donor.

Lysozyme catalyses the hydrolysis of p-nitrophenyl 2-acetamido-4-O-(2-acetamido-2-deoxy-β-D-glucopyranosyl)-2-deoxyglucopyranoside with a value of k_{cat} 20 times greater than p-nitrophenyl 4-O-(2-acetamido-2-deoxy-β-D-glucopyranosyl)-β-D-glucopyranoside [54]. These results suggest acetamido participation.

A concerted proton donor-proton acceptor or proton donor-nucleophilic catalysis is favoured, however, on the basis of kinetic results with small substrates [55]. Transglycosylation was demonstrated in lysozyme-catalysed reactions. This reaction has been used to synthesize p-nitrophenyl β-D-glucoside and a corresponding oligomer. Since lysozyme hydrolyses these glycosidic bonds and since they contain no acetylamino group, anchimeric assistance by the N-acetylamino group can be ruled out as a necessary pathway for the enzyme-catalysed cleavage of glycosidic bonds [56]. The lysozyme active site, while not completely characterized, must certainly be large, for the enzyme can interact with a hexasaccharide and there is strong kinetic dependence on the length of the saccharide. This implies the importance of non-productive complexes (complexes of enzyme and substrate that do not lead to reaction). Evidence for oxocarbonium ion formation and participation of glutamic acid-35 is particularly convincing [57]. Retention at the anomeric centre means that if an oxocarbonium ion is formed, it must be constrained by the enzyme.

12.4 Conclusion

From the mechanistic point of view enzyme-catalysed proton transfers are much less well characterized than their non-enzymatic counterparts. The cause for this has to be sought more in the complexity of enzyme structure and the lack of information concerning the groups constituting the active site than in any mechanistic peculiarities involved in enzymatic reactions. Enzymatic proton transfers could even be considered as lacking the mechanistic diversity of organic reactions. On the other hand, proton transfers are ubiquitous in enzyme-catalysed reactions, and it is remarkable in how many different ways one can accelerate organic reactions catalytically, by carrying out the chemical transformation through a pathway involving one or more proton transfers of the critical, rate-limiting steps.

A unique feature of enzyme-catalysed proton transfers is their stereospecificity. Again, this is not a result of the stereochemistry of the donor or acceptor groups or of the transition state alone, since the specificity is achieved by the judicious arrangement of the reaction partners in a stereospecific array, excluding the possibility of any favourable interaction between the catalytic groups and substrate in the 'wrong' orientation. In this sense, the enzymatic process can be conceived as a stereochemical selection in the ground state, rather than in the transition state. The role of water in the enzyme-catalysed reactions is still somewhat controversial. From the few examples considered, it is quite clear however, that the catalytically active proton donors/acceptors are covalently linked to the peptide chain. The catalytic advantage of such groups over water appears to stem from their restricted orientation and mobility, in comparison with the mobility of even very restricted 'structural water' molecules which sometimes can be detected by X-ray crystallography. We feel that it is this lack of rigidity and orientation which makes otherwise abundant water-protons unsuitable for enzymatic catalysis. The role of water as a polar prototropic solvent of enzymes is important, however, and most enzymatic proton donors/acceptors are in rapid equilibrium with water, their state of protonation being a direct function of the pH of the medium.

Much remains to be done before a clear, quantitative mechanistic picture will be available to describe enzymatic proton transfers. From the methodological point of view, the development of models for intramolecular proton transfers appear to be necessary for the quantification of the influence of H-bond strength and bond orientation on the catalytic rate. Also, new methods of experimentation will be required to assign the location of protons in rapidly equilibrating, extensively H-bonded structures.

From the theoretical point of view, quantitative theories are needed to correlate general acid-general base catalysed intramolecular reactions with the structure of the reagents.

Structural and mechanistic studies are only at their beginning for most enzyme classes other than hydrolases. Mechanistic studies of the latter group were greatly facilitated by the structural simplicity of the protein molecules involved and the reduced stoichiometry of only one substrate molecule and water. The task will be more arduous for multi-subunit and multi-substrate enzymes.

Looking back on the past twenty years which marked the development of enzyme-mechanistic studies, one cannot help but notice that the progress in the understanding of enzymatic reactions was always critically dependent upon the existence of appropriate model studies. From this point of view, the pioneering work of Professor R. P. Bell on proton transfer reactions is an invaluable asset without which enzyme mechanistic studies would probably still be in the exploratory stage.

Acknowledgment

This research was supported by grants from the National Institutes of Health and the National Science Foundation, U.S.A.

REFERENCES

[1] R. P. Bell, *Acid-Base Catalysis*, Oxford University Press, London (1941).

[2] W. P. Jencks, *Catalysis in Chemistry and Enzymology*, McGraw-Hill, New York, (1969).

[3] W. P. Jencks and M. Gilchrist, *J. Amer. Chem. Soc.*, **90**, 2622 (1968).

[4] I. A. Rose, in *The Enzymes*, 3rd Ed., Vol. 2, p. 281, P. D. Boyer, ed., Academic Press, N.Y. (1970).

[5] G. A. Hamilton, *Progress in Bioorganic Chemistry*, **1**, 83, (1971).

[6] M. W. Hunkapiller, S. H. Smallcombe, D. R. Whitaker, and J. H. Richards, *Biochemistry*, **12**, 4732 (1973).

[7] E. C. Lucas and A. Williams, *Biochemistry*, **8**, 5125 (1969).

[8] J. Drenth, J. N. Jansonius, R. Koekoek, H. M. Swen, and B. G. Wolthers, *Nature*, **218**, 929 (1968).

[9] M. L. Bender and L. J. Brubacher, *J. Amer. Chem. Soc.*, **88**, 5880 (1966).

[10] P. Valenzuela and M. L. Bender, *J. Amer. Chem. Soc.*, **93**, 3783 (1971).

[11] For a review, see W. P. Jencks, *Cold Spring Harbor Symposia on Quantitative Biology*, **36**, 1 (1971).

[12] C. D. Hubbard and J. F. Kirsch, *Biochem.* **11**, 2483 (1972).

[13] M. L. Bender and F. J. Kézdy, *Ann. Rev. Biochem.*, **34**, 49 (1965).

[14] G. P. Hess, in *The Enzymes*, 3rd Ed., Vol. 3, p. 231, P. D. Boyer, ed., Academic Press, N.Y. (1971).

[15] F. J. Kézdy, A. Thomson, and M. L. Bender, *J. Amer. Chem. Soc.*, **89**, 1004 (1967).

[16] I. A. Rose, E. L. O'Connell, and A. H. Mehler, *J. Biol. Chem.*, **240**, 1758 (1965).

[17] S. V. Reider and I. A. Rose, *J. Biol. Chem.*, **234**, 1007 (1959).

[18] E. J. Amshey Jr., S. P. Jindal, and M. L. Bender, *Arch. Biochem. Biophys.*, in press (1975).

[19] M. L. Bender, G. E. Clement, C. R. Gunter, and F. J. Kézdy, *J. Amer. Chem. Soc.*, **86**, 3697 (1964).

[20] M. L. Bender and F. J. Kézdy, *J. Amer. Chem. Soc.*, **86**, 3704 (1964).

[21] F. C. Wedler, F. J. Killian, and M. L. Bender; *Proc. Nat. Acad. Sci. U.S.A.*, **65**, 1120 (1970).

[22] A. L. Green and J. D. Nicholls, *Biochem. J.*, **72**, 70 (1959).

[23] W. Cohen and B. F. Erlanger, *J. Amer. Chem. Soc.*, **82**, 3928 (1960).

[24] P. W. Inward and W. P. Jencks, *J. Biol. Chem.*, **240**, 1986 (1965).

[25] M. L. Bender, *J. Amer. Chem. Soc.*, **84**, 2582 (1962).

[26] A. Frankfater and F. J. Kézdy, *J. Amer. Chem. Soc.*, **93**, 4039 (1971).

[27] H. Hirohara, M. L. Bender and R. Stark, *Proc. Nat. Acad. Sci. U.S.A.*, **71**, 1643 (1974).

[28] R. Henderson, *J. Mol. Biol.*, **54**, 341 (1970).

[29] F. J. Kézdy, G. E. Clement, and M. L. Bender, *J. Amer. Chem. Soc.*, **86**, 3690 (1964).

[30] (a) K. Tanizawa and M. L. Bender, *J. Biol. Chem.*, **249**, 2130 (1974); (b) H. Hirohara and M. L. Bender, *J. Biol. Chem.*, in press (1975).

[31] M. Caplow; *J. Amer. Chem. Soc.*, **95**, 2670 (1973).

[32] M. L. Bender and H. d'A. Heck, *J. Amer. Chem. Soc.*, **89**, 1211 (1967).

[33] A. Light, R. Frater, J. R. Kimmel, and E. L. Smith; *Proc. Nat. Acad. Sci. U.S.A.*, **52**, 1276 (1964).

[34] B. J. Finkle and E. L. Smith; *J. Biol. Chem.*, **230**, 669 (1958).

[35] M. L. Bender and J. R. Whitaker, *J. Amer. Chem. Soc.*, **87**, 2728 (1965).

[36] G. Lowe and A. Williams; *Proc. Chem. Soc.*, 140 (1964).

[37] M. L. Bender and L. J. Brubacher, *J. Amer. Chem. Soc.*, **86**, 5333 (1964).

[38] S. Warren, B. Zerner, and F. H. Westheimer; *Biochemistry*, **5**, 817 (1966).

[39] W. Tagaki and F. H. Westheimer; *Biochemistry*, **7**, 901 (1968).

[40] G. A. Hamilton and F. H. Westheimer, *J. Amer. Chem. Soc.*, **81**, 6332 (1959).

[41] F. H. Westheimer, *Proc. Chem. Soc.*, 253 (1963).

[42] R. Kluger and K. Nakaoka, *Biochemistry*, **13**, 910 (1974).

[43] E. T. Kaiser and B. L. Kaiser, *Accounts Chem. Res.*, **5**, 219 (1972).

[44] G. N. Reeke, J. A. Hartsuck, M. L. Ludwig, F. A. Quiocho, T. A. Steitz, and W. N. Lipscomb, *Proc. Nat. Acad. Sci. U.S.A.*, **58**, 2220 (1967).

[45] J. E. Coleman, *Progr. Bioorg. Chem.*, **1**, 159 (1971).

[46] D. Herbert, *Methods in Enzymol.*, **1**, 753 (1955).

[47] E. Grazi, T. Chang, and B. L. Horecker, *Biochem. Biophys. Res. Comm.*, **7**, 250 (1962).

[48] G. W. Kosicki and F. H. Westheimer, *Biochemistry*, **7**, 4303 (1968).

[49] G. D. Hegeman, E. Y. Rosenberg and G. L. Kenyon, *Biochemistry*, **9**, 4029 (1970).

[50] G. L. Kenyon and G. D. Hegeman, *Biochemistry*, **9**, 4036 (1970).

[51] D. C. Phillips, *Proc. Nat. Acad. Sci. U.S.A.*, **57**, 484 (1967).

[52] D. C. Phillips, *Scientific American*, **215**, No. **5**, 78 (1966).

[53] G. Lowe, *Proc. Roy. Soc. B*, **167**, 431 (1967).

[54] G. Lowe and G. Sheppard, *J. Chem. Soc. Chem. Comm.*, 529 (1968).

[55] G. Lowe, G. Sheppard, M. L. Sinnott, and A. Williams, *Biochem. J.*, **104**, 893 (1967).

[56] M. A. Raftery and T. Rand-Meir, *Biochemistry*, **7**, 3281 (1968).

[57] J. A. Rupley, V. Gates, and R. Bilbrey, *J. Amer. Chem. Soc.*, **90**, 5633 (1968).

13

Victor Gold, Colin J. Liddiard and Gareth D. Morgan

ACID CATALYSIS BY INSOLUBLE
MACROMOLECULAR ACIDS

13.1 Introduction

Insoluble materials containing acidic or basic functions are commonly known as ion exchangers, in recognition of their most useful property. Typical cation exchangers, for example, are polymeric acids. Their acidic hydrogen is exchangeable for metal ions which are retained whilst the displaced hydrogen ions pass into the solution phase. As they are acids, such cation exchangers, in the hydrogen form, can act as catalysts of reactions in solution that are accelerated by proton transfer from catalyst to substrate. Correspondingly, anion exchangers in their basic form can function as base or nucleophilic catalysts. Many examples of these phenomena, mostly concerned with polymeric organic sulphonic acids and far too numerous even to catalogue here, are known [1].

As long ago as 1911, Tacke and Süchting noted the inversion of sucrose in aqueous solution in the presence of acidic zeolites [2]. The effect was at first erroneously attributed to humic acids present in the soil; the correct interpretation was offered a few years later by Rice and Osugi [3]. The use of insoluble acid-base systems as buffers for solutions was considered by Bell and Prue [4].

From a purely practical point of view, the use of insoluble catalysts has important attractions, in particular the ease of separating the catalyst from reactants and products, which opens up the possibility of the use of flow reactors. Less obviously, greater selectivity in catalytic function can sometimes be achieved [4]. Set against this, there are some practical disadvantages, such as the high cost of the materials, and their thermal and mechanical instability, which is especially marked for organic ion exchange resins.

The study of catalysis by insoluble acids and bases is also of interest in several other contexts. Zeolites have already been mentioned, and clays and other mineral constituents of soil may exert similar effects. These are obviously relevant to soil sciences and agriculture. It seems likely that immobile parts of living organisms, such as bone or cell walls, may similarly function as catalysts. Less obviously perhaps, there is a connection between the common ion exchange resins and biological materials possessing enzymic activity, in that both can be looked upon as polyelectrolyte gels. Of course, it would be altogether wrong to ascribe enzyme activity solely to the presence of acidic and basic functions on a polymeric backbone. Enzymes are both more specific and more powerful as catalysts than ion exchange resins. Nevertheless an understanding of catalysis by chemically simple macromolecular acids and bases is clearly relevant to the proper description of the action of the chemically more complex macromolecular acid-base systems which we call enzymes. The study of these simplified macromolecular model systems therefore constitutes an approach to the enzyme problem which is complementary to the investigations on model compounds of low molecular weight, containing several specifically sited functional groups. This point of view has been emphasized in publications on catalysis by synthetic macromolecular acids and bases in solution, especially by Klotz, Morawetz and Overberger [6, 7], and their respective groups. Whilst there are clearly many parallelisms between work on soluble and work on insoluble macromolecular catalysts, some of which will be mentioned, there are also obvious differences. Both systems are, from the point of view of the solution kineticist (let alone from that of the gas kineticist), of staggering complexity which at present defies any completely general description. What is an important effect in one example may be relatively unimportant in an apparently closely related one.

The present chapter tries to set out some of the problems of catalysis by insoluble macromolecular acids and to establish its relationship to proton transfer processes in general. Of necessity most experimental work on catalysis by insoluble macromolecules has been concerned with practical and polymer aspects of the subject and therefore with the importance of factors other than the chemical reaction itself. By contrast, the aim of our contribution will be to work towards a discussion of reaction mechanisms, as that term is commonly used by chemists. We shall, in particular, examine progress in one relatively unexplored area, that of catalysis by polycarboxylic acids, and consequently the question whether *general acid catalysis* is an operationally significant concept for heterogeneous systems. In approaching this problem we have tried to learn from the development of reaction kinetics in homogeneous solution. Present-day views of proton transfer in solution rest largely on the quantitative study of general acid-base catalysis. R. P. Bell's classic book [8] *Acid-Base Catalysis* set out the development of the methodology required for this work. The possible adaptation of the methodology to heterogeneous acid-base catalysis and the usefulness of the solution kineticist's approach in that area will be considered below.

Much of the background to this contribution is described in detail in the excellent book on *Ion Exchange* by F. Helfferich [9], especially Chapter 11.

13.2 Types of cation exchangers

Cation exchangers can be either organic or inorganic. The latter are generally based on an alumino-silicate structure where the exchangeable cations are contained in the interstices of the silicate lattice. The important organic ion exchangers are based either on a condensation polymer (*e.g.*, a phenol-formaldehyde resin) or, more usually, on an addition polymer (*e.g.*, polystyrene) with incorporated acidic or basic groups. These are usually sulphonic acid groups for strong-acid resins, quaternary ammonium groups for strong-base resins, and amine groups, for weak-base resins, though other functional acidic or basic groups can also be attached. Commercially available weak-acid ion exchangers are manufactured almost exclusively by the addition polymerization of methacrylic acid. The ion exchangers of the addition polymer type are generally co-polymerized with varying amounts (ca. 2–10 per cent as a rule) of divinylbenzene to produce cross-linking and insolubility. Most previous kinetic work with cation exchangers has been carried out with cross-linked commercial polystyrenesulphonic acid resins in bead form – chiefly Amberlite IR-120 (Rohm & Haas), Dowex-50 (Dow) and the Russian product KU-2 – and with the sulphonated phenol formaldehyde resin Amberlite IR-100 (Rohm & Haas). The study of catalysis by weak-acid resins, to which reference will be made, was carried out by us with commercial specimens of cross-linked polymethacrylic acid, *viz.* Amberlite IRC-50 (Rohm & Haas), several variants of Zerolit 226 (Permutit), and with specimens of the same polymer prepared in the laboratory [11], and, to a smaller extent, with synthetic polystyrene carboxylic acids [12].

For catalytic purposes, ion exchange resins are commonly used either in columns of resin beads (for column operation) or in agitated suspensions (for batch operation). The study and interpretation of reactions in columns are more complicated and our attention will therefore be confined to batch reactions.

13.3 Terminology and units

The concentration of acid in a suspension of ion exchange resin ($\tilde{m}_c$) is usually expressed as gram-equivalents (or moles) of acid groups per kg of solvent added to the dry resin. This definition is similar to that used by Haskell and Hammett [13], who also defined the *efficiency* of the resin (q) as the ratio of the catalytic rate of an acid-catalysed reaction in the presence of ion exchanger to the catalytic rate of the same reaction without ion exchanger but in presence of the same bulk concentration ($\tilde{m}_c$) of hydrogen ion, added as hydrochloric acid.

A useful operational criterion for the existence of heterogeneous catalysis is the observation of a rate increase with increasing concentration of resin, when the chemical composition, including the hydrogen-ion concentration, in the supernatant (or external) phase is being kept constant.

Most of the early investigations [1] to which these concepts were applied related to polystyrenesulphonic acids. These are strong acids and catalysis must be due to solvated hydrogen ions. Being the counter-ions of the negatively charged polymer matrix, they are restrained by the electric potential from diffusion into the bulk solution. Helfferich [14], therefore summarized the position by stating that catalysis by ion-exchange resins 'is not a true case of heterogeneous catalysis, but may be described more adequately as homogeneous catalysis in the pore liquid'. We shall now examine some of the conclusions which can be drawn from studies of such systems before going on to the case of weak polymeric acids where, we submit, Helfferich's generalization is not wholly valid.

13.4 The role of diffusion

By analogy with other forms of non-homogeneous kinetics it is to be expected that diffusion can play an important part in reactions catalysed by solid acids. For reaction to occur, the substrate must diffuse to the catalytic sites where it reacts and the product must diffuse away. Only if the second of these steps is slow compared to the other two will the rate be controlled by a chemical process. The observed rate will then be a 'chemical reaction rate'. In fact, all types of rate control have been observed. For the sulphonic acid phenol-formaldehyde resin Amberlite IR-100, the rates of the inversion of sucrose [15] and of the esterification of oleic acid in 1-butanol [16] have been shown to be strongly dependent on surface area of the catalyst, whereas the hydrolysis of ethyl acetate, a substrate of smaller molecular size, on Amberlite, is not, provided the agitation of the suspension is maintained above a certain minimum rate [17]. Similarly, the inversion of sucrose, catalysed by the resin IR-120, is diffusion-controlled [18], whereas the hydrolysis of simple aliphatic esters is not [13]. The usual experimental criterion for concluding that chemical control operates is that the reaction velocity is independent of stirring rate and of size of catalyst bead (i.e. of the surface/volume ratio). As in one of the examples just cited, it is often found that the rates of reactions catalysed by an ion exchange resin, present as a suspension of fine beads, are not limited by diffusion when the solutions are well agitated. It is arguable whether use of the agitation-rate criterion is always justified. There are two phases of the diffusion process, namely diffusion through the Nernst layer surrounding the catalyst ('film diffusion') and diffusion through the pores of the catalyst. It does not seem at all certain that film diffusion (which is fairly rapid and thought to be only rarely completely rate-controlling) can be affected by the rates of agitation normally employed in the laboratory.

In work with weak-acid ion exchangers it may be possible to devise a different line of evidence regarding the role of diffusion (see Section 11).

Our present concern is with the chemical aspect of the catalysis and we shall therefore not pursue in detail the analysis of the situation when diffusion becomes partially or entirely rate-limiting. Catalysis by spheroidal particles is not the simplest case for a detailed mathematical analysis. Several contributions are of special interest in this connection [19]. Woermann and his group [19] present an interesting approach in which the catalytic effect of a sulphonic acid resin in the shape of a membrane is studied in two modes, first, when both sides of the membrane are exposed to the solution containing the substrate and, second, when only one side is exposed. The results allow both the chemical rate constant and the diffusion coefficients to be evaluated for the case where neither effect can be ignored (see also Section 5).

Cases are also known where a reaction product is firmly held to catalytic sites and diffuses out at a prohibitively low rate. The effect has been observed in several cases, including the esterification of alcohols, acting as reactant and solvent [16, 20]. It is thought that in these reactions water is the product which 'poisons' the acidic sites of the sulphonic acid catalyst. A more straightforward case is the hydrolysis of amides where one of the reaction products, the amine, forms an ammonium salt with the resin acid and thus decreases the number of free acid groups [21]. Similarly, cationic reaction products (formed from cationic reactants) may be retained in the resin by virtue of its ion exchanger function.

13.5 Partition effects and the two-phase model of resin catalysis

In a systematic formulation of heterogeneous catalysis under conditions where diffusion is rapid compared with the chemical process, Helfferich [14] considers that the reaction system can be divided into two homogeneous phases, a solution phase and a resin phase. The model is evidently over-simplified but has proved helpful in the chemical treatment of catalysis.

When diffusion effects are negligible, the resin phase is assumed to be freely permeable to solvent and substrate (R). The distribution of the latter between the two phases is expressed by a distribution coefficient λ_R, defined in equation 1, where $\bar{m}_R$ denotes the molal concentration of reactant R in the resin phase and m_R the corresponding concentration in the solution phase.*

$$\lambda_R = \bar{m}_R/m_R \qquad (1)$$

Material balance then requires equation 2,

$$\tilde{m}_R \tilde{V} = \bar{m}_R \bar{V} + m_R V \qquad (2)$$

where $\tilde{m}_R$ is the concentration of R in the total solution volume $\tilde{V}(\equiv \bar{V} + V)$. By combining equations 1 and 2, we can express the

---

* Following Helfferich [14], we use a bar over a symbol to denote the resin phase and a tilde to refer to average or total values relating to the system as a whole. A symbol not modified by either bar or tilde refers to the external phase.

concentration of R in the resin phase relative to the bulk concentration $\tilde{m}_R$, i.e.

$$\bar{m}_R = \frac{\lambda_R \bar{V}}{\lambda_R \bar{V} + V} \tilde{m}_R \qquad (3)$$

The stoicheiometric concentration of acid groups in the resin $(\bar{m}_c)$ phase can also be expressed as a bulk concentration $(\tilde{m}_c)$ over the entire solution volume, i.e.

$$\bar{m}_c = \tilde{m}_c \bar{V}/V \qquad (4)$$

For catalysis by the hydrogen ions of a sulphonic acid resin, we suppose that the reaction proceeds only in the resin phase where the concentration of hydrogen ion is equal to $\bar{m}_c$ and the rate coefficient $\bar{k}$ applies. The second-order rate equation for formation of product (P) (expressed in moles, n) by the resin-catalysed route is

$$\left(\frac{d\tilde{n}_P}{dt}\right)_{cat} = \bar{k}\bar{m}_R\bar{m}_c\bar{V} \qquad (5)$$

$$= \bar{k}\tilde{m}_R\tilde{m}_c\frac{\lambda_R\bar{V}^2}{(\lambda_R\bar{V} + V)} \qquad (6)$$

If the total chemical change in the system is expressed as a change in the average bulk concentration, we may write

$$\left(\frac{d\tilde{m}_P}{dt}\right)_{cat} = \frac{1}{\bar{V}}\frac{d\tilde{n}_P}{dt} = \bar{k}\tilde{m}_R\tilde{m}_c\frac{\lambda_R\bar{V}}{\lambda_R\bar{V} + V} \qquad (7)$$

When reaction in the external phase must also be taken into account, the full rate equation becomes

$$\left(\frac{d\tilde{m}_P}{dt}\right)_{total} = \tilde{m}_R\left(\bar{k}\tilde{m}_c\frac{\lambda_R\bar{V}}{\lambda_R\bar{V} + V} + k\frac{V}{\lambda_R\bar{V} + V}\right) \qquad (8)$$

where k is the rate constant for reaction of R in the external phase (at the pH, salt concentration, etc., of that phase). If $\bar{V} \ll \tilde{V}$, it follows that

$$\left(\frac{d\tilde{m}_P}{dt}\right)_{cat} = -\frac{d\tilde{m}_R}{dt} = \bar{k}\tilde{m}_R\tilde{m}_c\lambda_R \qquad (9)$$

It then follows that the efficiency q, defined above, measures the ratio

$$q = \frac{\bar{k}\lambda_R}{k_H} \qquad (10)$$

where k_H is the second order rate constant for the hydrogen ion-catalysed reaction in homogeneous solution. Equations 8 and 9, essentially due to Helfferich [14], are appropriate to the experimental procedures followed by Hammett [13, 22] (and in a number of subsequent investigations). As was shown by Bodamer and Kunin [18] the reaction may be more conveniently monitored by the appearance of product in the external phase. In that case we have

$$\left(\frac{dm_P}{dt}\right)_{cat} = \bar{k}\tilde{m}_R\tilde{m}_c\left(\frac{\lambda_R\tilde{V}}{\lambda_R\bar{V} + V}\right)\left(\frac{\tilde{V}}{\lambda_p\bar{V} + V}\right) \tag{11}$$

or, for the limiting case where $\bar{V} \ll \tilde{V}$, [cf. equation 9]

$$\left(\frac{dm_P}{dt}\right)_{cat} = \frac{d\tilde{m}_p}{dt} = \bar{k}\tilde{m}_R\tilde{m}_c\lambda_R \tag{12}$$

and hence again equation 10 for the efficiency of the resin.

It should be observed that some of the parameters $(m, \tilde{m}, \tilde{V})$ occurring in these equations are observables, whereas others $(\bar{m}, V)$ can be evaluated only with the aid of assumptions concerning the solution volume in the resin phase $(\bar{V})$. The product $\bar{k}\lambda_R$ is experimentally accessible through equations 7 and 12 which hold for the limit $\tilde{m}_R \to 0$.

A comparison of the catalytic coefficient of hydrogen ions in the resin phase $(\bar{k}_H)$ and in homogeneous solution (k_H) thus requires knowledge of the distribution coefficient λ_R. Tartarelli and co-workers [23] have attempted to evaluate this parameter for the hydrolysis of aliphatic carbonate esters in aqueous acetone solution in the presence of Dowex-50. The partition coefficient was determined by equilibration of the ester solution with the sodium form of the ion exchange resin, in the presence of which the esters do not hydrolyse. (The same stratagem has since been used by other authors.) The results obtained were consistent with the hypothesis that $q = \lambda_R$ and hence that $\bar{k}_H = k_H$. This evaluation of λ_R suffers from the possible objection that distribution coefficients may depend on the nature of the cation present in the resin, a point which requires further investigation. This assumption is likely to be even more suspect in the case of weak-acid resins. For such resins neutralization is accompanied by a large increase in ionic concentration. This in turn causes a substantial increase in gel volume (which would tend to increase λ) but may also exert an even more important salting-out effect from the resin phase (which would tend to decrease λ). It cannot be expected that these two, probably large effects will exactly compensate each other. Several reactions of macromolecular acid-base systems in solution have been found to be slowed down – by a larger factor than the reduction in the number of catalytically active acidic or basic groups – as the charge on the polymer backbone is increased [24]. This effect has been attributed to progressive exclusion of an organic substrate from the vicinity of the reactive groups of the macromolecule [25]. It can be reversed by addition of an organic co-solvent [26].

A further difficulty with the quantitative interpretation of Tartarelli's work [23] and of many other kinetic investigations lies in the use of mixed aqueous solvents. It is self-evident that the composition of the solvent in the resin phase may differ from that in the external phase owing to fractionation effects [27]. In such systems the effect of solvent change on the rate constant may be an additional factor influencing the value of the ratio $\bar{k}/k_H$ [15, 23]. The use of mixed solvents should therefore be avoided in kinetic investigations of heterogeneous catalysis.

All of the foregoing discussion assumes that the observed rate is that of a chemical reaction. Even when this is substantially correct, diffusion effects may not be entirely absent. This situation can be accommodated in the kinetic treatment [14, 19] by the introduction of a factor (the 'degree of catalyst utilization', a concept taken over from the treatment of heterogeneous catalysis of gas reactions) [28], which is given by equation 13 [29].

$$\eta_R = 3[(\omega_R \tanh \omega_R)^{-1} - \omega_R^{-2}] \tag{13}$$

Equation 13 involves the so-called [14] Thiele modulus of the substrate R, defined as

$$\omega_R = r_0(\bar{k}/\bar{D}_R)^{\frac{1}{2}} \tag{14}$$

where r_0 is the radius of the resin particle and $\bar{D}_R$ the diffusion co-efficient of R in the resin phase. The effect of including diffusion in this manner is to modify all rate equations in the section by the factor η_R. For example, equation 12 becomes

$$\left(\frac{dm_P}{dt}\right)_{\text{cat}} = \bar{k}\tilde{m}_R\tilde{m}_c\lambda_R\eta_R \tag{15}$$

The quantitative application of equations such as 15 depends on a knowledge of diffusion coefficients.

Whilst all these complications probably vitiate the detailed inter-pretations offered in some of the investigations, it appears that certain general conclusions may validly be drawn concerning the chemical step in catalysis by strong-acid ion exchangers:

1. It is not to be expected that $\bar{k}/k_H$ for a reaction which is first-order in hydrogen ion concentration and does not involve a very bulky substrate will differ from unity by a spectacular factor.

2. Because the chemical change proceeds in a region of high local concentration of hydrogen ions, reactions which are kinetically of second (or higher) order in hydrogen ion concentration will (other things being equal) be at an advantage under conditions of macro-molecular catalysis compared with a reaction at the same overall bulk concentration of hydrogen ion derived from a soluble mineral acid. The benzidine rearrangement is an example of such a reaction where this prediction [6] has been confirmed [30] for catalysis by soluble macromolecular acid (to which these considerations also apply). If a single substrate can react by several paths that are of different orders with respect to hydrogen ion concentration, different mechanisms may apply to resin and dilute homogeneous acid catalysis.

3. Dramatic effects on reaction rates (compared with homogeneous catalysis) may occur if there is severe fractionation of substrate between the phases. This is not pronounced with organic substrates of very low molecular weight unless they carry ionic charges. More marked partitioning effects are evidently to be expected if there are ionic charges in the substrate. The Donnan equilibrium conditions will result in virtual exclusion of mobile anions from a stationary anionic polymer phase. Cation exchangers are therefore expected to be ineffective as catalysts for anionic substrates. Conversely, cations will be preferentially sorbed. A cation exchanger may therefore create favourable conditions for a reaction between hydrogen ions and a cationic substrate [31]. Indeed the same charge effects can be observed with soluble polyelectrolytes of high molecular weight, which may similarly be regarded as constituting a 'two-phase' system. A good example is the reaction between two cations

$$Co(NH_3)_5Cl^{2+} + Hg^{2+} + H_2O \rightarrow Co(NH_3)_5H_2O^{3+} + HgCl^+$$

which is reported to be accelerated 176,000 times by addition of $5 \times 10^{-5}N$ polyvinylsulphonate [32]. One may look upon this catalysis as an extreme form of a special electrolyte effect. The attraction between an anionic polymer structure and a cationic substrate cannot, however, be divorced from the cation exchanger property of the resin. As reaction of a cationic substrate must also lead to a cationic product, it is possible for that product to be retained in the resin with release of hydrogen ion into the external phase.

Systematic studies of the acid-catalysed hydrolysis of esters in the presence of sulphonated polystyrene showed an increase in the value of q with decreasing solubility of the esters in water [17, 22]. Correspondingly, for soluble macromolecular acids the values of q have been found to decrease with increasing degree of sulphonation of the polymer. These effects are consistent with hydrophobic interaction between resin and substrate and its effect on the value of λ_R. The importance of hydrophobic interaction can be reduced if, for example, acetone is added to the solvent. Such solvent changes have the opposite effect on hydrophilic substrates [26].

For aromatic molecules of intermediate size these hydrophobic interactions seem to be more important than electrostatic charge effects, as has been concluded from work on soluble polymers [33]. However, it is also possible for hydrophobic interactions to become so strong that substrate and product molecules block the polymeric catalyst, and some observations of decreasing activity of catalyst with progressing reaction have been ascribed to this cause [1]. It has been noted that a similar effect may be caused by degradation (desulphonation) of the polysulphonic acid [1] or, as in Noller and Gruber's study of the decomposition of ethyl diazoacetate [19], by the accumulation of bubbles of a gaseous reaction product in the pores of the resin.

In their effect upon the distribution of hydrophobic solvents between phases, polymeric ion exchangers clearly bear a close resemblance to surfactants. The remarkable effects of these substances upon reaction velocities are generally referred to as 'micellar catalysis' and have been extensively reviewed [6, 34].

13.6 Cross-linking and swelling behaviour

Cation exchange resins manufactured by addition polymerization are generally co-polymerized with divinylbenzene in order to produce a cross-linked and water-insoluble structure. The cross-linking has some important effects on catalyst structure and behaviour, mainly by altering the size of the pores and the flexibility of the polymer network. It is generally not known whether such cross-linking takes the form of block polymers or of random co-polymers.

Dry ion exchange resins swell when they are added to water, and the volume of the beads is further dependent upon the nature of the cation present. The equilibrium state represents a minimization of

Gibbs free energy resulting from the balancing of solvation and electrostatic interactions between charges and of free energy of mixing and configurational entropy of the resin. The degree of swelling parallels the size of the *solvated* counter-ions ($H^+ > Li^+ > Na^+ > K^+ > Rb^+ > Cs^+$). In very highly cross-linked resins solvation of the counter-ions may not always be complete. Strong acid resins shrink on neutralization by base but weak-acid resins, being largely un-ionized in the acid form, swell upon conversion to a sodium (or other) salt. Important developments in the theory of these phenomena by Gregor, Katchalsky, and Rice and Harris were reviewed by Helfferich [35].

Although the volume change on addition of water decreases with increased divinylbenzene content of the resin, the effect clearly shows that even the cross-linked resins contain a fairly flexible structure. In itself, the swelling does not introduce any further complication into the theory, provided the composition (*i.e.* the degree of neutralization or cation loading) of the resin is kept constant. Comparisons between results for different cation compositions must be treated with greater circumspection. For example, specific effects on the catalyst efficiency were noted by Riesz and Hammett [22] in ester hydrolysis when the resin was partly neutralized by monovalent and divalent quaternary ammonium ions. It was noted that the efficiency of the catalyst (q) was increased by those ions which were structurally similar to the substrate ester. The effect is probably to be interpreted as an increase in the hydrophobic character of the resin phase, which results in an increased value of λ_R for a hydrophobic substrate (and a converse effect for a hydrophilic substrate, such as a sugar). This is of course related to our doubts concerning the applicability of values of λ_R obtained from measurements on a resin in the sodium form to the catalytically active acid form. Since the mobile aqueous phase in the resin is a fairly concentrated ionic solution, salting-out or salting-in effects must, in principle, always be present.

Resins with higher degrees of cross-linking tend to be less efficient catalysts, and Bernhard and Hammett noted that their efficiency q was lower the larger the substrate molecule [22]. As already mentioned, this is accompanied by a lower capacity for swelling and hence presumably a smaller pore size. If it is true that the ester hydrolyses are not diffusion-controlled in the case of the more highly cross-linked resins, then one must conclude that the catalyst contains volume regions which are totally inaccessible to substrate, whereas others retain their normal reactivity. There is some support for this picture of the cross-linked catalyst in Chen and Hammett's work [22]. However, one may also wonder whether the experimental criteria on the basis of which diffusion control was ruled out were sufficiently unequivocal.

13.7 Some observations concerning the two-phase model of a strong acid catalyst

The two-phase model [14] on which the preceding description of catalytic action in the absence of diffusion control was based, has the great virtue that it permits the ideas of hydrogen-ion catalysis in homogeneous solution to be directly transferred to the heterogeneous case. Not being dependent on any assumption concerning the structure of the catalyst (the term structure here being thought to embrace aspects of conformation and charge distribution), the model allows, at least in principle, the average reactivity of the hydrogen ion in the catalyst phase to be deduced. Conversely, of course, the analysis of kinetics in terms of the two-phase model does not provide much information on structural aspects. We wish to mention here a few of the problems which would arise if one were to devise a more detailed view of the reaction in the catalyst phase.

Even when we confine our attention to the case of unrestricted diffusion and a strong-acid resin, the assumption that only one type of chemical species is catalytically active within the resin particle may be questionable. If the pores of the resin are completely permeable to relatively large organic substrates, then they must be presumed to be wide enough for cations to move freely and to be fully solvated. Nevertheless most of the cations may be so close to the charged resin backbone as to be 'associated' in the sense in which Bjerrum defined the term. The problem is to give any quantitative expression to that statement, *i.e.* to assess the relative proportions of free and associated hydrogen ions [36]. Although the polyelectrolyte backbone of the resin is a particle carrying an enormous anionic charge, this can hardly be the relevant charge to consider in connection with ion association, since each sulphonate group acts largely as an independent centre of charge. In considerations of the physical chemistry of polyelectrolyte solutions, the idea of 'counter ion condensation' appears to be a useful concept [37]. According to this, the counter ions will associate with the surface of the poly-ion so as to maintain the net charge density at the surface of the polyelectrolyte below a certain critical charge density. Calculations based on this idea [38], treat the polyelectrolyte ion as an infinitely long cylinder with electric charge distributed uniformly over its surface. Whilst this is probably not too unrealistic for an infinitely dilute polyelectrolyte solution, the model would not appear to be readily applicable to the coiled-polyelectrolyte gel of an ion exchange resin.

If we, therefore, consider the polysulphonic acid of a strong-acid cation exchanger to be partially associated, the catalytic constant obtained by application of the two-phase model will be an average value for associated and free hydrogen ions in the resin phase. This possibility should be borne in mind when changes are made in the resin which affect the degree of association of hydrogen ions in the resin, in particular partial neutralization (replacement of hydrogen ion by a cation with a different ion-pairing tendency) or 'charge dilution' (*i.e.* chemical modification of the polymer so as to produce a lower frequency of sulphonate groups along the backbone). It is relatively easy to see how either of these changes could be introduced into the 'infinite-cylinder' model of polyelectrolytes. In the case of resins the situation is not so readily visualized, particularly since the structure of the polymer (as evidenced by the degree of swelling) alters when the density of free sulphonic acid groups in the resin is changed.

Apart from the ion-pairing aspect, the view that the activity of sulphonic acid groups does not markedly depend on local structure is perhaps supported by the observation by Sakurada *et al.* that a change in tacticity of the resin – which is expected to affect the geometrical distance between polymer groups – did not alter the catalytic behaviour of sulphonated polystyrene resins [26].

In recent years new types of ion exchangers, the so-called macroreticular resins, have become commercially available. They are based on a highly cross-linked matrix (which is resistant to swelling) but contain wide rigid pores. Their catalytic action has not so far been the subject of kinetic studies, but it seems improbable that the two-phase model would apply to them [39]. It is more likely that this catalysis would involve true surface reaction which may bear some analogies to reactions which are thought to occur at the interface of immiscible liquid solvents [40].

13.8 *General acid-base catalysis*

By constrast with the extensive studies on sulphonic acid resins, there has been virtually no work on catalysis by carboxylic acid resins. There are several reports [17, 18] of the ineffectiveness of such materials for reactions in which the use of more strongly acidic resins had proved successful. An exception is a report of the qualitative observation that the hydrolysis of several vinyl esters could be carried out at 80°C in the presence of Amberlite IRC-50 resin in the hydrogen form

[40]. As the experiments were carried out in unbuffered solution the significance of the observation (to which no further mechanistic inference was attached by the authors) is unclear. The hydrolysis of ethyl vinyl ether is, in fact, a well-studied reaction which is known to be subject to general acid catalysis by carboxylic acids [42] and was one of the substrates later selected by us to establish the possibility of general acid catalysis by solid catalysts.

Even in homogeneous solution there has been no work on macro-molecular catalysis in which an acid group which forms part of the polymer back-bone (rather than the hydrogen counter-ion) is impli-cated in the reaction as a simple proton donor. By contrast, an extensive literature has recently grown up on the catalysis of the hydrolysis of aryl acetates and related esters under the influence of polymeric bases in solution, especially polyvinylimidazole [43] and polyvinylpyridine [24]. By comparison with the catalytic power of the monomer bases, the polymers are sometimes much more effective, perhaps because of a pathway involving a second-order rate depend-ence on the catalyst moiety. The evidence points to the over-riding influence of hydrophobic interactions and sometimes of electrostatic forces between electrically charged groups in the polymer (even though they may not be the ones involved in the reaction) and substrates. In some cases the kinetic form of the catalysis obeys the Michaelis–Menten scheme [5, 33, 43–45], an observation which has been held to imply that the substrate-catalyst interactions resemble the inter-actions in enzyme systems [46], but this is an unwarranted inference. Ingenious modifications of both substrate and catalyst (including the introduction of two functional groups into the polymer) have succeeded in even closer simulation of the characteristics and powerful catalysis of enzyme action in these reactions [45, 47].

The chemical reaction involved in the catalytic process of these reactions is, however, not a proton transfer but an acyl group transfer so that in all these studies the mechanism is one of nucleophilic catalysis rather than of general base catalysis. Some of the experi-mental methods and theoretical considerations are nevertheless related to some of those relevant to the study of general acid-base catalysis with polycarboxylic acids.

The operational characteristic of general acid or base catalysis in homogeneous solution is the dependence of reaction velocity on total acid (or base) concentration when the concentration of hydrogen ion is being kept constant. Such variations in composition can be achieved by the use of buffer solutions. A constant buffer ratio (at constant and low ionic strength) ensures constancy of hydrogen ion concentration, although the buffer concentration (and hence the concentration of the buffer component – say, a carboxylic acid – the catalytic effect of which is to be tested) can be varied at will. In this case the contribution to the total rate made by hydrogen ion sets a practical limit to the detectability of catalysis by the carboxylic acid.

The procedure cannot immediately be transferred to the study of catalysis by solid macromolecular acids since the hydrogen ion concentration in the resin phase is not so easily regulated. Because of the Donnan equilibrium between internal and external phases, the internal and external hydrogen ion concentrations can differ by several powers of ten [48], and there is no straightforward method of relating buffer ratio and external or internal pH.

13.9 Experimental criteria for general acid catalysis and rate-limiting proton transfer

For reactions in homogeneous solution three main operational criteria are commonly employed to establish the occurrence of rate-limiting proton transfer to the substrate in the catalytic step. As just mentioned, the first two depend on the use of buffered solutions of weak acids and require (1) that at constant buffer ratio the rate of the catalysed reaction should increase linearly with the total concentration of buffer, and (2) that the dependence of the rate on buffer ratio can successfully be analysed in terms of independent contributions from any spontaneous reaction, catalysis by hydrogen ions and catalysis by the weak acid. On their own, these two criteria indicate the occurrence of general acid catalysis but do not establish the mechanism of the general acid catalysis. The third operational criterion is the effect of changing the solvent from water to deuterium oxide (and possibly also to H_2O-D_2O mixtures) [49]. It relies in part on the well-established rule that dissociation constants of weak acids (including those formed by protonation of the substrate) are reduced by an appreciable factor, which generally lies in the range 2–4, on going from protium oxide to deuterium oxide as solvent. In accordance with this, rate *increases* by a factor of this order are observed in D_2O for reactions involving pre-equilibrium proton (deuteron) transfer from aqueous hydrogen ions to substrate, but there is a cancellation of these effects when a buffer solution (at constant buffer ratio) of a weak acid is used. These findings apply even when the reaction exhibits general acid catalysis through intervention of the conjugate base of the catalysing acid in the second rate-limiting step. By contrast, the rate of a reaction governed by a rate-limiting

424

proton (deuteron) transfer from the aqueous hydrogen ion to the substrate *decreases* by a small factor as the solvent is changed from water to deuterium oxide and decreases by a larger factor (commonly in the range 5–8) for proton (deuteron) transfer from weak acids.

It is not easy to see how the first of these criteria could be applied to two-phase catalysis. The rate of the catalysed reaction depends upon the volume of the resin phase (for constant chemical composition of the resin) and hence on the total amount of catalyst per unit volume of suspension, quite irrespective of mechanism and even when the catalyst is a strong acid and catalysis is simply due to the higher concentration of hydrogen ions in the resin phase. The second criterion is, in principle, applicable, as discussed in the next section, but is complicated by the lack of an entirely satisfactory quantitative theory of the dissociation of poly-acids [38, 48, 50]. It should be possible to make some use of the third criterion, based on the solvent change to deuterium oxide.

13.10 Application of the two-phase model to weak-acid resins

The general equations 8 and 11 devised above are again applicable but the rate constant $\bar{k}$ for the catalysed reaction within the resin phase can now no longer be equated with the catalytic coefficient of hydrogen ions. It is a composite quantity and can be expressed as

$$\bar{k}\bar{m}_c = k_H\overline{H}^+] + k_{HA}\bar{m}_c(1 - \alpha) \tag{16}$$

where the first term on the right-hand side represents catalysis by hydrogen ions and the second term catalysis by undissociated resin acid, a fraction α of which has been neutralized by base.*

For small organic solutes (such as C_4H_8O and $C_4H_{10}O_2$, to which we shall apply this treatment) it is unlikely that a large error will be introduced by setting $\lambda_P = \lambda_R = 1$ in equation 11. On this assumption and on substituting for $\bar{k}$ from equation 16 we obtain

$$\left(\frac{dm_P}{dt}\right)_{cat} = \left[k_H\frac{[\overline{H}^+]}{\bar{m}_c(1 - \alpha)} + k_{HA}\right]\tilde{m}_R\tilde{m}_c(1 - \alpha) \tag{17}$$

The factor in square brackets corresponds to the value of $\bar{k}$ which is experimentally measurable as a function of α.

The difficulty about directly applying equation 17 to a reaction suspected of being catalysed by the resin acid is that it involves two concentration parameters relating to the resin phase – $[\overline{H}^+]$ and $\bar{m}_c$:

---

* $\bar{m}_c$ refers to the *sum* of resin acid and salt; $\bar{m}_c (1 - \alpha)$ is the concentration of the acid form.

the calculation of the former presents a formidable problem [48]. The evaluation of $\overline{m}_c$ requires knowledge of the effective volume of the resin phase which can, within the limitations of certain assumptions, be deduced from the water uptake of the dry resin. To obtain the concentration of hydrogen ion inside the resin from the concentrations in the external solution one must in addition balance the constraints imposed by the Donnan membrane conditions governing ionic distribution, osmosis and mechanical strain of the swelled polymer network. Although there is a considerable literature on this subject we are not aware of a quantitative and practical solution which could be applied to our problem. Various approximate procedures for the calculation of $[\overline{H}^+]$ could be discussed but we shall instead concentrate on an empirical approach, based on the results of studies of acid catalysis in homogeneous solution. The method derives from a suggestion by Brønsted and Grove [51] that the rate of hydrolysis of dimethyl acetal – a reaction uninfluenced by acid species other than the hydronium ion – could be used to measure low hydrogen ion concentrations in solution.

For the hydrolysis of dimethyl acetal under the influence of resin we may therefore set $\overline{k}_{HA}^{DMA}$ in equation 14 equal to zero. On the assumption that $\overline{k}_H^{DMA} = k_H^{DMA}$, an experimental determination of $\overline{k}^{DMA}$ will therefore allow us to calculate the ratio $[\overline{H}^+]/\overline{m}_c(1 - \alpha)$. The value of this ratio is a property of the resin and is independent of the substrate used (equation 15).

$$[\overline{H}^+]/\overline{m}_c(1 - \alpha) = \overline{k}^{DMA}/k_H^{DMA} \qquad (15)$$

The procedure seems of some interest not only to the problem of catalysis, which is our concern here, but also as an additional source of information which could be used to elucidate heterogeneous equilibria of ion exchange resins.

We next measure $\overline{k}$ for a reaction thought to be capable of being catalysed by the macromolecular acid, in the presence of the same resin and at the same value of α. If it is assumed that also for this reaction the same rate constant applies for hydrogen ion catalysis inside the resin and in homogeneous solution, it is then possible to evaluate $\overline{k}_{HA}$ from equation 14.

To illustrate the procedure, we have measured the rate of hydrolysis of ethyl vinyl ether, a reaction subject to general acid catalysis in homogeneous solution, under the same conditions as for the hydrolysis of dimethyl acetal. Both reactions took place at 25°C under the influence of the same sample of half-neutralized polymethacrylic acid (Zerolit 226 SRC 41) and in the presence of sodium chloride

426

([$\widehat{NaCl}$] = 0.1 molal) in the external solution. If we use the superscripts DMA and EVE to distinguish the two reactions, the evaluation of $\bar{k}_H^{EVE}$ is summarized in equation 16.

$$\bar{k}_{HA}^{EVE} = \bar{k}^{EVE} - k_H^{EVE}\bar{k}^{DMA}/k_H^{DMA} \qquad (16)$$

Taking $k_H^{EVE} = 1.75$ [41], $k_H^{DMA} = 0.289 \, dm^3 \, mol^{-1} \, s^{-1}$ [52]; $\bar{k}^{DMA} = 3.0 \times 10^{-6}$ [11], $\bar{k}^{EVE} = 7.3 \times 10^{-4} \, dm^3 \, mol^{-1} \, s^{-1}$ [11], we calculate $\bar{k}_{HA}^{EVE} = 7.0 \times 10^{-4} \, dm^3 \, mol^{-1} \, s^{-1}$. Using the same data in equation 15 we obtain $[\bar{H}^+]/\bar{m}_c(1 - \alpha) = 2 \times 10^{-5}$. The relative contributions of catalysis by hydrogen ions and resin acid to the hydrolysis of ethyl vinyl ether under these conditions can accordingly be assessed as 3 per cent and 97 per cent respectively. The catalytic intervention of undissociated macromolecular acid in the reaction therefore seems highly probable, even though the assumptions involved in the calculation must place rather wide error limits on the numerical values deduced. In particular, the limitations set by the

$$\left[-CH_2-\underset{\underset{CO_2H}{|}}{\overset{\overset{CH_3}{|}}{C}}-CH_2-\underset{\underset{CO_2^-}{|}}{\overset{\overset{CH_3}{|}}{C}}- \right]_n$$

$$1 \qquad Na^+$$

assumption $\lambda_P = \lambda_R = 1$ have yet to be established by experiment.

We should note that the value of $\bar{k}_{HA}^{EVE}$ derived in this manner applies only at a particular value of α. Experiments with samples of Zerolit SRC-45 resin neutralized to different degrees indicate that $\bar{k}_{HA}$ decreases as the resin is progressively converted into the sodium form. Since $\bar{k}_{HA}$ is evaluated on the basis of the free acid groups only, this implies – as has been pointed out on the basis of other evidence – that the microscopic acidity of the resin acids must decrease with progressive incorporation of carboxylate groups. Inductive, direct electrostatic, hydrogen-bonding and salting-out effects are all expected to operate in this direction. For diffusion-controlled reactions the opposite effect may be expected [19].

In comparing the derived value of $\bar{k}_{HA}^{EVE}$ with catalytic constants of other carboxylic acids we must note that the carboxylic acid groups in half-neutralized polymethacrylic acid 1 will be somewhat weakened relative to comparable monocarboxylic acids. Purely on the basis of inductive effects alone we estimate the appropriate microscopic dissociation constant as 4×10^{-6}. The corresponding value of k_{HA}^{EVE} calculated from the parameters of the catalysis law deduced by

Kresge [42] for the hydrolysis of ethyl vinyl ether in homogeneous solution is $5 \times 10^{-4}\,\mathrm{dm^3\,mol^{-1}\,s^{-1}}$, in close agreement with the deduced value $k_{HA}^{EVE} = 7.0 \times 10^{-4}\,\mathrm{dm^3\,mol^{-1}\,s^{-1}}$.

13.11 Solvent isotope effects

The two terms which, according to equation 14, may contribute to the value of k for a weak-acid resin should be differently affected by a change to deuterium oxide solvent. Catalysis by hydrogen ions in the resin phase is presumed not to differ in kind from catalysis in ordinary solutions. One should therefore expect the dichotomy of isotope effects on rate constants – which distinguishes rate-limiting proton transfers from catalysis involving proton-transfer pre-equilibria – to be maintained for the resin systems. In addition to the influence on rate constants, the isotopic solvent change affects acid dissociation constants and one would therefore expect the concentration of hydrogen ions inside the buffered resin phase to be approximately three times lower in D_2O than in H_2O. The following total solvent isotope effects on rate constants, determined for constant degrees of neutralization of the catalyst acid, are therefore expected. For a reaction involving a protonation (deuteronation) pre-equilibrium, as exemplified by the hydrolysis of dimethyl acetal, there should be hardly any effect (*i.e.* greater than ca. 50 per cent) since the overall pre-equilibrium $HA + S \rightleftharpoons A^- + SH^+$ does not involve formation of hydrogen ion, nor is a proton transfer, in general, involved in the subsequent rate-limiting step. The measured solvent isotope effect [11] $[(k^{DMA})_{D_2O}/(k^{DMA})_{H_2O} = 0.9]$ for the hydrolysis of dimethyl acetal in the presence of half-neutralized polymethacrylic acid is in accord with these conclusions.

For a rate-limiting proton transfer from the hydronium ion, large isotope effects $(k_{H_2O}/k_{D_2O} \sim 4\text{–}12)$ are expected, depending on the actual reaction studied. For rate-limiting proton transfer from a weak-acid resin, only the primary isotope effect $(k_{H_2O}/k_{D_2O} \sim 5\text{–}9)$ normally found for weak carboxylic acids in homogeneous solution should operate. However, it cannot be ruled out that steric constraints of the polymer structure may modify the manner of solvent access to the carboxylic acid/carboxylate grouping during proton transfer, and thus affect the result. In any event, the expectation that the ranges of solvent effects for rate-limiting proton transfers from hydrogen ions and from resin acid overlap, renders the diagnostic use of solvent isotope effects for distinguishing between these two mechanisms rather problematical.

Nevertheless it may always be useful to make observations of solvent isotope effects when it is uncertain whether the rate of a reaction is controlled by a chemical process.

13.12 Conclusions

In spite of the bewildering complexity of the system and of the phenomena reported, it is, in our view, possible to draw a few general, if tentative, conclusions which may serve as pointers to further work.

(1) Experimental conditions can be realized for which diffusion effects are not important and for which the two-phase model of catalysis by insoluble macromolecular acids represents a fair first-order theory.

(2) Weak-acid resins are capable of exhibiting dual catalysis, *viz.* by hydrogen ions, present as solute counter-ions in the resin phase, and by un-ionized acid groups attached to the polymer back-bone. The latter catalysis is expected to operate selectively, *viz.* in reactions which are subject to general acid catalysis in homogeneous solution.

(3) The degree of neutralization of the acid resin by base has a marked influence on the 'chemical' reaction velocity, for both strong- and weak-acid resins, but the cause or causes of this effect remain to be established. It is not known whether a reduction of the catalytic power of the remaining acid groups or salting-out of substrate from the resin phase is more important, though both these effects are presumed to be amongst the factors to be considered. These may also include ion association and changes in conformation and in volume of the resin phase.

(4) The rate of the hydrolysis of dimethyl acetal, a reaction catalysed specifically by hydrogen ions, appears promising as a probe for hydrogen ion concentration in the resin phase.

(4) Experimental conditions which give rise to simple kinetic behaviour are not the ones best suited to producing the largest catalytic effects. To achieve large rate enhancement and selectivity it is advisable to seek conditions under which the partitioning of the substrate favours the resin phase, by virtue of hydrophobic or electro-static interactions.

Acknowledgment

We thank Shell Research Ltd and the Science Research Council for grants in support of two of us.

REFERENCES

[1] N. G. Polyanskii, *Russ. Chem. Rev.*, **31**, 496 (1962); **39**, 244 (1970).

[2] B. Tacke and H. Süchting, *Landwirtsch. Jahrb.*, **41**, 717 (1911).

[3] F. E. Rice and S. Osugi, *Soil Sci.*, **5**, 333 (1918).

[4] R. P. Bell and J. E. Prue, *Trans. Faraday Soc.*, **46**, 5 (1950).

[5] P. Mastagli, A. Floc'h, and G. Durr, *Compt. rend*, **235**, 1402 (1952).

[6] I. M. Klotz, G. P. Royer, and I. S. Scarpa, *Proc. Nat. Acad. Sci. U.S.A.* **68**, 263 (1971). H. Morawetz and E. W. Westhead, Jr. *J. Polymer Sci.*, **16**, 273 (1955); C. G. Overberger and J. C. Salamone, *Accounts. Chem. Res.*, **2**, 217 (1969).

[7] H. Morawetz, *Adv. Catalysis*, **20**, 341 (1969).

[8] R. P. Bell, *Acid-Base Catalysis*, Clarendon Press, Oxford, 1941.

[9] F. Helfferich, *Ion Exchange*, McGraw-Hill, New York, 1962.

[10] Ref. 9, chapter 3.

[11] V. Gold and C. J. Liddiard, unpublished.

[12] V. Gold and G. D. Morgan, unpublished.

[13] V. C. Haskell and L. P. Hammett, *J. Amer. Chem. Soc.*, **71**, 1284 (1949).

[14] F. Helfferich, *J. Amer. Chem. Soc.*, **76**, 5567 (1954); ref. 9, chapter 11.

[15] E. Mariani, *Ann. Chim. applicata*, **39**, 283 (1949); **40**, 500 (1950).

[16] C. L. Levesque and A. M. Craig, *Ind. and Eng. Chem.*, **40**, 96 (1948).

[17] C. W. Davies and G. G. Thomas, *J. Chem. Soc.*, 1607 (1952).

[18] G. Bodamer and R. Kunin, *Ind. and Eng. Chem.*, **43**, 1082 (1951).

[19] N. L. Smith and N. R. Amundsen, *Ind. and Eng. Chem.*, **43**, 2156 (1951); H. Noller and A. Hässler, *J. Chim. phys.*, **55**, 255 (1958); *Z. phys. Chem. (Frankfurt)*, **11**, 267 (1957); H. Noller and P. E. Gruber, *ibid*, **38**, 184, 203 (1963); **41**, 353 (1964); J. Meyer, F. Sauer and D. Woermann, *Ber. Bunsengesellschaft Phys. Chem.*, **74**, 245 (1970); D. Woerman, *ibid.*, p. 385.

[20] M. B. Bochner, S. M. Gerber, R. W. Vieth, and A. J. Rodgers, *Ind. and Eng. Chem. (Fundamentals)*, **4**, 314 (1965); R. Tartarelli, G. Stoppato, F. Morelli, A. Lucchesi, and M. Giorgini, *J. Catalysis*, **13**, 108 (1969).

[21] P. D. Bolton and T. Henshall, *J. Chem. Soc.*, 1226, 3369 (1962).

[22] S. A. Bernhard and L. P. Hammett, *J. Amer. Chem. Soc.*, **75**, 1798, 5834 (1953); S. A. Bernhard, E. Garfield, and L. P. Hammett, *ibid.*, **76**, 991 (1954); P. Riesz and L. P. Hammett, *ibid.*, **76**, 992 (1954); C. H. Chen and L. P. Hammett, *ibid.*, **80**, 1329 (1958).

[23] R. Tartarelli, G. Nensetti, and M. Baccaredda, *Ann. Chim. (Italy)*, 1108 (1966); R. Tartarelli, F. Morelli, M. Giorgini, and A. Lucchesi, *Chimica e Industria (Milan)*, **50**, 528 (1968).

[24] E.g. H. Ladenheim and H. Morawetz, *J. Amer. Chem. Soc.*, **81**, 4860 (1959); R. L. Letsinger and R. J. Savereide, *ibid.*, **84**, 3122 (1962).

[25] H. Morawetz, *Macromolecules in Solution*, Interscience, New York, 1965, p. 419.

[26] S. Yoshikawa and O.-K. Kim, *Bull. Chem. Soc. Japan*, **39**, 1515, 1729 (1966); I. Sakurada, Y. Sakaguchi, T. Ono, and T. Ueda, *Makromol. Chem.*, **91**, 243 (1966).

[27] C. W. Davies and B. D. Owen, *J. Chem. Soc.*, 1676 (1956).

[28] A. Wheeler, *Adv. Catalysis*, **3**, 298 (1951).

[29] E. W. Thiele, *Ind. and Eng. Chem.*, **31**, 916 (1939).

[30] C. L. Arcus, T. J. Howard, and D. S. South, *Chem. and Ind.*, 1756 (1964).

[31] J. R. Whitaker and F. E. Deatherage, *J. Amer. Chem. Soc.*, **77**, 3360 (1955).

[32] B. Vogel and H. Morawetz, *J. Amer. Chem. Soc.*, **90**, 1368 (1969); H. Morawetz and B. Vogel, *ibid.*, **91**, 563 (1969).

[33] T. Kunitake, F. Shimoda, and C. Aso, *J. Amer. Chem. Soc.*, **91**, 2716 (1969); T. Kunitake and S. Shinkai, *ibid.*, **73**, 4247, 4256 (1971); Y. Okamoto and C. G. Overberger, *J. Polymer Sci.*, **10**, 3387 (1972).

[34] E. J. Fendler and J. H. Fendler, *Adv. Phys. Org. Chem.*, **8**, 271 (1970).

[35] F. Helfferich, *Ion Exchange*, McGraw-Hill, New York, (1962), Chapter 5.

[36] H. P. Gregor, *J. Amer. Chem. Soc.*, **73**, 3537 (1951); S. A. Rice and M. Nagasawa, *Polyelectrolyte Solutions*, Academic Press, New York, 1961; V. S. Sol'datov, *Russian J. Phys. Chem.*, **45**, 1629 (1971).

[37] N. Imai and T. Onishi, *J. Chem. Phys.*, **30**, 1115 (1959).

[38] See the review by G. S. Manning, *Ann. Rev. Phys. Chem.*, **23**, 117 (1972).

[39] N. W. Frisch, *Chem. Eng. Sci.*, **17**, 735 (1962).

[40] R. P. Bell, *J. Phys. Chem.*, **32**, 882 (1928); F. M. Menger, *J. Amer. Chem. Soc.*, **92**, 5965 (1970); *Chem. Soc. Rev.*, **1**, 229 (1972).

[41] M. F. Shostakovskii, A. S. Atavin, B. A. Trofimova, and A. V. Gusarov, *Zhur. Vsesoyuz Khim. Obshch. im. D. I. Mendeleeva*, **9**, 599 (1964).

[42] A. J. Kresge and Y. Chiang, *J. Chem. Soc. (B)*, 53, 58 (1967); A. J. Kresge, H. L. Chen, Y. Chiang, E. Murrill, M. A. Payne, and D. S. Sagatys, *J. Amer. Chem. Soc.*, **93**, 413 (1971); P. Salomaa, A. Kankaanperä, and M. Lajunen, *Acta Chem. Scand.*, **20**, 1790 (1966); P. Salomaa and A. Kankaanperä, *ibid.*, **20**, 1802 (1966); M. M. Kreevoy and R. Eliason, *J. Phys. Chem.*, **72**, 1313 (1969).

[43] For a review, see C. G. Overberger and J. C. Salamone, *Bul. Inst. Politeh Iasi*, **16**, 73 (1970).

[44] Yu. E. Kirsh, V. A. Kabasov, and V. A. Kargin, *Doklady Akad. Nauk U.S.S.R.*, **177**, 976 (1967).

[45] C. G. Overberger, M. Morimoto, I. Cho, and J. C. Salamone, *J. Amer. Chem. Soc.*, **93**, 3228 (1971).

[46] M. Dixon and E. C. Webb, *Enzymes*, Academic Press, New York, 1958, p. 62.

[47] C. G. Overberger and M. Morimoto, *J. Amer. Chem. Soc.*, **93**, 3222 (1971).

[48] A. Katchalsky, *Progr. Biophys. and Biophys. Chem.*, **4**, 1 (1954); I. Michaeli and A. Katchalsky, *J. Polymer Sci.*, **23**, 683 (1957).

[49] For a review see V. Gold, *Adv. Phys. Org. Chem.*, **7**, 2591 (1969).

[50] E. J. King, *Acid-Base Equilibria*, Pergamon, Oxford 1965, chapter 9.

[51] J. N. Brønsted and C. Grove, *J. Amer. Chem. Soc.*, **52**, 1394 (1930).

[52] M. Kilpatrick, *J. Amer. Chem. Soc.*, **85**, 1036 (1963).

AUTHOR INDEX

433

SUBJECT INDEX

446